Noise in High-Frequency Circuits and Oscillators

Noise in High-Frequency Circuits and Oscillators

Burkhard Schiek
Ilona Rolfes
Heinz-Jürgen Siweris

WILEY-INTERSCIENCE

A JOHN WILEY & SONS, INC., PUBLICATION

Published by John Wiley & Sons, Inc., Hoboken, New Jersey.
Published simultaneously in Canada.

For general information on our other products and services or for technical support, please contact our
Customer Care Department within the United States at (800) 762-2974, outside the United States at
(317) 572-3993 or fax (317) 572-4002.

Wiley also publishes its books in a variety of electronic formats. Some content that appears in print may
not be available in electronic format. For information about Wiley products, visit our web site at
www.wiley.com.

Library of Congress Cataloging-in-Publication Data:

Schiek, B. (Burkhard), 1938–
    Noise in high-frequency circuits and oscillators / B. Schiek, I. Rolfes, H.J. Siweris.
      p. cm.
    ISBN-13 978-0-471-70607-6
    ISBN-10 0-471-70607-8
    1. Electronic circuits—Noise.   2. Electronic circuit design.   3. Electromagnetic noise.   I.
Rolfes, I. (Ilona), 1973–   II. Siweris, H. J. (Heinz J.), 1953–   III. Title.

    TK7867.5.S35 2006
    621.382'24—dc22   2005056824

10 9 8 7 6 5 4 3 2 1

Contents

Preface

Electrical noise fundamentally limits the sensitivity and resolution of communication, navigation, measurement, and other electronic systems. This book introduces the reader to the most important noise mechanisms, the description of noise phenomena in electrical circuits by means of equivalent sources and analytical or numerical methods.

The consequences of noise in high-frequency systems are not always easy to understand. There might be complicated interactions between different circuit parameters. Furthermore, often not only one noise mechanism has to be considered, but an interaction of various different processes. One example is the noise in frequency converters or mixers, respectively. In order to obtain a sensitive input stage, the noise figure, which is a measure for the internal noise of the system, should be as low as possible. The noise figure of the complete circuit can be improved by inserting a low-noise amplifier in front of the down converter, for example. If this preamplifier does not provide sufficient gain, then the noise figure of the cascade connection of the amplifier and the mixer can still be deteriorated by the mixer according to the cascade formula. Nevertheless, the amplifier is often omitted for cost reasons or in order to improve the dynamic range. In this case, a low noise figure for the mixer is of great importance. If the intermediate frequency of the mixer is low, then the so-called flicker- or 1/f-noise can increase the noise figure significantly. A higher noise figure can also be caused by noise of the local oscillator. If the mixer is not perfectly balanced then the demodulated amplitude noise of the local oscillator increases the noise at the mixer output. An increase of the noise figure is also observed, if the phase or frequency noise of the local oscillator is unintentionally discriminated by frequency selective circuits. In order to minimize the disturbing effects of the noise in the whole system, knowledge of all these noise mechanisms, interactions, and dependencies is necessary.

One objective of this text is to convey a qualitative and quantitative comprehension of the noise phenomena in linear and non-linear high-frequency circuits. The book, however, does not claim to be complete in the sense of a reference work. For example, a detailed presentation of the physical origin of thermal noise and the available noise power of resistors is omitted.

The book contains a number of problems with a varying degree of difficulty. The solutions are given at the end of the book.

This book is the result of many years of research and education. It is based on the course "Noise in High Frequency Circuits and Oscillators", which has been presented regularly at the Ruhr-University Bochum, Germany. The notes for that course were the basis for this book.

This book addresses graduate students but should also be useful for academics, engineers, and physicists.

We wish to express our gratitude to Bianca Will for her aid in the technical editing of the manuscript, especially of the problem solutions and the proofreading.

We wish to give special thanks to Dr. Steve Nightingale for proofreading and his many valuable suggestions for the improvement of the text.

Appreciation is expressed for the steady encouragement and support by our colleagues Dr. Thomas Musch, Prof. Dr. Edgar Voges, Dr. Reinhard Stolle, Dr. Michael Gerding, Prof. Dr. Heinz Chaloupka, Prof. Dr. Volkert Hansen, Prof. Dr. Hermann Eul, Prof. Dr. Holger Heuermann.

We are particularly grateful to Prof. Dr.-Ing. Dr.h.c.mult. Ulrich L. Rohde for his encouragement to write this book and his helpful suggestions.

Comments from our reviewers were very much appreciated.

Finally, we wish to give special thanks to the many students over the years who have attended the course "Noise in High Frequency Circuits and Oscillators" at the Ruhr-University Bochum, Germany, for their many fruitful questions and discussions. Special thanks go to Nils Pohl for proofreading.

B. SCHIEK
I. ROLFES
H.-J. SIWERIS

Bochum, Germany

1

Mathematical and System-oriented Fundamentals

In this chapter the most important mathematical, statistical and system-oriented theoretical fundamentals are presented as they will be needed in the following chapters. However, a certain knowledge of the theory of probabilities and statistics will be anticipated.

In this book, noise signals continuous in time will be considered predominantly. These are the kind of noise signals that normally appear in high-frequency circuits.

The time-dependent behavior of noise signals cannot be predicted in general. It is only possible to characterize their properties with the help of mean values, as for example the mean square value, i.e. the mean power. The noise power per unit bandwidth, the so-called spectrum, will turn out to be a particularly important quantity for the description of noise signals. This spectral representation is particularly useful and common for high-frequency techniques. Therefore, it is one aim of a noise description to determine the spectral distribution of the noise power, i.e. the spectrum, at any point of a circuit or a system quantitatively. For this purpose, one can either attempt to calculate the spectrum of the circuit or to measure it. However, both procedures have to be performed with care. The best way is to make measurements as well as calculations and to bring both results into agreement. Having achieved this agreement it can be expected to some degree that the noise phenomena of the circuit or the system are understood quite well. On this basis it is often possible to take some measures to reduce the noise. Whether a noise reduction is necessary or not depends on the magnitude of the signal of interest in the circuit and furthermore on the required signal-to-noise ratio. These questions will be answered in depth in the following pages.

However, some fundamentals are needed, which will be discussed in the first chapter. One significant result of this first chapter will be that the spectrum is transmitted within a circuit according to the magnitude squared value of the complex system transfer function. This statement will be of great importance for the noise analysis, as will be seen later on.

1.1 INTRODUCTION

1.1.1 Technical relevance of noise

The term noise stems from the perception a person has when electrical jitter effects in the audio frequency range are amplified sufficiently and then are passed to a loudspeaker. This phenomenon is generally known from a broadcast receiver, for example. Later, the term noise was extended to frequencies outside the audible range. In general, the electrical noise originates from current or voltage fluctuations in electronic circuits. One disturbing effect of noise is that it limits the sensitivity of receivers of communication systems or reduces their transfer capacity. Furthermore, it limits the accuracy of measurement systems. Without noise the transmitter power could be reduced down to a limit set by interference from communication channels. As a consequence the electrical noise has a large influence on the system design and thus on the costs.

On the other hand, noise, considered as a physical phenomenon, often contains useful information. Temperature measurements can be performed with the help of thermal noise over large distances by means of antennas (Remote Sensing). In radio-astronomy weak noise signals can give information, e.g. on molecules in distant galaxies. For electronic devices the frequency behavior and the amplitude of the noise is often useful to evaluate the functionality and the quality of the device.

1.1.2 Physical origins of noise

This text mainly deals with the effects of noise in electronic circuits, particularly in high-frequency circuits. Thus, the physical origins of noise will be discussed less extensively. One focus will be on the **thermal noise** in electrical conductors. Thermal noise, which always exists at non-zero temperatures, originates from vibrations of the lattice atoms, which are transferred to the free electrons. The electrons are thus performing an unsteady movement, being interrupted by collisions. This unsteady movement leads to an irregularly fluctuating voltage between both ends of the conductor. It will be seen that the available noise power of a resistor only depends on the absolute temperature of the resistor. The thermal noise is thus especially well suited to serve as a reference noise source. Other kinds of noise phenomena can be compared with it to advantage. Thermal noise is a relatively weak noise phenomenon,

which can be further reduced by cooling. For many systems it is often sufficient, if the overall noise, referred to the input of the system, is of a similar
level as the thermal noise.

Another noise mechanism, which is particularly important for electronic
devices, is the so-called **shot noise**. The transition of electrical potential
barriers is a statistical process, because the charge of the carriers, electrons or
ions, is always an integer multiple of the elementary charge. Consequently, the
current emitted by a cathode at constant temperature and voltage is not a pure
d.c. current, rather it fluctuates around a time average. Because the emitted
electrons arrive irregularly at the anode, the term shot noise became accepted
for this phenomenon. A similar situation can be found for potential barriers in
solid-state devices, i.e. junctions between semiconductors or between metals
and semiconductors. Therefore, such junctions also show shot noise for the
current flowing. Later on, it will be shown that a comparison between shot
noise and thermal noise will lead to the conclusion that, in general, shot noise
has the smaller available noise power.

Very high noise is generated by a semiconductor junction operated at breakdown. This noise mechanism is called **avalanche noise**. Accelerated electrons generate new electron-hole pairs by collision. In particular, the electrons
are able to generate further electron-hole pairs after an acceleration, so that
the current increases very rapidly. In low-noise devices breakdown must be
strictly avoided. On the other hand calibrated noise sources utilize the breakdown mechanism of a pn- or Schottky-junction in order to achieve a high
well-defined noise power for measurement purposes.

The electrical properties of surfaces or boundary layers are influenced ener-
getically by so-called boundary layer states, which are also subject to statistical fluctuations. This leads to the so-called **flicker noise** or **1/f-noise** for the
current flow. This kind of noise is especially observable at low frequencies and
generally decreases with increasing frequency f according to a $1/f$ law until
it will be covered by frequency independent noise mechanisms, e.g. thermal
noise or shot noise. It will be seen, that flicker noise is of great importance
for high-frequency oscillators, because it can modulate the carrier frequency
by non-linear processes. Unwanted amplitude and phase fluctuations with a
flicker noise characteristic will, therefore, be impressed on the oscillator signal.

Oscillators can be interpreted as amplifiers with a high-Q (quality factor)
feedback network. The noise signals, which always exist, are amplified after the supply voltage is turned on until finally saturation effects lead to an
approximately sinusoidal oscillation with constant amplitude.

1.1.3 General characteristics of noise signals

The time characteristic of noise signals can be displayed with the help of an
oscilloscope, if, for example, the thermal noise voltage of a resistor is amplified
sufficiently as shown in Fig. 1.1. The irregular time behavior of the signal is
a characteristic property of electrical noise. A mathematical description in

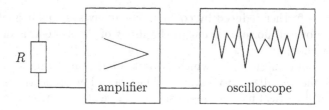

Fig. 1.1   Noise signal displayed by an oscilloscope.

the time domain obviously is not possible. Also, the future behavior of the signal cannot be predicted. In contrast to sinusoidal signals, the description of noise signals is thus restricted to different mean values. For this purpose, the methods of statistical signal theory are applied.

In the frequency domain, the power of a sinusoidal signal is concentrated at a single frequency. In contrast, for a noise signal the power at a single frequency is always zero. Power can only be measured for a non-zero bandwidth. This offers the possibility of a measurement to discriminate between sinusoidal signals and noise signals.

1.2   MATHEMATICAL BASICS FOR THE DESCRIPTION OF NOISE SIGNALS

In this section the most important mathematical fundamentals for the description of noise signals will be discussed. More detailed introductions to the theory of probability and of stochastic processes can be found in the literature.

1.2.1   Stochastic process and probability density

Figure 1.2 shows N equal resistors with $R_1 = R_2 = \ldots = R_N = R$ at the same temperature T. For each resistor the time dependent open circuit voltage is measured and recorded.

All noise voltages are recorded simultaneously. Together, they form a **stochastic process**. Each single curve is a so-called **realization** of the stochastic process. If a specified time, for example t_1, is considered, then the different realizations provide a sequence of voltage values $U_{1i}(t_1)$, $i = 1, 2, \ldots, N$. Herein $U_1(t_1)$ is called a **random variable**. For general considerations a random variable will be denoted by Y in the following pages. If the random variable can only adopt a limited number of values, then Y is called a discrete random variable. If Y can adopt all values within a given interval, then Y is a continuous random variable. This text will deal predominantly with continuous random processes. The random variable Y can designate a voltage, a current, a noise wave or similar signals. The continuous random

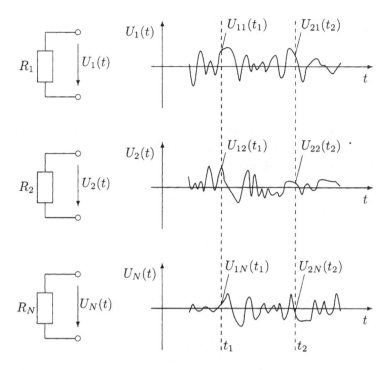

Fig. 1.2   Set of resistors with noise voltages $U_{1i}(t_1)$ and $U_{2i}(t_2)$, respectively, with $i = 1, 2, \ldots, N$.

variable $Y = Y(t_1)$, that results, if the number N of realizations tends towards infinity, shows a certain amplitude distribution. The probability to find an amplitude value of the random variable Y at a time t_1 in the interval from y to $(y + \Delta y)$ is described by $p(Y = y) \cdot \Delta y = p(y) \cdot \Delta y = p(t_1, y) \cdot \Delta y$. Here, $p(y)$ is called the **probability density** or **amplitude distribution density** or **distribution density** or briefly **density**. The normalization condition requires

$$\int\limits_{-\infty}^{+\infty} p(y) \, dy = 1 \ . \tag{1.1}$$

Looking at the probability W_k that the amplitude is lower or equal to y_1' at the time t_1, then the following relation holds:

$$W_k\{Y \leq y_1'\} = \int\limits_{-\infty}^{y_1'} p(t_1, y) \, dy \ . \tag{1.2}$$

In the same way the following equation applies for the probability that the amplitude lies within the boundaries y_1' and y_1''

$$W_k\{y_1'' \leq Y \leq y_1'\} = \int_{y_1''}^{y_1'} p(t_1, y)\, dy \quad \text{with} \quad y_1'' < y_1' \ . \tag{1.3}$$

Electrical noise phenomena often show an amplitude distribution, which is called **normal distribution** or **Gaussian distribution**. The normal distribution is given by the following probability density, which is also illustrated in Fig. 1.3

$$p(y) = \frac{1}{\sqrt{2\pi \cdot \sigma^2}} \cdot e^{-(y - \mu)^2/(2 \cdot \sigma^2)} \ . \tag{1.4}$$

The normal distribution is an even function with respect to μ.

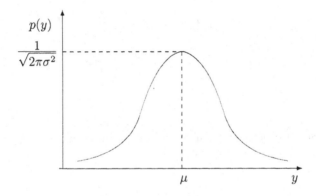

Fig. 1.3   Probability density of a normal distribution.

Problem

1.1   Show that the normal distribution satisfies the normalization condition of Eq. (1.1).

The parameter σ is also called statistical spread. The thermal noise and the shot noise show a normal amplitude distribution. This is due to the following theorem: the central limit theorem of statistics states that, under certain assumptions, the sum of a large number of independent and random variables with an arbitrary distribution is again a random variable but with a normal or Gaussian distribution. For example, thermal noise arises from the not predictable thermal movement of many single electrons, which move independently from each other. From this it follows that the open-circuit voltage of a resistor with thermal noise is normally distributed.

1.2.2 Compound probability density and conditional probability

The term $p(t_1, y_1; t_2, y_2)dy_1 dy_2$ stands for the compound probability density. This is the probability to find, for the same realization, at a time t_1 the amplitude in the interval dy_1 around y_1 and at a time t_2 the amplitude in the interval dy_2 around y_2.

The probability densities $p(t_1, y_1)$ and $p(t_2, y_2)$ can be calculated from the compound probability density with

$$p(t_1, y_1) \quad = \quad \int_{-\infty}^{+\infty} p(t_1, y_1; t_2, y_2)\, dy_2 \ , \tag{1.5}$$

$$p(t_2, y_2) \quad = \quad \int_{-\infty}^{+\infty} p(t_1, y_1; t_2, y_2)\, dy_1 \ . \tag{1.6}$$

A further quantity, which is commonly used for statistics is the so-called conditional probability. The term $p(t_2, y_2 \,|\, t_1, y_1)dy_2$ represents the probability, that the amplitude at a time t_2 can be found in an interval dy_2 around y_2, under the condition that it was in the interval dy_1 around y_1 at a time t_1. For the conditional probability the normalization applies again:

$$\int_{-\infty}^{+\infty} p(t_2, y_2 \,|\, t_1, y_1)\, dy_2 = 1 \ . \tag{1.7}$$

The compound probability density and the conditional probability density are related by the following equation:

$$p(t_2, y_2; t_1, y_1) = p(t_1, y_1) \cdot p(t_2, y_2 \,|\, t_1, y_1) \ . \tag{1.8}$$

If the amplitude at the time t_2 does not depend on the amplitude at the time t_1, then the random variables $Y(t_1)$ and $Y(t_2)$ are called statistically independent. In this case, the conditional probability density is equal to the probability density

$$p(t_2, y_2 \,|\, t_1, y_1) = p(t_2, y_2) \ . \tag{1.9}$$

Because of Eq. (1.9) the compound probability of statistically independent variables is given as follows:

$$p(t_1, y_1; t_2, y_2) = p(t_1, y_1) \cdot p(t_2, y_2) \ , \tag{1.10}$$

i.e. the product of the probability densities of the single variables.

1.2.3 Mean value and moments

For noise processes the main interest is focused on mean values and their transfer characteristics in circuits. The **ensemble average** $\langle Y(t_1) \rangle$ or the

expected value $E\{Y(t_1)\}$ of the random variable Y at the time t_1 is defined by the following expression:

$$\langle Y(t_1) \rangle = E\{Y(t_1)\} = \int_{-\infty}^{+\infty} y_1 \cdot p(t_1, y_1) \, dy_1 \ . \tag{1.11}$$

The expected values of integer powers of Y are called moments of the distribution and are defined by

$$E\{Y^n(t_1)\} = \int_{-\infty}^{+\infty} y_1^n \cdot p(t_1, y_1) \, dy_1 \quad \text{with} \quad n = 1, 2, 3, \ldots \ . \tag{1.12}$$

The first moment or the expected value $E\{Y\}$ is equal to zero for many noise processes, as for example for thermal noise. This is due to the fact that the probability density $p(t_1, y_1)$ of these processes is an even function of y_1. The second moment $E\{Y^2\}$ is of particular interest, because, if Y represents currents, voltages or waves, then it is a measure for the mean noise power.

The distribution of the random variables Y around their expected values is denoted by the **central moments**. This means that effectively only the alternating part is considered:

$$E\{(Y(t_1) - E\{Y(t_1)\})^n\} = \int_{-\infty}^{+\infty} (y_1 - E\{Y(t_1)\})^n \cdot p(t_1, y_1) \, dy_1 \ . \tag{1.13}$$

The second moment is called **variance** σ^2 and σ itself is called **statistical spread** or **standard deviation**.

The electrical fluctuation phenomena are mostly **stationary processes**, this means it can be assumed that the various densities, probabilities, mean values and moments do not change with time. A simple example for a non-stationary process can be imagined for the set of resistors of Fig. 1.2, if they are stored in a thermal bath, the temperature of which is varied over time. For a stationary process, time can be omitted and Eq. (1.11) can be replaced by

$$E\{Y\} = \int_{-\infty}^{+\infty} y \cdot p(y) \, dy \tag{1.14}$$

and for Eq. (1.12) and Eq. (1.13) the results are

$$E\{Y^n\} = \int_{-\infty}^{+\infty} y^n p(y) \, dy \ , \tag{1.15}$$

$$E\{(Y - E\{Y\})^n\} = \int_{-\infty}^{+\infty} (y - E\{Y\})^n \cdot p(y) \, dy \ . \tag{1.16}$$

Moreover, this book will essentially deal with **ergodic** processes. An ergodic process is a stationary process where the time average is equal to the ensemble average. Then, in fact, it will generally be possible to observe noise processes with only one measurement curve (e.g. the curve of R_1 in Fig. 1.2). In the following, the main interest will be focused on the time average. Every ergodic process is stationary, whereas the reversal is not necessarily true. The assumption that the time and the ensemble average are equal is a useful means to perform calculations, because it is often possible to predict the characteristics of densities in advance. By denoting the time average with a bar, Eq. (1.11) can be rewritten as follows for an ergodic and thus stationary process:

$$\overline{y(t)} = \lim_{T \to \infty} \frac{1}{2 \cdot T} \cdot \int_{-T}^{T} y(t)\, dt = \mathrm{E}\{Y\} = \int_{-\infty}^{+\infty} y \cdot p(y)\, dy \ . \tag{1.17}$$

In Eq. (1.17) the variable t describes the time dependence of the stochastic amplitude $y(t)$. For the quadratic time average the result is

$$\overline{y^2(t)} = \lim_{T \to \infty} \frac{1}{2 \cdot T} \cdot \int_{-T}^{T} y^2(t)\, dt = \mathrm{E}\{Y^2\} = \int_{-\infty}^{+\infty} y^2 \cdot p(y)\, dy \tag{1.18}$$

and the higher moments

$$\overline{y^n(t)} = \mathrm{E}\{Y^n\} = \int_{-\infty}^{+\infty} y^n \cdot p(y)\, dy \ . \tag{1.19}$$

We get for the central moments of ergodic processes:

$$\overline{(y(t) - \overline{y(t)})^n} = \mathrm{E}\{(Y - \mathrm{E}\{Y\})^n\} \ . \tag{1.20}$$

In order to emphasize it again: this book will deal nearly solely with continuous, stationary and ergodic noise processes. The quadratic time average will thus be of quite predominant interest. However, the relations above allow one to calculate time averages also by means of probability densities which might sometimes be advantageous.

1.2.4   Auto- and cross-correlation function

The autocorrelation function $\rho(t_1, t_2)$ defines the averaged product of the amplitude values $y(t_1)$ at a time t_1 and $y(t_2)$ at a time t_2. For a stationary process this mean value only depends on the time difference $\theta = t_2 - t_1$. Considering the compound probability density with $t_1 = t$ and $t_2 = t + \theta$ one can write:

$$\rho(\theta) = \overline{y(t) \cdot y(t+\theta)} = \lim_{T \to \infty} \frac{1}{2 \cdot T} \cdot \int_{-T}^{T} y(t) \cdot y(t+\theta)\, dt$$

$$= \mathrm{E}\{Y(t_1) \cdot Y(t_2)\} = \int\!\!\int_{-\infty}^{+\infty} y_1 \cdot y_2 \cdot p(t_1, y_1; t_2, y_2)\, dy_1 dy_2$$

$$= \int\!\!\int_{-\infty}^{+\infty} y_1 y_2\, p(y_1, y_2, \theta)\, dy_1 dy_2 \ . \tag{1.21}$$

For the last expression on the right side it was again made use of the fact that, for a stationary process, the compound probability density only depends on the time difference. The determination of the autocorrelation function with the help of a probability density will be a useful calculation method.

The autocorrelation function is always an even function of θ for stationary processes. This is obvious, if the substitution $t + \theta = \tau$ is introduced:

$$\rho(\theta) = \overline{y(t)y(t+\theta)} = \overline{y(\tau)y(\tau-\theta)} = \rho(-\theta) \ . \tag{1.22}$$

For $\theta = 0$ the autocorrelation function is identical to the mean square value.

The autocorrelation function is a measure of how strongly the value of the random variable at the time t is influenced by the values it had before. In other words, a large value of $\rho(\theta)$ means that, with the knowledge of $y(t)$, the values $y(t \pm \theta)$ can be calculated with a higher probability than for a small or even vanishing correlation.

If the averaging in Eq. (1.21) is not performed with two values of the same process but with two values of different processes X and Y, then the cross-correlation function ρ_{xy} of these processes is obtained. For stationary processes it can be written, similar to Eq. (1.21):

$$\rho_{xy}(\theta) = \overline{x(t)y(t+\theta)} = \lim_{T \to \infty} \frac{1}{2 \cdot T} \int_{-T}^{T} x(t)y(t+\theta)\, dt$$

$$= \mathrm{E}\{X(t_1)Y(t_2)\} = \int\!\!\int_{-\infty}^{+\infty} x_1 y_2\, p(t_1, x_1; t_2, y_2)\, dx_1 dy_2$$

$$= \int\!\!\int_{-\infty}^{+\infty} x_1 y_2\, p(x_1, y_2, \theta)\, dx_1 dy_2 \ . \tag{1.23}$$

In contrast to the autocorrelation function, the cross-correlation function is, in general, not an even function.

The cross-correlation function of two signals X and Y describes how similar both signals are. The maximum magnitude is obtained, if both signals are identical except for a common factor, and if, for example, they are originating from the same noise source. Then the signals are completely correlated. On the other hand, the cross-correlation function is always equal to zero, if the signals originate from two physically completely independent sources, such as, for example, from two separate resistors with thermal noise. In this case, the signals are uncorrelated.

1.2.5   Description of noise signals in the frequency domain

In the time domain noise signals are described by the auto- or cross-correlation functions. The auto correlation function for the time shift $\theta = 0$ is a measure for the quadratic time average and thus the signal power. In the frequency domain the description of the noise signals is performed with the help of the **power spectral density** or the **power spectrum** or briefly the **spectrum** W(f) for the frequency f. Here, W(f)df is the contribution of the frequency interval df at the frequency f to the mean square value or the signal power. The following relation thus applies:

$$\overline{y^2(t)} = \int_0^\infty W(f)\, df \ .$$
(1.24)

Apart from the so-called **one-sided spectrum** W(f), which only exists for positive frequencies, the **two-sided spectrum** $W(f)$ is defined for positive as well as for negative frequencies as the Fourier transform of the autocorrelation function. Conversely, the autocorrelation function $\rho(\theta)$ is the inverse Fourier transform of the spectrum $W(f)$.

$$W(f) = \int_{-\infty}^{+\infty} \rho(\theta) e^{-j2\pi f\theta}\, d\theta$$

$$\rho(\theta) = \int_{-\infty}^{+\infty} W(f) e^{+j2\pi f\theta}\, df \ .$$
(1.25)

The autocorrelation function and the spectrum form a pair of Fourier transforms. These are the so-called Wiener-Khintchine-relations. The derivation is omitted here. These relations will be used extensively in the following chapters. Because $\rho(\theta)$ is a real and even function in θ, it follows that the two-sided spectrum $W(f)$ is also a real and even function in f. The proof is easy. According to Eq. (1.25), the spectrum can be written as

$$W(f) = \int_{-\infty}^{+\infty} \rho(\theta) e^{-j2\pi f\theta}\, d\theta = \int_{-\infty}^{+\infty} \rho(-\theta) e^{-j2\pi f\theta}\, d\theta \ ,$$
(1.26)

because $\rho(\theta) = \rho(-\theta)$ is an even function. The substitution $\theta = -\tau$ and an interchange of the integral boundaries leads to:

$$W(f) = \int_{-\infty}^{+\infty} \rho(\tau)e^{j2\pi f\tau}\, d\tau = W^*(f) = W(-f) \ . \qquad (1.27)$$

Thus $W(f)$ is a real and even function in f.

The two-sided spectrum is often more convenient to use because of the symmetry of the transformations to and from the autocorrelation function. However, only the one-sided spectrum $\mathsf{W}(f)$, which is defined for positive frequencies only, has a physical relevance. The one-sided and the two-sided spectra, $\mathsf{W}(f)$ and $W(f)$, are related in a simple way:

$$\mathsf{W}(f) = W(f) + W(-f) = 2 \cdot W(f) \ . \qquad (1.28)$$

The difference in the factor 2 results from the different definitions of the integral boundaries. For example, the mean square value is given by

$$\overline{u^2(t)} = \rho(\theta = 0) = \int_{-\infty}^{+\infty} W(f)\, df = 2 \cdot \int_{0}^{\infty} W(f)\, df = \int_{0}^{\infty} \mathsf{W}(f)\, df \ . \qquad (1.29)$$

By analogy with Eq. (1.25), the cross-spectrum W_{12} and the cross-correlation function $\rho_{12}(\theta)$ also form a pair of Fourier transforms:

$$W_{12}(f) \;=\; \int_{-\infty}^{+\infty} \rho_{12}(\theta)e^{-j2\pi f\theta}\, d\theta$$

$$\rho_{12}(\theta) \;=\; \int_{-\infty}^{+\infty} W_{12}(f)e^{+j2\pi f\theta}\, df \ . \qquad (1.30)$$

In general, the cross-correlation function is not an even function, but it is a real valued function. Hence,

$$W_{12}(f) = \int_{-\infty}^{+\infty} \rho_{12}(\theta)e^{-j2\pi f\theta}\, d\theta = W_{12}^*(-f) \ . \qquad (1.31)$$

Thus, the real part of the cross-spectrum is an even function in f and the imaginary part is an odd function in f. Furthermore, using the definition (1.30), it can be shown that

$$\rho_{12}(\theta) = \rho_{21}(-\theta) \ . \qquad (1.32)$$

With the substitution $\tau = -\theta$, an interchange of the integral boundaries leads to

$$
\begin{aligned}
W_{12}(f) &= \int_{-\infty}^{+\infty} \rho_{12}(\theta) e^{-j2\pi f\theta}\, d\theta \\
&= \int_{-\infty}^{+\infty} \rho_{21}(-\theta) e^{-j2\pi f\theta}\, d\theta \\
&= \int_{-\infty}^{+\infty} \rho_{21}(\tau) e^{j2\pi f\tau}\, d\tau = W_{21}^*(f) \ .
\end{aligned}
\tag{1.33}
$$

Furthermore, a comparison of Eq. (1.31) and Eq. (1.33) yields

$$
W_{21}(f) = W_{21}^*(-f) \ . \tag{1.34}
$$

Note that, in general, the cross-spectrum is a complex valued function.

1.2.6 Characteristic function and the central limit theorem

The inverse Fourier transform of the probability density function $p(x)$ is called the characteristic function $C(u)$. This is equivalent to the statement that the characteristic function is the expected value of the function e^{jux}:

$$
C(u) = E(e^{jux}) = \int_{-\infty}^{+\infty} e^{jux} p(x)\, dx = F^{-1}(p(x)) \ . \tag{1.35}
$$

The characteristic function always exists, because $|e^{jux}| = 1$ and $p(x) \geq 0$ and real. Thus, it is given by

$$
|C(u)| \leq \int_{-\infty}^{+\infty} p(x)\, dx = C(0) = 1 \ . \tag{1.36}
$$

Since $p(x)$ is real, the complex conjugate value of C is

$$
C^*(u) = C(-u) \ . \tag{1.37}
$$

This means that $\mathrm{Re}\{C\}$ is an even and $\mathrm{Im}\{C\}$ is an odd function of u. If $p(x)$ is an even function, then $C(u)$ is also an even and thus a real function. If the characteristic function is known, then the corresponding probability density can be calculated by a Fourier transformation:

$$
p(x) = \frac{1}{2\pi} \int_{-\infty}^{+\infty} e^{-jux} \cdot C(u)\, du \ . \tag{1.38}
$$

The characteristic function and the probability density form a pair of Fourier transforms.

As an important example as well as an application of the characteristic function, the probability density of the sum of two **independent** random variables X and Y will be determined. The probability density $p_s(s)$ for the sum $s = x + y$ of the two variables with the associated probability densities $p_1(x)$ and $p_2(y)$ is given by:

$$
\begin{aligned}
p_s(s) &= \int_{-\infty}^{+\infty} p_1(x) \cdot p_2(s-x)\,dx \\
&= \int_{-\infty}^{+\infty} p_2(y) \cdot p_1(s-y)\,dy \; .
\end{aligned}
\tag{1.39}
$$

The integral in Eq. (1.39) is a convolution integral. The appropriate extension to more than two independent random variables leads to a multiple convolution integral for $p_s(s)$

$$
p_s(s) = p_1 \star p_2 \star p_3 \star p_4 \ldots \; .
\tag{1.40}
$$

The order of the convolution can be chosen arbitrarily. According to a theorem of Fourier transformation, the characteristic function of $p_s(s)$, namely, $C_s(u)$, is obtained as the product of the characteristic functions C_i of the single probability densities p_i, $i = 1, 2, 3 \ldots$:

$$
C_s(u) = \prod_{i=1}^{n} C_i(u)
\tag{1.41}
$$

The central limit theorem of statistics states that, under quite general conditions, the sum of a large number of statistically independent random variables shows a Gaussian distribution. This result is independent of the distributions of the single variables. In problem 1.2 it shall be shown with the help of the convolution theorem that, even for the sum of only three variables with rectangular distributions, a distribution quite similar to the Gaussian distribution results.

Problem

1.2   Calculate the distribution $p_s(s) = p_s(x + y + z)$ for three rectangular distributions $p_1(x), p_2(y), p_3(z)$, as shown in the figure.

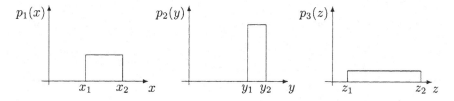

The probability densities are assumed to be independent of each other.
Numerical example: $x_1 = 2; x_2 = 4; y_1 = 3; y_2 = 4; z_1 = 1; z_2 = 5$

For a very large number n of equal and independent distribution densities with a rectangular shape

$$p(x) = \begin{cases} \dfrac{1}{\beta} & \text{for} \quad -\dfrac{\beta}{2} \leq x \leq \dfrac{\beta}{2} \\[3mm] 0 & \text{for} \quad |x| > \dfrac{\beta}{2} \end{cases} \tag{1.42}$$

it can be shown analytically by using the characteristic function $C_s(u)$, that the probability density of the sum $p_s(s)$ adopts a Gaussian distribution. The characteristic function of the single random variable $C_1(u)$ is calculated as

$$C_1(u) = \frac{1}{\beta} \int_{-\beta/2}^{\beta/2} e^{jux} \, dx = \frac{\sin \dfrac{\beta u}{2}}{\dfrac{\beta u}{2}} \tag{1.43}$$

and $C_s(u)$ with Eq. (1.41) as

$$C_s(u) = C_1^n(u) \ . \tag{1.44}$$

An inverse Fourier transformation yields the probability density $p_s(s)$ for the sum of n random variables with a rectangular distribution:

$$p_s(s) = \frac{1}{2\pi} \int_{-\infty}^{+\infty} \left(\frac{\sin(\beta u)/2}{(\beta u)/2} \right)^n e^{-jus} \, du \ . \tag{1.45}$$

For large numbers of n the function $\mathrm{si}^n(\beta u/2)$ is different from zero only in a small region of $u = 0$. Therefore, a series expansion for $\mathrm{si}^n(\beta u/2)$ may be truncated beyond the first two elements:

$$\text{si}^n \left(\frac{\beta u}{2} \right) \simeq \left(1 - \frac{(\beta u)^2}{24} \right)^n . \tag{1.46}$$

However, for large n the last term is identical to a Gaussian function, as can be shown by comparing the coefficients of the associated series expansions.

$$\left(1 - \frac{(\beta u)^2}{24} \right)^n = \exp \left(-\frac{n}{24} (\beta u)^2 \right) \quad \text{for large } n . \tag{1.47}$$

Thus, for $p_s(s)$ the following result is obtained by employing an integral table:

$$
\begin{aligned}
p_s(s) &= \frac{1}{2\pi} \int_{-\infty}^{+\infty} \exp \left(-\frac{n}{24} (\beta u)^2 \right) \cdot \exp(-jus) \, du \\
&= \frac{\exp \left(\frac{-s^2}{2\sigma_n^2} \right)}{\sqrt{2\pi \cdot \sigma_n^2}} \quad \text{with } \sigma_n = \frac{\beta \cdot \sqrt{n}}{\sqrt{12}} .
\end{aligned}
\tag{1.48}
$$

This is a Gaussian distribution with the standard deviation σ_n. The standard deviation σ_1 for a single random variable with a rectangular distribution according to Eq. (1.42) is given by:

$$\sigma_1 = \frac{\beta}{\sqrt{12}} . \tag{1.49}$$

This leads to:

$$\sigma_n = \sqrt{n} \cdot \sigma_1 . \tag{1.50}$$

It will be shown next, that the result of Eq. (1.50) could have been expected due to general relations which hold for the sum of independent variables. For this purpose, the variance σ_s^2 of $S = X + Y$ is considered. The expected value of S, $E\{S\} = \bar{s}$, is equal to the sum of the expected values of X and Y, i.e. $\bar{s} = \bar{x} + \bar{y}$. This leads to:

$$
\begin{aligned}
\sigma_s^2 &= \int\!\!\int_{-\infty}^{+\infty} (s - \bar{s})^2 \cdot p_1(x) \cdot p_2(y) \, dx \, dy \\
&= \int\!\!\int_{-\infty}^{+\infty} [(x - \bar{x}) + (y - \bar{y})]^2 \cdot p_1(x) \cdot p_2(y) \, dx \, dy \\
&= \int_{-\infty}^{+\infty} (x - \bar{x})^2 p_1(x) \, dx + \int_{-\infty}^{+\infty} (y - \bar{y})^2 p_2(y) \, dy
\end{aligned}
$$

$$+ \quad 2 \cdot \underbrace{\int_{-\infty}^{+\infty} (x - \bar{x}) p_1(x)\, dx}_{= 0} \cdot \underbrace{\int_{-\infty}^{+\infty} (y - \bar{y}) p_2(y)\, dy}_{= 0} \quad . \tag{1.51}$$

Thus it follows:

$$\sigma_s^2 = \sigma_x^2 + \sigma_y^2 \quad . \tag{1.52}$$

Obviously, the result of Eq. (1.51) can be extended to n variables:

$$\sigma_s^2 = \sum_{i=1}^{n} \sigma_i^2 \quad . \tag{1.53}$$

The variance of the random variable S, which results from the sum of the independent random variables X and Y, is equal to the sum of the variances of the individual random variables. Also, the mean value of the sum is equal to the sum of the mean values:

$$\bar{s} = \sum_{i=1}^{n} \bar{x}_i \quad . \tag{1.54}$$

The only requirement the single random variables must meet is that they are statistically independent from each other. However, the result does not depend on the individual distribution densities. Eq. (1.50) resulted from the summation of n equal distributions with the standard deviation σ_1. Hence, the standard deviation σ_n of the sum of the variables is larger by a factor $\sqrt{n}$. The equations (1.52) and (1.53) are also valid for differences of random variables. For independent random variables with a Gaussian distribution the validity of Eq. (1.52) can be shown by direct calculation with the help of the characteristic function (problem 1.3).

Problem

1.3   The independent random variables possess a Gaussian distribution. Show by direct calculation utilizing the characteristic function, that the variance of the sum of the random variables is obtained from the sum of the individual variances.

If the characteristic function is known, then the probability density can be calculated by a Fourier transform (Eq. (1.35)). Differentiating Eq. (1.35) once leads to

$$\frac{dC(u)}{du} = j \int_{-\infty}^{+\infty} e^{jux} \cdot x \cdot p(x)\, dx \quad . \tag{1.55}$$

From this it follows for $u = 0$

$$\frac{1}{j} \cdot \frac{dC(u)}{du}\bigg|_{u=0} = \int_{-\infty}^{+\infty} x \cdot p(x)\, dx = \mathrm{E}\{X\} \ . \tag{1.56}$$

Repeating the differentiation process leads to

$$\frac{1}{j^n} \cdot \frac{d^n C(u)}{du^n}\bigg|_{u=0} = \mathrm{E}\{X^n\} \ . \tag{1.57}$$

For a known characteristic function, the moments can thus be determined by differentiation. Quite often, the result proves to be very useful.

The definition of the characteristic function as the inverse Fourier transform of the density function can be extended to more variables. The characteristic function for two variables is given by a double Fourier integral:

$$C(u, v) = \int\!\!\!\int_{-\infty}^{+\infty} e^{jux + jvy} p(x, y)\, dx\, dy \ . \tag{1.58}$$

Here, the bivariate probability density of the random variables X and Y is denoted by $p(x, y)$. By inversion the bivariate probability density results from Eq. (1.58)

$$p(x, y) = \frac{1}{(2\pi)^2} \int\!\!\!\int_{-\infty}^{+\infty} e^{-jux - jvy} \cdot C(u, v)\, du\, dv \ . \tag{1.59}$$

Calculating the kth and lth derivative of Eq. (1.58) with respect to u and v, the following relation for the mixed moments results:

$$\mathrm{E}\{X^k \cdot Y^l\} = \frac{1}{j^{k+l}} \cdot \frac{\partial^{k+l} C(u, v)}{\partial u^k \cdot \partial v^l}\bigg|_{u=v=0} \ . \tag{1.60}$$

Furthermore, Eq. (1.58) yields:

$$\begin{aligned}
C(u, 0) &= C(u) \\
C(0, v) &= C(v)
\end{aligned}$$

and

$$|C(u, v)| \le C(0, 0) = 1 \ . \tag{1.61}$$

If the random variables X and Y are statistically independent, then $p(x, y) = p(x) \cdot p(y)$ and

$$C(u, v) = C(u) \cdot C(v) \ . \tag{1.62}$$

1.2.7   Interrelationship between moments of different orders

In general, moments of different orders cannot be converted into one another. However, if the random variables $X(t_i)$ of a process are normally distributed for all times t_i and, if they possess a normally distributed multivariate density, then the higher order moments can be calculated from those of the first and second order. Processes of this type are called Gaussian processes. In the following, two random variables $X_1(t_1)$ and $X_2(t_2)$ of different ergodic Gaussian processes with normally distributed densities $p_1(x_1)$ at a time t_1 and $p_2(x_2)$ at a time t_2 are considered. The expected values are assumed to be zero. As will be shown in problem 1.4, the corresponding characteristic function is given by

$$
\begin{aligned}
C(u_1, u_2) &= \int\!\!\!\int_{-\infty}^{+\infty} e^{ju_1x_1 + ju_2x_2} p(x_1, x_2)\, dx_1 dx_2 \\
&= \exp\left(-\frac{1}{2}\left(\rho_{11} \cdot u_1^2 + \rho_{22} \cdot u_2^2 + 2\rho_{12} \cdot u_1 \cdot u_2\right)\right) . \quad (1.63)
\end{aligned}
$$

Here, $p(x_1, x_2)$ is the Gaussian distributed bivariate density.

Problem

1.4   Derive equation Eq. (1.63).

Using Eq. (1.60) it can be shown that ρ_{11} is the variance of $X_1(t_1)$, ρ_{22} is the variance of $X_2(t_2)$, and ρ_{12} is the so-called covariance of $X_1(t_1)$ and $X_2(t_2)$. When applied to a single process with $X_1 = X_2 = X$ for $t_1 = t_2$, then $\rho_{12} = \rho_{12}(\theta)$ is the autocorrelation function depending on $\theta = t_2 - t_1$ and $\rho_{11} = \rho_{22} = \rho_{12}(\theta = 0) = \rho(0)$ is the autocorrelation function at $\theta = 0$. Equation (1.60) is useful to derive the following relation between the moment of 4th order $E\{X^2(t) \cdot X^2(t+\theta)\}$ and the moment of 2nd order (problem 1.5):

$$
E\{X^2(t) \cdot X^2(t + \theta)\} = \rho^2(0) + 2\rho^2(\theta) . \quad (1.64)
$$

Problem

1.5   Derive equation (1.64) using the equations (1.60) and (1.63).

Equation (1.64) will be required later to calculate the standard deviation of noise power measurement results for a finite measurement time and a finite bandwidth.

1.3   TRANSFER OF NOISE SIGNALS BY LINEAR NETWORKS

1.3.1   Impulse response and transfer function

A linear two-port is considered with an impulse response function $h(t)$.

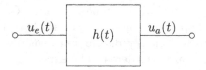

Fig. 1.4   Two-port network, characterized by its impulse response function.

The voltage at the output $u_a(t)$ is related to the voltage at the input $u_e(t)$ by a convolution integral.

$$u_a(t) = \int\limits_{-\infty}^{+\infty} h(t')u_e(t-t')\,dt' \tag{1.65}$$

For physical reasons, a property of the weighting function $h(t)$ is:

$$h(t) = 0 \quad \text{for} \quad t < 0 \tag{1.66}$$

because an effect cannot occur prior to its cause. This means that the system is causal. Choosing a sinusoidal input voltage in complex form with U_e, U_a as complex phasors

$$u_e(t) = \mathrm{Re}\left\{|U_e|e^{j\phi}\cdot e^{j\omega t}\right\} = \mathrm{Re}\left\{U_e\cdot e^{j\omega t}\right\} \ , \tag{1.67}$$

then the output signal has a sinusoidal form

$$u_a(t) = \mathrm{Re}\left\{U_a\cdot e^{j\omega t}\right\} \tag{1.68}$$

with

$$
\begin{aligned}
U_a\cdot e^{j\omega t} &= \int\limits_{-\infty}^{+\infty} h(t')\cdot U_e\cdot e^{j\omega t}e^{-j\omega t'}\,dt' \\
&= U_e\cdot e^{j\omega t}\cdot \int\limits_{-\infty}^{+\infty} h(t')e^{-j\omega t'}\,dt' \ .
\end{aligned}
\tag{1.69}
$$

With the definition

$$\int\limits_{-\infty}^{+\infty} h(t')e^{-j\omega t'}\,dt' = V(\omega) \ , \tag{1.70}$$

where $V(\omega)$ is the complex voltage amplification or complex transfer function of the two-port, the following complex notation can be applied

$$U_a = V(\omega) \cdot U_e \ . \tag{1.71}$$

Obviously, $h(t)$ and $V(\omega)$ or $V(f)$, respectively, are a pair of Fourier transforms

$$V(f) = \int_{-\infty}^{+\infty} h(t)e^{-j2\pi ft} \, dt \ ,$$

$$h(t) = \int_{-\infty}^{+\infty} V(f)e^{+j2\pi ft} \, df \ . \tag{1.72}$$

The function $h(t)$ is used for calculations in the time domain, whereas $V(f)$ is preferred for calculations in the frequency domain. Since $h(t)$ is real, we have

$$V^*(f) = \int_{-\infty}^{+\infty} h(t) \cdot e^{j2\pi ft} \, dt = V(-f) \tag{1.73}$$

or

$$V^*(-f) = V(f) \ . \tag{1.74}$$

Thus, the real part of $V(f)$ is an even function of the frequency, whereas the imaginary part is an odd function.

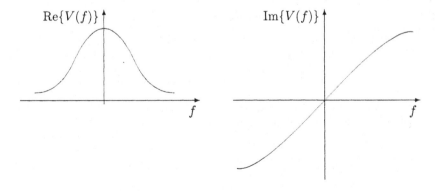

Fig. 1.5   Real and imaginary part of a complex transfer function.

1.3.2   Transformation of the autocorrelation function and the power spectrum

As already discussed in the previous section, it will be assumed in the following, that the noise processes under investigation are stationary. This means that there is no time dependence of mean values. Furthermore, it will be assumed that the noise processes are ergodic, which means that a mean value at a fixed time over a large ensemble of similar noise processes leads to the same result as the time average of a single noise process.

The autocorrelation function $\rho_e(\theta)$ of the input voltage $u_e(t)$ is thus given by

$$\rho_e(\theta) \;\; = \;\; \overline{u_e(t) \cdot u_e(t + \theta)} = \mathrm{E}\{u_e(t) \cdot u_e(t + \theta)\}$$

$$= \;\; \lim_{T \to \infty} \frac{1}{2T} \int\limits_{-T}^{T} u_e(t) \cdot u_e(t + \theta)\, dt \; . \tag{1.75}$$

A corresponding definition holds for ρ_a, the autocorrelation function, at the output. The two autocorrelation functions ρ_a and ρ_e are related by a double convolution integral as will be shown in the following. First, we have

$$u_a(t) \cdot u_a(t + \theta)$$

$$= \int\limits_{-\infty}^{+\infty} h(t') \cdot u_e(t - t')\, dt' \cdot \int\limits_{-\infty}^{+\infty} h(t'') \cdot u_e(t + \theta - t'')\, dt''$$

$$= \int\limits_{-\infty}^{+\infty}\!\!\!\int h(t')h(t'') \cdot u_e(t - t')u_e(t - t'' + \theta)\, dt'dt'' \; . \tag{1.76}$$

Now the mean value with respect to t is formed on both sides, making use of the fact that the order of integration and the calculation of the mean value can be exchanged. The substitution $\tau = t - t'$ yields

$$\overline{u_e(t - t') \cdot u_e(t - t'' + \theta)}$$
$$= \overline{u_e(\tau) \cdot u_e(\tau + t' - t'' + \theta)} = \rho_e(\theta + t' - t'') \; . \tag{1.77}$$

In this way, a relation between ρ_a and ρ_e in the form of a double convolution integral finally results, which will be used later:

$$\rho_a(\theta) = \int\limits_{-\infty}^{+\infty}\!\!\!\int h(t')h(t'') \cdot \rho_e(\theta + t' - t'')\, dt'dt'' \; . \tag{1.78}$$

ρ_e as well as ρ_a are even functions of θ.

On the basis of the transformation of the autocorrelation function between the input and the output of the two-port (Eq. (1.78)), the transformation of the power spectrum can also be calculated, i.e. the relation between the power spectrum W_e at the input and W_a at the output. W_a is the power spectrum corresponding to the autocorrelation function ρ_a. The relation (1.78) leads to

$$
\begin{aligned}
W_a &= \int_{-\infty}^{+\infty} \rho_a(\theta) e^{-j2\pi f\theta}\, d\theta \\[2mm]
&= \iiint_{-\infty}^{+\infty} h(t')h(t'') \cdot \rho_e(\theta + t' - t'') e^{-j2\pi f\theta}\, d\theta\, dt'\, dt'' \\[2mm]
&= \iiint_{-\infty}^{+\infty} h(t')h(t'') \cdot \rho_e(\theta + t' - t'') \\[2mm]
&\qquad \cdot e^{-j2\pi f(\theta + t' - t'')} \cdot e^{j2\pi ft'} e^{-j2\pi ft''}\, d\theta\, dt'\, dt'' \quad . \quad (1.79)
\end{aligned}
$$

The last step is just an extension of the previous expression. Next, the order of integration is changed. First, an integration over $\tau = \theta + t' - t''$ is performed, where $t' - t''$ is kept constant. The result is

$$
\begin{aligned}
W_a(f) &= W_e(f) \left\{ \int_{-\infty}^{+\infty} h(t') e^{+j2\pi ft'}\, dt' \right\} \left\{ \int_{-\infty}^{+\infty} h(t'') e^{-j2\pi ft''}\, dt'' \right\} \\[2mm]
&= W_e(f) \cdot V^*(f) \cdot V(f) \\[2mm]
&= |V(f)|^2 \cdot W_e(f) \qquad\qquad\qquad\qquad\qquad (1.80)
\end{aligned}
$$

$V(f)$ is the complex transfer function of the two-port, i.e. the Fourier transform of the impulse response $h(t)$. Thus, the power spectrum is shaped according to the magnitude squared value of the corresponding transfer function.

1.3.3 Correlation between input and output noise signals

A possible correlation between the input and the output noise signals of a two-port network is described by the cross-correlation function $\rho_{ea}(\theta)$:

$$
\begin{aligned}
\rho_{ea}(\theta) &= \overline{u_e(t) \cdot u_a(t + \theta)} \\[2mm]
&= \mathrm{E}\left\{ \int_{-\infty}^{+\infty} h(t') \cdot u_e(t) \cdot u_e(t + \theta - t')dt' \right\} \\[2mm]
&= \int_{-\infty}^{+\infty} h(t') \cdot \rho_e(\theta - t')dt' \quad . \qquad\qquad (1.81)
\end{aligned}
$$

Again, the order of the calculation of the mean value and the integral have been changed. Generally, the cross-correlation function is not an even function. By a Fourier transformation of Eq. (1.81) to the frequency domain the corresponding cross-spectrum $W_{ea}(f)$ is obtained:

$$
\begin{aligned}
W_{ea}(f) &= \int_{-\infty}^{+\infty} \rho_{ea}(\theta) e^{-j2\pi f\theta} d\theta \\
&= \int\!\!\!\int_{-\infty}^{+\infty} h(t')\rho_e(\theta - t') e^{-j2\pi f(\theta - t')} e^{-j2\pi ft'} \, d\theta \, dt' \\
&= W_e \cdot V(f) \ .
\end{aligned}
\tag{1.82}
$$

As before, use was made of an extension of the equation and an interchange of the order of integration.

In contrast to W_e and W_a the cross-spectrum W_{ea} is complex in general. Due to $W_a = |V|^2 \cdot W_e$ we also have

$$
W_{ea}(f) = \frac{W_a}{|V|^2} \cdot V = \frac{W_a}{V^*} \ .
\tag{1.83}
$$

Furthermore, it can be shown that

$$
W_{ae} = W_{ea}^* = W_e \cdot V^* = \frac{W_a}{V} \ ,
\tag{1.84}
$$

because

$$
\begin{aligned}
\rho_{ae} &= \overline{u_a(t) \cdot u_e(t + \theta)} \\
&= \mathrm{E}\left\{ \int_{-\infty}^{+\infty} h(t') \cdot u_e(t - t') \cdot u_e(t + \theta) \, dt' \right\} \\
&= \int_{-\infty}^{+\infty} h(t') \cdot \rho_e(\theta + t') \, dt'
\end{aligned}
\tag{1.85}
$$

and furthermore

$$
W_{ae} = \int\!\!\!\int_{-\infty}^{+\infty} h(t')\rho_e(\theta + t') \cdot e^{-j2\pi f(\theta + t')} e^{+j2\pi ft'} \, d\theta \, dt' \ .
\tag{1.86}
$$

Thus $W_{ae} = W_e \cdot V^* = W_{ea}^*$, which proves Eq. (1.84). The normalized cross-spectrum of the input and output signals of a two-port is defined by

$$
k_{ea} = \frac{W_{ea}}{\sqrt{W_e \cdot W_a}} = \frac{W_e \cdot V}{W_e \cdot |V|} = \frac{V}{|V|} \ .
\tag{1.87}
$$

The magnitude of the normalized cross-spectrum is equal to 1. This means that the input and output signals are completely correlated. This is not surprising since both signals stem from the same origin.

1.3.4 Superposition of partly correlated noise signals

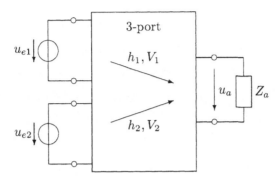

Fig. 1.6 Superposition of noise signals.

By means of a three-port (Fig. 1.6) two noise voltages, u_{e1} and u_{e2}, which are partly correlated, are superimposed at the load impedance Z_a. The auto-correlation function of the output voltage $u_a(t)$ will be calculated.

$$\overline{u_a(t) \cdot u_a(t + \theta)}$$

$$= \mathrm{E}\left\{ \int\limits_{-\infty}^{+\infty}\!\!\!\int [h_1(t') \cdot u_{e1}(t - t') + h_2(t')u_{e2}(t - t')]\, dt' \right.$$

$$\left. \cdot [h_1(t'')u_{e1}(t + \theta - t'') + h_2(t'') \cdot u_{e2}(t + \theta - t'')]\, dt'' \right\} \quad . \quad (1.88)$$

This equation consists of four parts:

$$\overline{u_a(t) \cdot u_a(t + \theta)} = \rho_a(\theta)$$

$$= \int\limits_{-\infty}^{+\infty}\!\!\!\int h_1(t') \cdot h_1(t'') \cdot \rho_{e1}(\theta + t' - t'')\, dt'\, dt''$$

$$+ \int\limits_{-\infty}^{+\infty}\!\!\!\int h_2(t') \cdot h_2(t'') \cdot \rho_{e2}(\theta + t' - t'')\, dt'\, dt''$$

$$+ \int\limits_{-\infty}^{+\infty}\!\!\!\int h_1(t') \cdot h_2(t'') \cdot \rho_{e1e2}(\theta + t' - t'')\, dt'\, dt''$$

$$+ \int\limits_{-\infty}^{+\infty}\!\!\!\int h_1(t'') \cdot h_2(t') \cdot \rho_{e2e1}(\theta + t' - t'')\, dt'\, dt'' \quad . \quad (1.89)$$

Here, the first two parts describe the autocorrelation function and the last two parts represent the cross-correlation function. Finally, the Fourier transform of this expression is calculated and the order of integrations is changed. This leads to

$$
\begin{aligned}
W_a &= \int_{-\infty}^{+\infty} \rho_a(\theta) e^{-j2\pi f\theta}\, d\theta \\
&= |V_1|^2 \cdot W_{e1} + |V_2|^2 \cdot W_{e2} \\
&\quad + V_1^* \cdot V_2 \cdot W_{e1e2} + V_2^* \cdot V_1 \cdot W_{e2e1} \ .
\end{aligned}
\tag{1.90}
$$

Now, the noise voltages will be replaced by sinusoidal signals of the same frequency. According to the symbolic complex phasor notation the output voltage is

$$
U_a = V_1 \cdot U_{e1} + V_2 \cdot U_{e2}
\tag{1.91}
$$

or

$$
\begin{aligned}
|U_a|^2 &= |V_1|^2 \cdot |U_{e1}|^2 + |V_2|^2 \cdot |U_{e2}|^2 \\
&\quad + V_1^* \cdot V_2 \cdot U_{e1}^* \cdot U_{e2} + V_1 \cdot V_2^* \cdot U_{e1} \cdot U_{e2}^* \ .
\end{aligned}
\tag{1.92}
$$

A comparison with Eq. (1.90) shows that there is a simple correspondence between the calculation with power and cross-spectra and the calculation with complex phasors. One simply has to replace $|U_e|^2$ by W_e, $|U_a|^2$ by W_a and $U_{e1}^* U_{e2}$ by W_{e1e2}. However, it should be mentioned that, generally, the cross-spectrum cannot be calculated from $U_{e1}^* U_{e2}$. This equivalence between spectra and complex phasors establishes a method to perform calculations with noise signals as comfortable as with sinusoidal signals. The main difference is that the correlation between the noise signals has to be taken into account.

The substitution of spectra by the product of complex phasors will frequently be used in the following. Two signals can be completely correlated, if, for example, one originates from the other. They can be completely uncorrelated or, which is the most general case, be partly correlated. If two signals are completely uncorrelated, then the powers or spectra can simply be added. It is not easy to determine the correlation between two noise signals. However, if the correlation is known, then, linear circuit calculations with noise signals are no more difficult than with sinusoidal signals.

Generally, the cross-correlation function $\rho_{12}(\theta)$ is not an even function, but certainly a real function. Because of Eq. (1.33), Eq. (1.90) can also be written in the following form:

$$
W_a = |V_1|^2 \cdot W_{e1} + |V_2|^2 \cdot W_{e2} + 2\mathrm{Re}\{V_1^* V_2 \cdot W_{e1e2}\} \ .
\tag{1.93}
$$

Thus, W_a always is a real valued function, a necessary condition for a power spectrum.

Problems

1.6   Two non-overlapping frequency bands at different frequencies are to be filtered out of a white noise signal. What is the correlation between these two noise signals?

1.7   Calculate the autocorrelation function of a band limited noise signal with a rectangularly shaped spectrum.

1.8   Consider the low-pass filter in the figure below. Calculate the autocorrelation function of the output noise signal, if the input signal is white noise, generated by the resistor R.

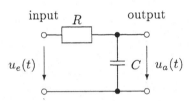

2

Noise of Linear One- and Two-Ports

Thermal noise is one of the most fundamental noise phenomena. It is present in nearly every electronic circuit. Therefore, analytical methods for the calculation of thermal noise and its effects in electronic circuits are of fundamental interest.

The noise behavior of thermally noisy electronic devices can be described with the help of equivalent circuits. Typically, the thermal noise of one- or two-ports is represented by equivalent sources, for example current- and voltage sources and by noiseless two-ports or noiseless impedance networks. However, if different representations are to be transformed into one another one has to take into account that in addition to the magnitudes it is also necessary to know the correlation between the different equivalent noise sources. For thermally noisy networks it will be seen that it is always possible to calculate the correlation of arbitrarily chosen sets of equivalent sources.

A representation by equivalent noise sources and noiseless networks can also be adopted for non-thermally noisy linear networks. For example, linear amplifiers can be described with the help of noise current and noise voltage sources at the input and output. For this noise model the amplifier is assumed to be noiseless while its gain and impedances remain unchanged.

Another important representation of amplifier noise is based on the noise factor or noise figure. The noise figure describes the deterioration of the signal-to-noise ratio when a signal passes through the amplifier. The noise figure depends on the source impedance of the generator. Thus it is possible to minimize the noise figure of a two-port circuit by transforming the source impedance with the help of a lossless and noiseless network. However, such a so-called noise matching is successful only if the loss of the transformation network is sufficiently low.

2.1 NOISE OF ONE-PORTS

2.1.1 Thermal noise of resistors

The noise of resistors is caused by the thermal movement of the electrons or holes in metals or semiconductors. This noise phenomenon is also called *Johnson noise* or *thermal noise*. Experiments yield the following expression for the time average of the magnitude squared of the short circuit current in a frequency bandwidth Δf:

$$\overline{i^2(t)} = \frac{4kT}{R} \cdot \Delta f = 4kTG \cdot \Delta f \ . \tag{2.1}$$

The resistance is denoted by R and the conductance by $G = 1/R$, respectively. T is the absolute temperature in K and k is Boltzmann constant with

$$k = 1.38 \cdot 10^{-23} \text{Ws/K} \ . \tag{2.2}$$

Similarly, it can be found by a voltage measurement of the open circuit noise voltage:

$$\overline{u^2(t)} = 4kTR \cdot \Delta f \ . \tag{2.3}$$

The spectral density function $\mathsf{W}(f)$ represents the mean square value of the voltage or current, respectively, in 1 Hz bandwidth. Thus, we have

$$\mathsf{W}_u(f) = 4kTR \ , \tag{2.4}$$

$$\mathsf{W}_i(f) = 4kTG \ . \tag{2.5}$$

For thermal noise the spectral density function does not depend on the frequency, if the frequency is not too high and if the temperature is not too low, as will be seen in the following. The time average of the squared voltage and the squared current can be calculated via $\mathsf{W}(f)$. In Eq. (2.6) f_2 is the upper and f_1 the lower frequency boundary.

$$\overline{u^2(t)} = \int_{f_1}^{f_2} \mathsf{W}_u(f) df$$

$$\overline{i^2(t)} = \int_{f_1}^{f_2} \mathsf{W}_i(f) df \ . \tag{2.6}$$

The **spectral density function** is also called the **spectral distribution** or **spectrum** or **power spectrum**.

(a)   (b)

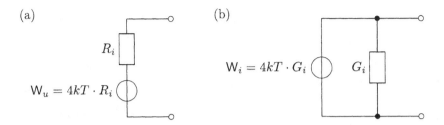

Fig. 2.1 Noise equivalent circuits of a thermally noisy resistor with (a) a voltage source and (b) a current source.

For a thermally noisy resistor the circuits shown in Fig. 2.1 are equivalent. In the equivalent circuits, the internal resistance R_i and the internal conductance G_i are noiseless. The voltage source is assumed to have zero internal resistance and the current source has infinite resistance.

2.1.2 Networks of resistors of identical temperature

If several resistors at the same temperature are combined, then an equivalent circuit can be defined for the resulting circuit. Whether the overall resistance is determined first and then an equivalent noise source is calculated or whether the equivalent noise source of all individual resistors are determined first and subsequently are combined, will lead to the same result. The same holds for a network of resistors. However, one necessary condition for this approach is that the noise sources are uncorrelated, i.e. that the root mean square values can be added. Later on, some examples will follow where this is not a valid assumption.

Problem

2.1 Two resistors in series are connected to a third resistor in parallel, as depicted in the figure below. Show that the same overall noise equivalent circuit results, if 1) an overall resistance is calculated first or if 2) the noise equivalent source is determined first. Assume that $T_1 = T_2 = T_3 = T_0$.

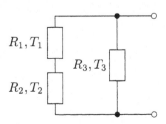

Apparently, a resistor cannot be divided into arbitrarily small parts. The assumption of statistical independence might not be valid, for example, if the dimensions become smaller than the mean free path length of the electrons. However, under these extreme conditions, it is also no longer possible to define a resistor in the usual way.

2.1.3   The RC-circuit

As was shown in section 1.3.4, calculations of noisy linear networks can be performed by means of the well-known rules for sinusoidal signals.

The noise spectra at the input and output of a device are related according to the magnitude squared of the transfer function $V(f)$. As an example, the spectrum W_{uc} at the output of the capacitor of the circuit in Fig. 2.2 will be calculated. Only the resistor R is assumed to generate thermal noise.

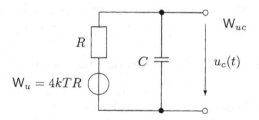

Fig. 2.2   Thermally noisy resistor with a capacitor in parallel.

Applying simple voltage divider relations leads to

$$W_{uc} = |V(f)|^2 \cdot W_u = \left|\frac{1/(j\omega C)}{R + 1/(j\omega C)}\right|^2 \cdot 4kTR = \frac{4kTR}{1 + (\omega CR)^2} \ . \qquad (2.7)$$

The spectral density W_{uc} becomes frequency dependent because of the capacitor. The mean square value of the voltage at the capacitor can be calculated by integration over the entire frequency range:

$$
\begin{aligned}
\overline{u_c^2(t)} &= \int_0^\infty W_{uc}(f)df = \int_0^\infty \frac{4kTR}{1 + (\omega CR)^2}df \\
&= \frac{2kT}{\pi \cdot C} \int_0^\infty \frac{1}{1 + \eta^2}d\eta \quad \text{with} \quad \eta = \omega CR \\
&= \frac{2kT}{\pi C} \cdot \arctan \eta \Big|_0^\infty \\
&= \frac{kT}{C} \ . \qquad\qquad\qquad\qquad\qquad (2.8)
\end{aligned}
$$

Thus, the mean square voltage at the capacitor is finite, although the frequency range was supposed to be unlimited. The result is independent of R, which might be interpreted physically.

2.1.4 Thermal noise of complex impedances

In a thought experiment a real-valued resistor R' and a complex impedance $Z(f)$ are connected by a band-pass filter (BPF). R' and $Z(f)$ are assumed

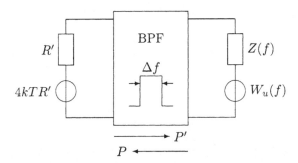

Fig. 2.3 For the explanation of the thermal noise of a complex impedance.

to be at the same temperature T. The band-pass filter is assumed to be lossless, therefore, it does not contribute to the noise of the setup. In the thermodynamic equilibrium the noise power P', which is transmitted by the resistor R' to the load $Z(f)$, must be equal to the noise power P, which is transmitted by the impedance $Z(f)$ to the load R', i.e. $P = P'$. The noise powers are given by

$$P' = \frac{4kTR'}{|R' + Z(f)|^2} \cdot \text{Re}\{Z\} \cdot \Delta f \ , \tag{2.9}$$

$$P = \frac{\mathsf{W}_u}{|R' + Z(f)|^2} \cdot R' \cdot \Delta f \ , \tag{2.10}$$

and, since $P' = P$, one can determine

$$\mathsf{W}_u = 4kT \cdot \text{Re}\{Z\} \ . \tag{2.11}$$

Similarly, with $Y = 1/Z$, it is concluded that

$$\mathsf{W}_i = 4kT \cdot \text{Re}\{Y\} \ . \tag{2.12}$$

The noise sources for thermally noisy complex impedances or admittances are thus also known. Both representations or equivalent circuits in Fig. 2.4 are equivalent. For example, for the right-hand circuit in Fig. 2.4 the spectrum

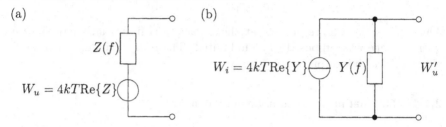

(a)   (b)

$Z(f)$

$W_i = 4kT\mathrm{Re}\{Y\}$ $Y(f)$ W'_u

$W_u = 4kT\mathrm{Re}\{Z\}$

Fig. 2.4   Equivalent representation of complex thermally noisy impedances.

of the open circuit voltage W'_u is calculated as

$$
\begin{aligned}
\mathsf{W}'_u &= \mathsf{W}_i \frac{1}{|Y(f)|^2} \\
&= 4kT \cdot \mathrm{Re}\{Y\} \cdot \frac{1}{Y(f) \cdot Y^*(f)} \\
&= 4kT \cdot \frac{Y + Y^*}{2 \cdot Y \cdot Y^*} \\
&= 4kT \cdot \frac{1}{2} \cdot \left(\frac{1}{Y} + \frac{1}{Y^*} \right) = 4kT \cdot \mathrm{Re}\{Z(f)\} \\
&= \mathsf{W}_u \ .
\end{aligned}
\tag{2.13}
$$

The equivalent circuits of Fig. 2.4 for complex impedances and admittances remain valid, even for combinations of lumped elements and transmission lines. The lines can either be lossless or lossy.

2.1.5   Available noise power and equivalent noise temperature

The available noise power P_{av} is obtained if a circuit is terminated by the complex conjugate of the generator source impedance. In this case, the power

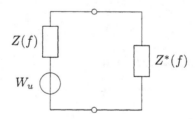

$Z(f)$

$Z^*(f)$

W_u

Fig. 2.5   For the explanation of the available noise power.

P_l transmitted to $Z^*(f)$ is given by

$$
P_l = \frac{\mathsf{W}_u}{|Z + Z^*|^2} \cdot \mathrm{Re}\{Z^*\} \cdot \Delta f
$$

$$\begin{aligned}
&= \frac{W_u}{4 \cdot \mathrm{Re}^2\{Z\}} \cdot \mathrm{Re}\{Z\} \cdot \Delta f \\
&= \frac{4kT \cdot \mathrm{Re}\{Z\} \cdot \mathrm{Re}\{Z\} \cdot \Delta f}{4 \cdot \mathrm{Re}^2\{Z\}} \\
&= kT \cdot \Delta f = P_{av} \ .
\end{aligned} \tag{2.14}$$

The available noise power only depends on the temperature T. It is independent of the resistance value. The noise temperature can thus be used as a quantity to describe the noise behavior of a general lossy one-port network. The noise temperature T_n is then called the equivalent noise temperature of a one-port. The definition can also be extended to non-thermally noisy one-ports.

The so-called Nyquist-relation in Eq. (2.14) is not valid for all frequencies and temperatures, because it is derived from statistical thermodynamics. For high frequencies and/or low temperatures a quantum mechanical correction factor has to be introduced. This correction term results from Planck's Radiation Law which applies to blackbody radiation. In the general case, $P_{av} = kT\Delta f$ must be replaced by

$$P_{av} = kT\Delta f \cdot p(f,T) \tag{2.15}$$

with

$$p(f,T) = \frac{hf/kT}{\exp(hf/kT) - 1} \tag{2.16}$$

and

$$h = 6.626 \cdot 10^{-34} \mathrm{Ws}^2 \quad \text{(Planck's constant)} \ . \tag{2.17}$$

At room temperature and up to 10 GHz, $p(f,T) \approx 1$. The Planck correction of the Nyquist formula also prevents that the noise power becomes infinite for arbitrarily large bandwidths.

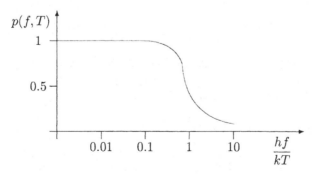

Fig. 2.6 Normalized radiation power as a function of frequency and temperature.

2.1.6   Networks with inhomogeneous temperature distribution

A network with impedances $Z_1, Z_2, Z_3, \ldots, Z_j$, which are at different temperatures $T_1, T_2, T_3, \ldots, T_j$, is now considered. Fig. 2.7 shows an example with three impedances. The network could be more complex than shown here. Furthermore, it might also include transmission line elements. The open cir-

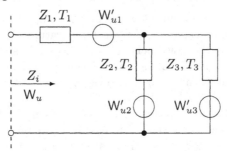

Fig. 2.7   Noisy one-port with three temperatures.

cuit noise source W_u at the external terminals will be calculated. For this purpose, the superposition principle is applied, which is valid for all linear circuits. The internal noise sources W'_{uj} are consecutively transformed to the input. In this way, the equivalent circuit in Fig. 2.8 is obtained. Here, Z_i is

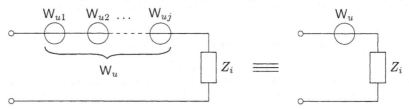

Fig. 2.8   Equivalent sources to Fig. 2.7.

the input impedance of the one-port network, which is calculated by shorting all noise sources. It can be assumed that the $W_{u1}, W_{u2}, \ldots, W_{uj}$ are all uncorrelated, because they originate from different, independent impedances. Thus

$$W_u = \sum_j W_{uj} = 4kT_n \cdot \mathrm{Re}\{Z_i\} \; . \tag{2.18}$$

The W_{uj} are related to the W'_{uj} by the magnitude squared of a transfer function, i.e. by a real coefficient. Consequently, the equivalent temperature T_n of the one-port is a linear function of the individual T_j:

$$T_n = \sum_j \beta_j \cdot T_j \; . \tag{2.19}$$

An obvious constraint for the real coefficients β_j is that

$$\sum_j \beta_j = 1 \ , \tag{2.20}$$

because if all T_j are equal, then $T_n = T_j$ must also hold. The noise temperature T_n can also be interpreted as the result of an averaging operation, which means that

$$T_{j,min} \leq T_n \leq T_{j,max} \ . \tag{2.21}$$

2.1.7 Dissipation theorem

For a reciprocal network the coefficients β_j can also be interpreted as the fraction of power absorbed by the impedance $\text{Re}\{Z_j\}$ when unity power is fed into the network. This so-called dissipation theorem will be proven first. Next, its application will be explained for some examples.

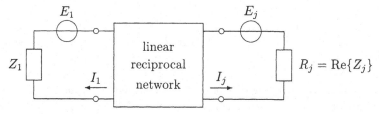

Fig. 2.9   For the explanation of the dissipation theorem.

According to Fig. 2.9, a resistor R_j with the temperature T_j is extracted from the linear and reciprocal one-port network. This resistor R_j can thus be considered as an external termination of a linear and reciprocal two-port. The impedance Z_1 and the noise voltage generator E_1 describe the input side, the resistor R_j and the equivalent noise source E_j represent the output side. Due to reciprocity, the current I_j, which is generated by E_1 for $E_j = 0$, and the current I_1, which is generated by E_j for $E_1 = 0$, are related as

$$\left.\frac{E_1}{I_j}\right|_{E_j=0} = \left.\frac{E_j}{I_1}\right|_{E_1=0} = \frac{1}{y} \ , \tag{2.22}$$

where y is a complex constant. The impedance Z_1 is chosen in order to achieve a power match to the input impedance Z_{in}, i.e. $Z_1 = Z_{in}^*$. In this case, the network should supply the available power P_{av} to the impedance Z_1 according to Eq. (2.14). The noise power P_{1n}, transmitted from the thermally noisy resistor R_j to the real part of Z_1, is equal to:

$$\begin{aligned} P_{1n} &= |I_{1n}|^2 \cdot \text{Re}\{Z_1\} = |E_{jn}|^2 \cdot |y|^2 \cdot \text{Re}\{Z_1\} \\ &= 4kT_j R_j \Delta f \cdot |y|^2 \cdot \text{Re}\{Z_1\} = kT_j \beta_j \Delta f \ . \end{aligned} \tag{2.23}$$

With the index n a noise value will be denoted. From Eq. (2.23) it follows that

$$\beta_j = 4 \cdot R_j \cdot |y|^2 \cdot \text{Re}\{Z_1\} \ . \tag{2.24}$$

This calculation can be performed for each thermally noisy element of the network. The sum of all power contributions must be equal to the overall available noise power P_{av}:

$$P_{av} = k \cdot \Delta f \cdot \sum_j (T_j \beta_j) \ . \tag{2.25}$$

The β_j are identical to those of Eq. (2.19). It can also be shown that the term β_j represents the ratio of the power P_j, absorbed by R_j, to the incident (available) power P_1:

$$\frac{P_j}{P_1} = |I_j|^2 \cdot R_j \frac{4Re\{Z_1\}}{|E_1|^2} = 4 \cdot R_j \cdot |y|^2 \cdot \text{Re}\{Z_1\} = \beta_j \ . \tag{2.26}$$

In other words, the relative contribution of a resistor R_j at the temperature T_j to the effective noise temperature T_n of the one-port, expressed by the coefficient β_j, just amounts to the power absorbed in the resistor R_j, normalized to the power fed into the one-port. Thermal noise power is very closely related to power dissipation. A component, which cannot dissipate real power, cannot emit thermal noise power. The dissipation theorem was derived for circuits with lumped elements. However, it can be extended to circuits including distributed elements.

Problems

2.2 Calculate the equivalent noise temperature T_n of the circuit in Fig. 2.7 with the help of the dissipation theorem. The impedances Z_1, Z_2, Z_3 are complex.

2.3 Determine the input noise temperature T_n of the circuit in Fig. 2.10 by using the dissipation theorem.

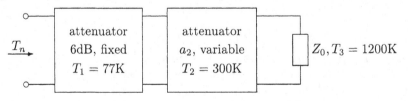

Fig. 2.10 A noise source with variable temperature consisting of a fixed and a variable attenuator and a hot one-port device.

2.4 Calculate the equivalent noise temperature using the dissipation theorem for the antenna setup in Fig. 2.11.

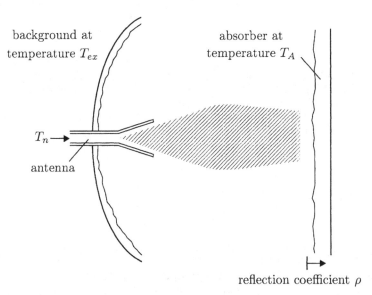

background at
temperature T_{ex}

absorber at
temperature T_A

$T_n \rightarrow$

antenna

reflection coefficient ρ

Fig. 2.11   Determination of the noise temperature of an antenna.

2.2 NOISE OF TWO-PORTS

2.2.1 Description of the internal noise by current and voltage sources

The noise of linear two-ports can be described with the help of equivalent current or voltage sources at the input and/or output of the circuit. The two-port itself is assumed to be noiseless and it is described in the usual way by a matrix which linearly relates the currents and voltages at the input and output. In Fig. 2.12 an equivalent circuit of a noisy two-port with noise current sources at the input and output is shown. The noiseless two-port itself is represented by an admittance matrix $[Y]$. The currents and voltages are represented in a symbolic description by phasors which depend on the frequency f:

$$U, I = U(f), I(f) \ . \tag{2.27}$$

Apart from the admittance matrix Y, various noise parameters must be known in order to perform noise calculations. In the symbolic description the squared magnitudes of the noise current sources $|I_{n1}|^2, |I_{n2}|^2$ are equal to the equiva-

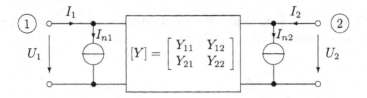

Fig. 2.12   Noise equivalent circuit of a two-port with noise current sources.

lent two-sided spectral densities:

$$|I_{n1}|^2 = W_{n1}, \quad |I_{n2}|^2 = W_{n2} . \tag{2.28}$$

Furthermore, for the complete description of the noise behavior the cross-spectrum W_{n12} must be known. In the symbolic description, W_{n12} corresponds to:

$$W_{n12} = I_{n1}^* \cdot I_{n2} . \tag{2.29}$$

Equation (2.29) does not normally serve as a rule for the calculation of W_{n12}. The cross-spectrum, in general, is a complex quantity and thus both its real and imaginary part must be known or its amplitude and phase, respectively. It is often not easy to determine the cross-spectrum. However, several examples will follow where it is possible to calculate the cross-spectrum. If the magnitudes of the noise equivalent sources as well as their cross-spectra are known, then all noise parameters of interest of linear circuits can be calculated in principle.

In another description of a noisy two-port, noise voltage sources at the input and output of the circuit are used and the noiseless two-port is described by an impedance matrix $[Z]$ (Fig. 2.13). Different representations for the same

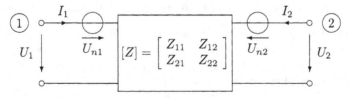

Fig. 2.13   Noise equivalent circuit of a two-port with noise voltage sources.

noisy two-port can be converted into one another. This will be demonstrated for the circuit arrangements in Figs. 2.12 and 2.13 as an example. For such transformations or other noise calculations it is convenient to define directions or oriented arrows for the equivalent noise currents and voltages as well as the terminal currents and voltages, respectively. Although, they can be chosen arbitrarily at first, it is necessary to adhere strictly to the chosen directions of the arrows in the following calculations. If a resulting solution has a negative sign, this means that the actual directions of the currents or voltages are opposite to the direction defined as positive. Although the final aim is to calculate a noise spectrum, that is the magnitude squared of a current or voltage, which is certainly positive, the current or voltage of the entire circuit often result from a superposition of single currents or voltages and this superposition has to be performed with the correct signs.

For the representation with current sources in Fig. 2.12 and taking into account the chosen orientation of the arrows, the following two-port equations hold:

$$I_1 = Y_{11} \cdot U_1 + Y_{12} \cdot U_2 + I_{n1}$$
$$I_2 = Y_{21} \cdot U_1 + Y_{22} \cdot U_2 + I_{n2}$$

or, in matrix form,

$$\begin{bmatrix} I_1 \\ I_2 \end{bmatrix} = \begin{bmatrix} Y_{11} & Y_{12} \\ Y_{21} & Y_{22} \end{bmatrix} \cdot \begin{bmatrix} U_1 \\ U_2 \end{bmatrix} + \begin{bmatrix} I_{n1} \\ I_{n2} \end{bmatrix} , \qquad (2.30)$$

which may be written more compactly as

$$[I] = [Y] \cdot [U] + [I_n] . \qquad (2.31)$$

For the description with voltage sources according to Fig. 2.13, the following two-port matrix equations apply, if first an admittance representation is chosen. Here, $[Y] = [Z]^{-1}$, because the transformation for the two-port parameters are independent of the noise sources. Therefore,

$$I_1 = Y_{11} \cdot (U_1 - U_{n1}) + Y_{12} \cdot (U_2 - U_{n2})$$
$$I_2 = Y_{21} \cdot (U_1 - U_{n1}) + Y_{22} \cdot (U_2 - U_{n2})$$

or, in a matrix form,

$$[I] = [Y] \cdot [U] - [Y] \cdot [U_n] . \qquad (2.32)$$

A comparison of Eq. (2.31) with Eq. (2.32) leads to a relationship between the noise sources:

$$[I_n] = -[Y] \cdot [U_n] \qquad (2.33)$$

or

$$[U_n] = -[Z] \cdot [I_n] , \qquad (2.34)$$

respectively. With the help of Eq. (2.34) the spectra $|U_{n1}|^2$, $|U_{n2}|^2$ and the cross-spectrum $U_{n1}^* U_{n2}$ of the noise sources can be calculated, provided that the cross-spectrum $I_{n1}^* I_{n2}$ and the power spectra $|I_{n1}|^2$ and $|I_{n2}|^2$ of the noise current sources are known. In detail, we have

$$W_{u1} = U_{n1}^* \cdot U_{n1}$$

$$
\begin{aligned}
&= (Z_{11} \cdot I_{n1} + Z_{12} \cdot I_{n2})^* \cdot (Z_{11} \cdot I_{n1} + Z_{12} \cdot I_{n2}) \\
&= |Z_{11}|^2 \cdot |I_{n1}|^2 + |Z_{12}|^2 \cdot |I_{n2}|^2 \\
&\quad + Z_{11}^* \cdot Z_{12} \cdot (I_{n1}^* \cdot I_{n2}) + Z_{12}^* \cdot Z_{11} \cdot (I_{n1}^* \cdot I_{n2})^* \\
&= |Z_{11}|^2 \cdot W_{i1} + |Z_{12}|^2 \cdot W_{i2} + Z_{11}^* \cdot Z_{12} \cdot W_{i12} \\
&\quad + Z_{12}^* \cdot Z_{11} \cdot W_{i12}^* \ ,
\end{aligned}
\tag{2.35}
$$

$$
\begin{aligned}
W_{u2} &= U_{n2}^* \cdot U_{n2} \\
&= (Z_{21} \cdot I_{n1} + Z_{22} \cdot I_{n2})^* \cdot (Z_{21} \cdot I_{n1} + Z_{22} \cdot I_{n2}) \\
&= |Z_{21}|^2 \cdot |I_{n1}|^2 + |Z_{22}|^2 \cdot |I_{n2}|^2 \\
&\quad + Z_{21}^* \cdot Z_{22} \cdot (I_{n1}^* \cdot I_{n2}) + Z_{22}^* \cdot Z_{21} \cdot (I_{n1}^* \cdot I_{n2})^* \\
&= |Z_{21}|^2 \cdot W_{i1} + |Z_{22}|^2 \cdot W_{i2} + Z_{21}^* \cdot Z_{22} \cdot W_{i12} \\
&\quad + Z_{22}^* \cdot Z_{21} \cdot W_{i12}^* \ ,
\end{aligned}
\tag{2.36}
$$

$$
\begin{aligned}
W_{u12} &= U_{n1}^* \cdot U_{n2} \\
&= (Z_{11} \cdot I_{n1} + Z_{12} \cdot I_{n2})^* \cdot (Z_{21} \cdot I_{n1} + Z_{22} \cdot I_{n2}) \\
&= Z_{11}^* \cdot Z_{21} \cdot |I_{n1}|^2 + Z_{12}^* \cdot Z_{22} \cdot |I_{n2}|^2 \\
&\quad + Z_{11}^* \cdot Z_{22} \cdot (I_{n1}^* \cdot I_{n2}) + Z_{12}^* \cdot Z_{21} \cdot (I_{n1}^* \cdot I_{n2})^* \\
&= Z_{11}^* \cdot Z_{21} \cdot W_{i1} + Z_{12}^* \cdot Z_{22} \cdot W_{i2} + Z_{11}^* \cdot Z_{22} \cdot W_{i12} \\
&\quad + Z_{12}^* \cdot Z_{21} \cdot W_{i12}^* \ .
\end{aligned}
\tag{2.37}
$$

A more compact representation for the description results from the definition of a correlation matrix $[C_u]$

$$
\begin{aligned}
[C_u] &= [U_n]^* \cdot [U_n]^T = \begin{bmatrix} U_{n1}^* \\ U_{n2}^* \end{bmatrix} \cdot \begin{bmatrix} U_{n1} & U_{n2} \end{bmatrix} \\
&= \begin{bmatrix} |U_{n1}|^2 & U_{n1}^* U_{n2} \\ U_{n1} U_{n2}^* & |U_{n2}|^2 \end{bmatrix} = \begin{bmatrix} W_{u1} & W_{u12} \\ W_{u12}^* & W_{u2} \end{bmatrix} \\
&= (-[Z] \cdot [I_n])^* \cdot (-[Z] \cdot [I_n])^T = [Z]^* [I_n]^* [I_n]^T [Z]^T \\
&= [Z]^* [C_i] [Z]^T
\end{aligned}
\tag{2.38}
$$

with

$$
[C_i] = [I_n]^* [I_n]^T = \begin{bmatrix} W_{i1} & W_{i12} \\ W_{i12}^* & W_{i2} \end{bmatrix} \ .
\tag{2.39}
$$

Here, the superscripts denote the complex conjugate matrix ($*$) and the transposed matrix (T).

For a two-port there are six possibilities to combine the currents and voltages and thus there are six different corresponding matrix descriptions. Accordingly, six different possibilities exist to arrange the equivalent noise sources. In addition to the two configurations in Fig. 2.12 and 2.13, the four configurations in Fig. 2.14 are possible. Consequently, there are $6 \cdot 6 = 36$ ways to combine the equivalent noise sources with the algebraic two-port matrix descriptions. Which one of these combinations is the most appropriate depends on the particular problem. As an example, a configuration based on current

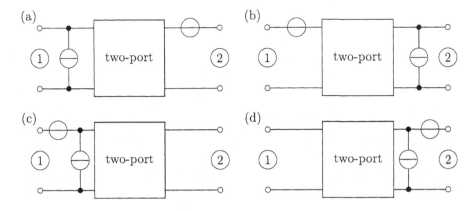

Fig. 2.14 Further configurations for the noise equivalent sources.

sources is favorably combined with an admittance description whereas a representation with voltage sources may be combined with an impedance matrix description. The configuration in Fig. 2.14c is well suited for the calculation of the noise figure as will be seen later.

The 36 different representations can be converted into each other, where, in general, the correlation of the noise sources will change as well. Instead of a description by noise currents and voltages it is also possible to perform the calculations on the basis of noise waves (section 2.2.3). This is often advantageous for high frequency circuits.

Problems

2.5 Convert the equivalent noise circuit below into the circuit shown in Fig. 2.14c. Calculate the correlation of the new noise current and voltage sources.

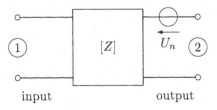

2.6 A noisy two-port, which is described by an equivalent noise circuit as shown in Fig. 2.12 is terminated at its output by a complex admittance Y_2 which is noise-free. At the input of the circuit the complex admittance Y_1 is connected, the thermal noise of which is represented by a parallel current source I_g. What is the magnitude of the noise power and the noise spectrum W_2, respectively, at the output admittance Y_2?

2.2.2   Noise equivalent sources for thermally noisy two-ports at homogeneous temperature

For the special case of thermally noisy two-ports at homogeneous temperature, the noise equivalent sources can be calculated in a general way on the basis of the two-port parameters. The term homogeneous temperature means that all losses of the two-ports network, for example resistances, dielectric losses of capacitors, the losses of eddy currents in iron, the cable losses and suchlike, are related to the same physical temperature T. The two-port is described by the admittance matrix $[Y]$ and noise current sources at the input and output as shown in Fig. 2.12. The two-port considered here could as well characterize any two selected ports of a N-port network, where all other ports are assumed to be terminated by an impedance, also at the temperature T.

The squared magnitudes of I_{n1} and I_{n2} can be calculated in a simple way by terminating the other port with a short and by considering the Nyquist-relation (two-sided spectrum):

$$\begin{aligned} |I_{n1}|^2 &= W_{n1} = 2kT \cdot \text{Re}\{Y_{11}\} \ , \\ |I_{n2}|^2 &= W_{n2} = 2kT \cdot \text{Re}\{Y_{22}\} \ . \end{aligned} \qquad (2.40)$$

However, this does not give any information about the cross-spectrum $W_{i12} = I_{n1}^* \cdot I_{n2}$. In order to determine the cross-spectrum, the input admittance Y_{in} at port 1 will be calculated under the condition that port 2 is terminated by an open circuit, $I_2 = 0$. The resulting two-port will generate thermal noise at the temperature T. This two-port can be described with the help of an equivalent noise current source I'_{n1} (Fig. 2.15), where Y_{in} is the input admittance at port 1. Therefore, we have

$$|I'_{n1}|^2 = 2kT \cdot \text{Re}\{Y_{in}\} \ . \qquad (2.41)$$

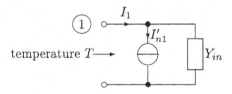

Fig. 2.15   Equivalent circuit of a one-port with an open circuit at port 2.

On the other hand, the noise current I'_{n1} for an open circuit at port 2 in Fig. 2.12 can be calculated as well on the basis of the two noise sources I_{n1}

and I_{n2}. For this purpose, the noise current I_{1k} for a short circuit at the input, which means $U_1 = 0$, and an open circuit at the output, which means $I_2 = 0$, is determined. The matrix relation (2.31)

$$[I] = [Y] \cdot [U] + [I_n]\Big|_{U_1=0,I_2=0} \tag{2.42}$$

yields

$$\begin{aligned} I_{1k} &= Y_{12} \cdot U_2 + I_{n1} \\ 0 &= Y_{22} \cdot U_2 + I_{n2} \ . \end{aligned} \tag{2.43}$$

In Eq. (2.43) U_2 can be eliminated. This leads to

$$I_{1k} = I_{n1} - \frac{Y_{12}}{Y_{22}} \cdot I_{n2} \ . \tag{2.44}$$

It can be assumed that $|I'_{n1}|^2$ is equal to $|I_{1k}|^2$:

$$\begin{aligned} |I'_{n1}|^2 &= 2kT \cdot \mathrm{Re}\{Y_{in}\} = 2kT \cdot \mathrm{Re}\left\{ Y_{11} - \frac{Y_{12} \cdot Y_{21}}{Y_{22}} \right\} \\ &= |I_{1k}|^2 = 2kT \cdot \mathrm{Re}\{Y_{11}\} + \left|\frac{Y_{12}}{Y_{22}}\right|^2 \cdot 2kT \cdot \mathrm{Re}\{Y_{22}\} \\ &\quad - I^*_{n1} \cdot I_{n2} \cdot \frac{Y_{12}}{Y_{22}} - I_{n1} \cdot I^*_{n2} \cdot \frac{Y^*_{12}}{Y^*_{22}} \ . \end{aligned} \tag{2.45}$$

In Eq. (2.45) the term $2kT \cdot \mathrm{Re}\{Y_{11}\}$ can be eliminated so that finally the following relationship results:

$$\begin{aligned} &I^*_{n1} \cdot I_{n2} \cdot \frac{Y_{12}}{Y_{22}} + I_{n1} \cdot I^*_{n2} \cdot \frac{Y^*_{12}}{Y^*_{22}} \\ &= \left|\frac{Y_{12}}{Y_{22}}\right|^2 \cdot 2kT \cdot \mathrm{Re}\{Y_{22}\} + 2kT \cdot \mathrm{Re}\left\{\frac{Y_{12} \cdot Y_{21}}{Y_{22}}\right\} \ . \end{aligned} \tag{2.46}$$

A very similar equation, however with exchanged indices 1 and 2, can be derived, if the same calculation is performed for port 2, that means that an open circuit is connected to port 1. Both resulting equations are linear with respect to the parameters $I^*_{n1} \cdot I_{n2}$ and $I_{n1} \cdot I^*_{n2}$. After some manipulations the following equation for the term $I^*_{n1} \cdot I_{n2}$ results:

$$I^*_{n1} \cdot I_{n2} = W_{n12} = kT \cdot (Y^*_{12} + Y_{21}) \ . \tag{2.47}$$

Similarly, the following relation is valid for a N-port considering the ports i and j:

$$I^*_{ni} \cdot I_{nj} = W_{nij} = kT \cdot (Y^*_{ij} + Y_{ji}) \ . \tag{2.48}$$

These simple relations for the cross-spectrum of the noise current sources at the input and output of the circuit are valid for the case of thermally noisy two-ports at a homogeneous temperature. It is valid for reciprocal passive circuits, that means $Y_{ij} = Y_{ji}$, as well as for non-reciprocal passive circuits. The relation (2.47) can be transformed into the other 35 description possibilities. If, on the other hand, the noise characteristics of a two-port or N-port are in accordance to Eqs. (2.40) and (2.47) or (2.48), respectively, then the device can be described by a homogeneous temperature. This often leads to a notably easy characterization of the noise behavior of a two-port. In Chapter 5 this property will be used for the calculation of the noise behavior of a frequency converter. A passive thermally noisy two-port at a homogeneous temperature is an example, where the calculation of the two-port equivalent noise sources including the correlation is possible.

Problems

2.7 Derive the relation for the cross-spectrum as given by Eq. (2.47).

2.8 Describe a thermally noisy two-port at a homogeneous temperature T by the noise current sources I_{n1} and I_{n2} at the input and output. The input is terminated by a thermally noisy complex admittance Y_1, which is also at the temperature T and which is represented by the noise current source I_g. Derive a noise equivalent one-port circuit for the output port with the help of Eq. (2.40) and Eq. (2.47).

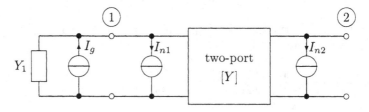

2.2.3 Noise description by waves

A matrix representation very common for high-frequency circuits, which is particularly suited to transmission line structures, is the scattering matrix. For this representation it is also possible to introduce noise equivalent sources. Apart from the noise waves $A_{1,2}$ propagating towards the two-port and the waves $B_{1,2}$ leaving the two-port, the noise equivalent waves $X_{1,2}$ are introduced for the description of the intrinsic noise of the two-port. The waves, illustrated in Fig. 2.16, are defined according to

$$B_1 = S_{11} \cdot A_1 + S_{12} \cdot A_2 + X_1$$
$$B_2 = S_{21} \cdot A_1 + S_{22} \cdot A_2 + X_2$$

or in a matrix description

$$[B] = [S] \cdot [A] + [X] \ . \tag{2.49}$$

$[S]$ denotes the scattering matrix of the two-port. A special symbol is not in common use for the noise equivalent waves X_1, X_2. In Fig. 2.16 oriented

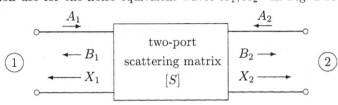

Fig. 2.16 Representation of a noisy two-port with a scattering matrix and noise equivalent waves X_1 and X_2.

arrows are used instead. In this context, the squared magnitudes of the noise waves A, B, X are supposed to have the dimension of a spectral noise power density, that means power per bandwidth.

The wave description can also be transformed into a representation with current or voltage sources. The following equations are used for this purpose:

$$\frac{U_{1,2}}{\sqrt{Z_0}} = A_{1,2} + B_{1,2} \ ,$$
$$I_{1,2} \cdot \sqrt{Z_0} = A_{1,2} - B_{1,2} \ . \tag{2.50}$$

The real-valued reference impedance is denoted by Z_0 and U, I are the currents and voltages of the current-voltage-representation. Problem 2.9 will clarify this transformation.

Problem

2.9 Transform the configuration in Fig. 2.12 with current sources into the representation with equivalent waves of Fig. 2.16.

2.2.4 Noise of circulators and isolators

An ideal circulator, for example a 3-port-circulator, is perfectly matched at all of its 3 ports and it transmits without any losses from port I to port II, from II to III and from III to I. The scattering matrix of an ideal 3-port circulator with the transfer phase φ is thus given by

$$[S] = \begin{bmatrix} 0 & 0 & e^{j\varphi} \\ e^{j\varphi} & 0 & 0 \\ 0 & e^{j\varphi} & 0 \end{bmatrix} \ . \tag{2.51}$$

An ideal circulator is lossless and thus noiseless. Such an ideal circulator can be used advantageously for thought experiments. A real circulator using ferrites is generally not perfectly matched, it has finite transmission losses, and a finite isolation. But still a real circulator based on ferrites is passive and generates thermal noise only.

If one port of the circulator is terminated by the reference impedance Z_0, the circuit works as an isolator as depicted in Fig. 2.17.

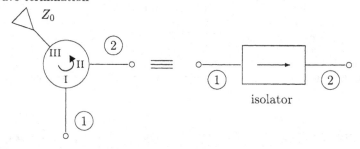

Fig. 2.17 Isolator on the basis of a circulator with one port being terminated by a wave impedance Z_0.

The scattering matrix of the isolator can be written as follows, if φ is assumed to be an arbitrary phase:

$$[S] = \begin{bmatrix} 0 & 0 \\ e^{j\varphi} & 0 \end{bmatrix} . \tag{2.52}$$

In general, a real isolator based on ferrites has finite losses in a forward direction, but also a finite isolation in a backward direction. Under usual laboratory conditions, circulators as well as isolators will be at a homogeneous temperature.

It should be noted that isolators are usually not realized on the basis of circulators. Nevertheless, their noise behavior with respect to the ports can accurately be represented by a circulator with one port terminated by a wave impedance Z_0. This model will be used with advantage later.

2.2.5   Noise waves for thermally noisy two-ports at a homogeneous temperature

In the following the thermally noisy two-port is assumed to be at the homogeneous temperature T. It is the aim to derive a similar equation for the equivalent noise waves and scattering matrices as in Eq. (2.40) and (2.47) for the current sources. For this purpose, the two-port is terminated at its input and output with impedances $Z_1 = Z_2 = Z_0$, which also have the same temperature T (Fig. 2.18).

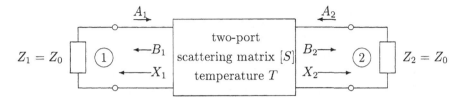

Fig. 2.18 Two-port with scattering matrix $[S]$ at a homogeneous temperature T, which is terminated on both sides by the impedance Z_0.

Because of the thermodynamic equilibrium and because of the matched terminations at the input and output, i.e.

$$Z_1 = Z_2 = Z_0 \ , \tag{2.53}$$

we get

$$|A_1|^2 = |B_1|^2 = |A_2|^2 = |B_2|^2 = kT \ . \tag{2.54}$$

The noise waves A_1 and A_2 originate from the terminating impedances Z_1 and Z_2. Therefore, they are uncorrelated, that is $A_1^* A_2 = 0$, and A_1 and A_2 are also uncorrelated with the equivalent noise waves X_1 and X_2 of the two-port, because they are generated in different parts of the circuit. Therefore, it is $A_{1,2}^* \cdot X_{1,2} = 0$. Under these conditions the following relationship can be derived for a two-port of a homogeneous temperature:

$$
\begin{aligned}
|B_1|^2 &= B_1 \cdot B_1^* \\
&= (S_{11} \cdot A_1 + S_{12} \cdot A_2 + X_1)(S_{11}^* \cdot A_1^* + S_{12}^* \cdot A_2^* + X_1^*) \\
&= |S_{11}|^2 \cdot |A_1|^2 + |S_{12}|^2 \cdot |A_2|^2 + |X_1|^2 \ .
\end{aligned}
\tag{2.55}
$$

From this and together with Eq. (2.54) it follows that

$$
\begin{aligned}
|X_1|^2 &= kT \left(1 - |S_{11}|^2 - |S_{12}|^2\right) \ , \\
|X_2|^2 &= kT \left(1 - |S_{22}|^2 - |S_{21}|^2\right) \ .
\end{aligned}
\tag{2.56}
$$

The squared magnitudes of the equivalent noise waves X_1 and X_2 are thus determined. Finally, the cross-spectrum between X_1 and X_2 is needed, that is the term $X_1^* \cdot X_2$. For its calculation it is advantageous to choose a method which is very similar to the solution based on the admittance matrix. For example, in Fig. 2.18 the termination of port 2 with an open and short circuit can be considered and then $|B_1|^2$ can be calculated for both cases. Under the condition of the thermodynamic equilibrium and for the termination of port 1 with Z_0, the parameter $|B_1|^2$ must be equal for both of the cases, namely either open or short connected to port 2. It is, in fact, $|B_1|^2 = kT$. This leads to a system of equations for $X_1^* \cdot X_2$ and $X_1 \cdot X_2^*$. The details of the calculation are left to problem 2.10.

Problem

2.10   Derive the cross-spectrum $X_1^* \cdot X_2$ in the representation with equivalent noise waves and the scattering matrix for a thermally noisy two-port at a homogeneous temperature. For this purpose, each port has to be terminated with an open or a short circuit.

In the following, a more illustrative and shorter way will be presented in order to calculate the cross spectrum $X_1^* \cdot X_2$. From Eq. (2.47) it can be concluded that the noise waves of an isolated two-port, that is a two-port with $Y_{12} = Y_{21} = 0$, are also uncorrelated, as long as the noise is of thermal origin at a homogeneous temperature. As will be seen for some examples, this often is not a trivial assertion. Assuming that this two-port is not only isolated but also matched at both sides, then we have

$$B_1^* \cdot B_2 = 0 \ , \tag{2.57}$$

because of $B_1 \sim I_{n1}$ and $B_2 \sim I_{n2}$. According to Fig. 2.19 a two-port will be considered which is embedded between two ideal passive circulators. The

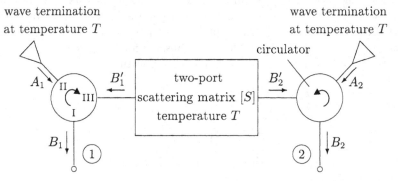

Fig. 2.19   Thermally noisy mismatched two-port with the scattering matrix $[S]$ and the temperature T, embedded between two ideal circulators.

circuit in Fig. 2.19 is assumed to be at a homogeneous temperature T. This assumption, in fact, applies to the whole circuit, that is the two-port and the terminations of the circulators. The circulators are assumed to be free of losses and they are thus noiseless. Eq. (2.57) can be applied to the circuit between port 1 and port 2 in Fig. 2.19, because the circuit between port 1 and 2 is isolated due to the ideal circulators and, furthermore, it is at a homogeneous temperature. As a consequence, Eqs. (2.49) and (2.57) yield

$$\begin{aligned}
B_1^* \cdot B_2 &= 0 \\
&= (S_{11}^* \cdot A_1^* + S_{12}^* \cdot A_2^* + X_1^*)(S_{21} \cdot A_1 + S_{22} \cdot A_2 + X_2)
\end{aligned}$$

and hence

$$X_1^* \cdot X_2 = -\left(S_{11}^* \cdot S_{21} + S_{12}^* \cdot S_{22}\right) \cdot kT \ . \tag{2.58}$$

The reasoning which led to Eq. (2.58) can directly be extended to a N-port. Therefore, the following relation for the cross-spectrum at the ports i, j is obtained:

$$X_i^* \cdot X_j = kT \left([1] - [S^*][S]^T\right)_{i,j} \ . \tag{2.59}$$

Here, the unity matrix is denoted by $[1]$ and $[S]^T$ is the transpose of the scattering matrix of the N-port. In Eq. (2.59) the element i, j of the matrix in the brackets is chosen. For $i = j$, Eq. (2.59) is identical to the already known result of Eq. (2.56).

The correlation of the equivalent noise waves $X_1^* \cdot X_2$ becomes zero according to Eq. (2.58), if the ports of the two-port are isolated. This result could have been anticipated, because it was already implied as a precondition. However, an interesting new result arises, i.e. the correlation also disappears for the case that both ports are matched, that is for $S_{11} = S_{22} = 0$. As will be seen later, this characteristic of matched, passive two-ports at a homogeneous temperature is of great interest for noise measurement techniques and hence is often utilized.

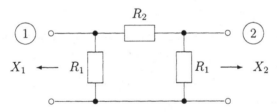

Fig. 2.20 π-attenuator with concentrated resistors.

An example of such two-ports are attenuators, which are matched at both sides and which possess a homogeneous temperature distribution under laboratory conditions. It is not relevant, how the attenuators work in detail. The correlation of a π-attenuator with lumped resistors, as shown in Fig. 2.20, will be calculated explicitly in problem 2.11. As expected, the correlation becomes zero for a homogeneous temperature.

Problem

2.11   Calculate explicitly the cross-spectrum $X_1^* X_2$ of a matched π-attenuator as shown in Fig. 2.20. All three resistors are assumed to have the same temperature.

It might be astonishing that the correlation of the equivalent noise waves disappears, although the noise of the resistors R_1 or R_2, respectively, is transmitted to port 1 as well as to port 2.

A further example, which seems to contradict the theory, is shown in Fig. 2.21. The signal divider with $\lambda/4$-lines is terminated at port 1 by a matched impedance. Also for this configuration, one might assume at first

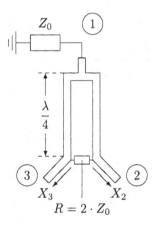

Fig. 2.21   Signal divider with $\lambda/4$-lines and a $R = 2\,Z_0$ resistor for isolation.

glance that there must be a correlation of the noise signals at port 2 and port 3 due to the common noise source at port 1. In fact, port 2 and port 3 are not coupled and they are matched at the center frequency and, therefore, the noise equivalent waves of port 2 and 3 are not correlated. This can also be proven by a direct calculation, as shown in problem 2.12.

Problem

2.12   Show for the signal divider of Fig. 2.21 that the cross-spectrum $X_3^* \cdot X_2$ is equal to zero, if the resistor R is at the same temperature T_0 as the input resistor Z_0 at port 1.

As a further example, two antennas are considered, the radiation patterns of which are directed in such a way that the antennas receive thermal noise from the same area of an absorber (Fig. 2.22). The noise waves which the antennas receive from the absorber are uncorrelated, if the absorber is at a homogeneous temperature and if the antennas and the absorber, respectively, are well matched. Furthermore, the antennas will be well isolated, if the side lobe attenuation is high enough. In the relation for the cross-spectrum in Eq. (2.58) the reflections and the isolation enter multiplicatively, so that the correlation will be especially low. Another possible explanation is the

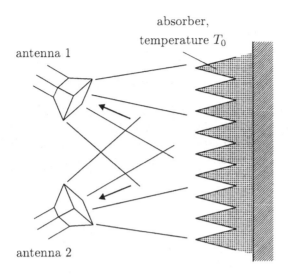

Fig. 2.22 Two antennas receive noise from the same area of an absorber.

following: the absorber radiates into different directions in space, whereby the correlation is destroyed. Both antennas must be placed at different positions for geometrical reasons.

2.2.6 Equivalent noise waves for linear amplifiers

An amplifier is not a passive two-port at a homogeneous temperature. Therefore, in general linear amplifiers of arbitrary design show a correlation between the input and output noise waves. For example, preamplifiers with low-noise bipolar transistors have a typical magnitude of the normalized cross spectrum of about 0.5.

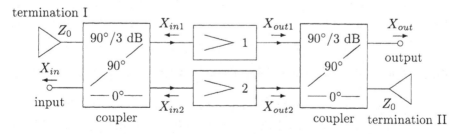

Fig. 2.23 Uncorrelated amplifier, based on a pair of amplifiers and two $90°$-$3\,\mathrm{dB}$ couplers (balanced amplifier).

In a variety of noise measurement problems it is desirable to use amplifiers with uncorrelated input and output noise waves, because the measurements often are simplified in this way. A possible realization of an amplifier with uncorrelated input and output noise waves is shown in Fig. 2.23. This setup consists of two matched amplifiers of the same kind, connected in parallel with the help of two 90°- 3 dB couplers.

For the following consideration the couplers are assumed to be perfectly matched and lossless. The cross-spectrum $X_{in}^* X_{out}$ will be calculated and it will be shown that the cross-spectrum becomes zero for identical amplifiers and perfect couplers. Because of the 90°-couplers, which show a phase shift of $+90° \hat{=} j$, the following relationships hold, where $X_{in1,2}$ and $X_{out1,2}$ denote the input and output equivalent noise waves of the amplifiers:

$$X_{in} = \frac{1}{\sqrt{2}}(j X_{in1} + X_{in2}) \ ,$$

$$X_{out} = \frac{1}{\sqrt{2}}(X_{out1} + j X_{out2}) \ . \tag{2.60}$$

Hence, the cross-spectrum is given by:

$$X_{in}^* X_{out} = \frac{1}{2}(-j X_{in1}^* X_{out1} + j X_{in2}^* X_{out2} - j X_{in1}^* j X_{out2} + X_{in2}^* X_{out1}) \ . \tag{2.61}$$

X_{in1} is uncorrelated with X_{out2}, and X_{in2} is uncorrelated with X_{out1}, which means $X_{in1}^* X_{out2} = 0$ and $X_{in2}^* X_{out1} = 0$, because the noise waves originate from two separate amplifiers. If, furthermore, the amplifiers have equal properties, then

$$X_{in1}^* X_{out1} = X_{in2}^* X_{out2} \ . \tag{2.62}$$

With these assumptions, Eq. (2.61) yields

$$X_{in}^* X_{out} = 0 \ . \tag{2.63}$$

The noise from the termination I in Fig. 2.23 is not transferred to the output. Hence, the overall noise figure of the uncorrelated amplifier is not changed as long as the couplers are lossless.

2.3   NOISE FIGURE OF LINEAR TWO-PORTS

The noise factor or noise figure F of a two-port is a measure for the additional noise that results if a signal, which can also be a noise signal, passes through a two-port. Some examples of such two-ports are amplifiers, attenuators, transmission lines or filters. If the two-port is loss-free, then the noise figure is $F = 1$ or 0 dB. There are several equivalent definitions for the noise figure. Two of them will be explained in the following in more detail.

2.3.1   Definition of the noise figure

The one-sided power spectrum at the output or load impedance Z_l of the two-port is denoted by W_2. The output impedance Z_l is assumed to be noiseless in this definition. Moreover, it will be shown that the impedance value of Z_l does not influence the noise figure. This is not necessarily true for a practical circuit, where the output impedance will be noisy in general. For the measurement of the noise figure, it will be necessary to guarantee a negligible influence of the output resistance in order to be consistent with the definition of the noise figure.

As already stated, W_2 denotes the power spectrum at the output resistor Z_l of the two-port at the frequency f_0 (Fig. 2.24). W_{20} represents the power spectrum, if the considered two-port is assumed to be noiseless.

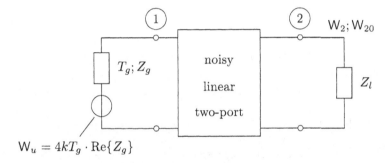

Fig. 2.24   Basic setup for the definition of the noise figure.

The noise figure is defined as the ratio of the power spectra at the output resistor for a noisy two-port, W_2, and for a noiseless two-port, W_{20},

$$F = \frac{W_2}{W_{20}} \; . \tag{2.64}$$

For the definition of the noise figure the source resistor Z_g is assumed to generate thermal noise at the ambient temperature $T_g = T_0$. Hence, it generates the spectrum W_{20} which contributes to the spectrum W_2. In general, it is assumed that T_g is at room temperature and T_0 is 290 K. Describing by ΔW_2 that part of the power spectrum at the output resistor which is solely caused by the noisy two-port, then

$$W_2 = W_{20} + \Delta W_2 \; , \tag{2.65}$$

because both parts of the spectrum originate from different regions and are thus uncorrelated. With Eq. (2.65) the relation for the noise figure can also be written in the form

$$F = 1 + \frac{\Delta W_2}{W_{20}} \; . \tag{2.66}$$

As the noise figure can be described by the ratio of spectra, the representation on the basis of one-sided spectra can just as well be replaced by two-sided spectra. With the same meaning of the indices, the noise figure is also given by

$$F = \frac{W_2}{W_{20}} = 1 + \frac{\Delta W_2}{W_{20}} \ . \tag{2.67}$$

Instead of a description by spectra, the noise figure can as well be defined by means of the associated noise powers P in the frequency band Δf. If the spectrum W is constant within the bandwidth Δf, then

$$P = W \cdot \Delta f \ , \tag{2.68}$$

otherwise

$$P = \int_{f_1}^{f_2} W(f) df \ . \tag{2.69}$$

Hence, if the indices are again the same as for the spectra, the noise figure may be defined by

$$F = \frac{P_2}{P_{20}} = 1 + \frac{\Delta P_2}{P_{20}} \ . \tag{2.70}$$

For a further discussion of the noise figure, some definitions regarding the amplification of a linear two-port have to be introduced.

The power gain G of a two-port represents the ratio of the power P_2 delivered to the load resistance and the power P_1 delivered to the two-port at its input.

power gain: $G = \dfrac{P_2}{P_1} \ .$ (2.71)

The gain G_p denotes the ratio of the power P_2 delivered to the load resistance to the available generator power P_g of the signal source:

transducer power gain: $G_p = \dfrac{P_2}{P_g} \ .$ (2.72)

The available gain describes the ratio of the available output power P_{2av} at the two-port's output to the available generator power. The available output power P_{2av} results from choosing Z_l such that a complex conjugate match is achieved at the output, i.e. a power match.

available power gain: $G_{av} = \dfrac{P_{2av}}{P_g} \ .$ (2.73)

As can be seen, these definitions of the power gain are not pure two-port quantities. They also depend on the circuitry around the two-port.

In contrast, the maximum available power gain G_m is a pure two-port quantity. It can be achieved, if a power match is provided at the input as well as at the output of the two-port.

With the definition of the gain G_p according to Eq. (2.72), the spectrum W_{20} or the power P_{20} for a noiseless two-port can also be written as follows:

$$W_{20} = G_p \cdot W_g = G_p \cdot kT_0 \qquad (2.74)$$

or

$$P_{20} = G_p \cdot P_g = G_p \cdot kT_0 \cdot \Delta f \ . \qquad (2.75)$$

Here, $W_g = kT_0$ is the available spectral power density of the generator. For the noise figure F, another notation can thus be derived

$$
\begin{aligned}
F \ &= \ \frac{W_2}{G_p kT_0} = 1 + \frac{\Delta W_2}{G_p kT_0} \\
&= \ \frac{P_2}{G_p kT_0 \Delta f} = 1 + \frac{\Delta P_2}{G_p kT_0 \Delta f} \ .
\end{aligned}
\qquad (2.76)
$$

A further definition of the noise figure results, if the numerator and the denominator in Eq. (2.70) are multiplied by S_g. S_g denotes a signal power being available at the input and S_2 denotes the associated signal power at the output or at the load resistance Z_l, respectively. Using the following relation of the signal power and the gain,

$$S_2 = G_p \cdot S_g \ , \qquad (2.77)$$

the noise figure can be written as

$$F = \frac{P_2}{P_{20}} = \frac{S_g}{S_g} \cdot \frac{P_2}{G_p \cdot P_g} = \frac{S_g/P_g}{S_2/P_2} \ . \qquad (2.78)$$

From Eq. (2.78) it follows that the noise figure is equal to the quotient of the signal-to-noise ratio at the input to the signal-to-noise ratio at the output. The noise figure can thus be interpreted as a quantity which describes the deterioration of the signal-to-noise ratio when a signal passes through a two-port.

2.3.2 Calculation of the noise figure based on equivalent circuits

The calculation of the noise figure of a linear two-port, represented by an equivalent circuit, can be performed on the basis of the symbolic notation. For this purpose, the correlation between the individual equivalent noise sources has to be known.

For a general discussion, the equivalent circuit with a current and voltage source at the input (Fig. 2.25) is quite convenient. As the two-port itself is assumed to be noise-free, the noise figure can be determined at the plane $a - a'$ in front of the two-port, because a noiseless two-port connected in cascade does not change the noise figure of the whole setup. The calculation of the noise figure F of this circuit will be presented in the following. First, the

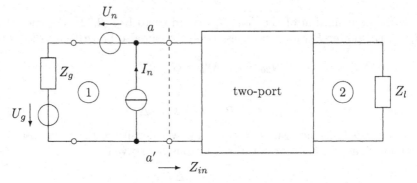

Fig. 2.25   Equivalent circuit of a linear two-port for the calculation of the noise figure.

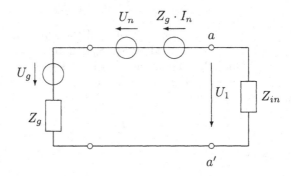

Fig. 2.26   Equivalent circuit of Fig. 2.25 with a voltage source instead of the current source.

current source I_n will be converted into an equivalent voltage source according to the basic rules for electrical networks (Fig. 2.26). The oriented arrows of the current and voltage sources can be chosen arbitrarily at first. However, for the subsequent transformations and calculations, the rules concerning the orientation of the arrows have strictly to be obeyed, otherwise, the resulting sign of the cross spectra may be wrong. From Fig. 2.26 the voltages U_1 or U_{10} for the noiseless case are obtained as

$$U_1 = (U_g + U_n + Z_g \cdot I_n) \cdot \frac{Z_{in}}{Z_{in} + Z_g} \tag{2.79}$$

$$U_{10} = U_g \cdot \frac{Z_{in}}{Z_{in} + Z_g} \tag{2.80}$$

and thus the noise figure F is

$$F = \frac{|U_1|^2}{|U_{10}|^2} = \frac{|U_g + U_n + Z_g \cdot I_n|^2}{|U_g|^2}$$

$$= \frac{|U_g|^2 + |U_n|^2 + |Z_g|^2 \cdot |I_n|^2 + 2 \cdot \mathrm{Re}\{Z_g \cdot U_n^* \cdot I_n\}}{|U_g|^2} \quad . \quad (2.81)$$

By introducing spectra instead of current and voltage phasors according to

$$W_g = |U_g|^2 = 2kT_0 \cdot \mathrm{Re}\{Z_g\} \quad (2.82)$$

$$W_u = |U_n|^2 , \quad W_i = |I_n|^2, \quad W_{ui} = U_n^* \cdot I_n , \quad (2.83)$$

the final result for the noise figure is

$$F = 1 + \frac{W_u + |Z_g|^2 \cdot W_i + 2\mathrm{Re}\{Z_g \cdot W_{ui}\}}{2 \cdot kT_0 \cdot \mathrm{Re}\{Z_g\}} \quad . \quad (2.84)$$

It can be seen that the noise figure is independent of the load resistance Z_l or the input impedance Z_{in}, while, on the other hand, it strongly depends on the source impedance Z_g. Consequently, the noise figure is not a pure two-port quantity. It is, however, independent of the load resistance.

As a further example, the noise figure of the circuit in Fig. 2.27 with current sources at both the input and the output will be calculated.

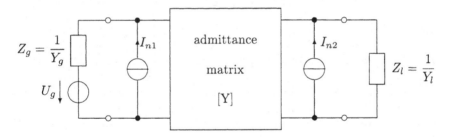

Fig. 2.27 Calculation of the noise figure for an equivalent circuit with two current sources.

According to problem 2.13, the following relationship can be derived for the noise figure of the circuit in Fig. 2.27 with $W_{n1} = |I_{n1}|^2$, $W_{n2} = |I_{n2}|^2$ and $W_{n12} = I_{n1}^* \cdot I_{n2}$:

$$F = 1 + \frac{|Y_{21}|^2 W_{n1} + |Y_{11} + Y_g|^2 W_{n2} - 2 \cdot \mathrm{Re}\{Y_{21}^* \cdot (Y_{11} + Y_g) W_{n12}\}}{|Y_{21}|^2 \cdot 2kT_0 \cdot \mathrm{Re}\{Y_g\}} \quad (2.85)$$

Problem

2.13 Derive equation (2.85).

2.3.3   Noise figure of two-ports with thermal noise

Very simple relations result for thermally noisy passive two-ports at a homogeneous temperature. As an example, a matched attenuator with a homogeneous temperature T_1 and the power attenuation $\kappa_1 = |S_{21}|^2$ shall be considered. The source resistance Z_g is assumed to be matched to the real reference impedance Z_0, i.e. $Z_g = Z_0$. Since the noise figure does not depend on the load resistance Z_l, the output load is also chosen to be matched with $Z_l = Z_0$ for simplicity (Fig. 2.28).

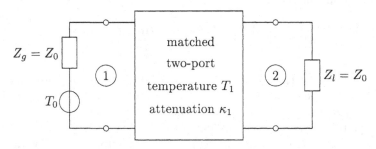

Fig. 2.28   For the calculation of the noise figure of a passive two-port at a homogeneous temperature T_1 being matched at both sides.

If, for the moment, the temperature of the two-port is assumed to be T_0, then the noise power P_2 at the load resistance Z_l is the available noise power for T_0 with $P_2 = kT_0\Delta f$. Here, a part $P_{20} = \kappa_1 kT_0\Delta f$ originates from the generator and a second part $\Delta P_2 = (1 - \kappa_1)kT_0\Delta f$ stems from the two-port. This latter part changes to $\Delta P_2 = (1 - \kappa_1)kT_1\Delta f$, if the temperature of the two-port is assumed to be at T_1. Thus, the following relationship results for the noise figure:

$$
\begin{aligned}
F &= \frac{P_2}{P_{20}} = 1 + \frac{\Delta P_2}{P_{20}} \\
&= 1 + \frac{(1 - \kappa_1)\, kT_1\Delta f}{\kappa_1 kT_0\Delta f} = 1 + \frac{1 - \kappa_1}{\kappa_1} \cdot \frac{T_1}{T_0} \ .
\end{aligned}
\tag{2.86}
$$

If the temperature of the matched two-port is equal to the reference temperature T_0, then Eq. (2.86) yields

$$
F = \frac{1}{\kappa_1} \quad \text{for} \quad T_1 = T_0 \ .
\tag{2.87}
$$

This means that an attenuator at ambient temperature T_0 with an attenuation of, for example 6 dB, has a noise figure of 6 dB. The relationships (2.86) and (2.87) are also valid for a matched non-reciprocal passive two-port, as for example an isolator or a circulator with a matched termination.

The noise figure of a passive and reciprocal two-port can also be determined by means of the dissipation theorem. This is a particularly useful approach if the two-port consists of more than one temperature region. The part ΔP_2 of the two-port at the load impedance is equal to

$$\Delta P_2 = kT_1\beta_1\Delta f \ . \tag{2.88}$$

Here, the coefficient β_1 represents the fraction of the power absorbed in the two-port if a signal is fed in from the load side. Obviously, due to the assumptions of reciprocity and output match, we have

$$\beta_1 + \kappa_1 = 1 \ , \tag{2.89}$$

which for the noise figure immediately leads to the result of Eq. (2.86).

In a similar way, the calculation of the noise figure of a passive two-port at the homogeneous temperature T_1 can be performed if the two-port is not matched at both ports and if, additionally, the generator and load resistances are not equal to the real reference impedance Z_0. As the noise figure still does not depend on the load impedance Z_l, again for simplicity a complex conjugate match or a power match, respectively, can be assumed for the load impedance at the output.

If the temperature of the two-port is equal to T_0 for the time being, then the noise power P_2 at the load impedance Z_l is equal to $P_2 = kT_0\Delta f$. Let the available power gain be denoted by G_{av}, then the part $P_{20} = G_{av}kT_0\Delta f$ originates from the generator and the part $\Delta P_2 = P_2 - P_{20} = (1 - G_{av})kT_0\Delta f$ stems from the two-port. The latter part becomes $\Delta P_2 = (1 - G_{av})kT_1\Delta f$, if the temperature of the two-port is T_1. Thus, the noise figure is given by

$$F = \frac{P_2}{P_{20}} = 1 + \frac{\Delta P_2}{P_{20}} = 1 + \frac{(1 - G_{av})}{G_{av}} \cdot \frac{T_1}{T_0} \ . \tag{2.90}$$

Apparently, G_{av} replaces the term κ_1 in Eq. (2.86). For a match on all sides and thus $\kappa_1 = G_{av}$, the equations (2.86) and (2.90) are equal. Eq. (2.90) is also valid for a non-reciprocal passive two-port.

Also for the mismatched case, it is instructive to determine the noise figure of a passive but reciprocal two-port with the help of the dissipation theorem. Equation (2.88) is again valid, and because of the reciprocity and the power match at the output, the absorption coefficient β_1 is apparently related to the available gain G_{av} by

$$G_{av} = 1 - \beta_1 \ . \tag{2.91}$$

Here, use was made of the fact that for a reciprocal two-port the gain from port 1 to port 2 is equal to the gain from port 2 to port 1, i.e. $G_{p12} = G_{p21}$, as will be shown in problem 2.14.

The calculation of the noise figure of a passive and reciprocal two-port with the help of the dissipation theorem is advantageous if the two-port consists of more than one temperature region. This will be demonstrated in problem 2.15.

If the temperature of a passive two-port, reciprocal or not, is equal to the reference temperature $T_1 = T_0$, then the noise figure according to Eq. (2.90) is

$$F = \frac{1}{G_{av}} \; . \tag{2.92}$$

In problem 2.16, Eq. (2.92) will be verified for an example by direct calculation.

Problems

2.14   Prove that the gain of a reciprocal network does not depend on the direction.

2.15   Two attenuators, being matched on both sides, with an insertion loss of 3 dB and 6 dB, respectively, are cascaded. The temperatures are T_1 and T_2. What is the noise figure of the combination?

2.16   Calculate the noise figure for the following equivalent circuit. Prove that the noise figure does not depend on the load impedance and that Eq. (2.92) is valid. R_1, R_2, Z_g are assumed to be real.

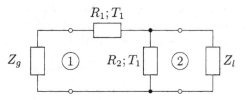

2.3.4   Noise figure of cascaded two-ports

For several two-ports in cascade connection with individual power gains or attenuations $\kappa_1, \kappa_2, \kappa_3, \ldots$ and noise figures $F_1, F_2, F_3, \ldots$ the noise figure of the entire network can be calculated (Fig. 2.29). At first, it will be assumed

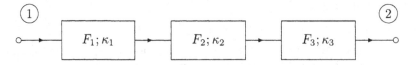

Fig. 2.29   The noise figure of two-ports in cascade connection.

that each two-port is terminated on both sides by the reference impedance Z_0, thus no reflections occur. The source and the load impedances are also assumed to be equal to Z_0. This is a common design goal for high-frequency circuits since it usually results in a smooth frequency response of the gain. For two-ports, matched at both sides and connected to matched terminations, the following relation for the output noise contribution of the two-port can be derived from Eq. (2.76) and $G_p = \kappa$:

$$\Delta W = (F - 1) \cdot \kappa \cdot kT_0 \; . \tag{2.93}$$

Thus the total or overall noise figure is given by

$$
\begin{aligned}
F_t &= 1 + \frac{(F_1 - 1)\kappa_1\kappa_2\kappa_3 kT_0}{\kappa_1\kappa_2\kappa_3 kT_0} \\
&\quad + \frac{(F_2 - 1)\kappa_2\kappa_3 kT_0}{\kappa_1\kappa_2\kappa_3 kT_0} + \frac{(F_3 - 1)\kappa_3 kT_0}{\kappa_1\kappa_2\kappa_3 kT_0} \\
&= F_1 + \frac{F_2 - 1}{\kappa_1} + \frac{F_3 - 1}{\kappa_1\kappa_2} \; . \tag{2.94}
\end{aligned}
$$

This form of the so-called cascade formula is valid only, if the same impedance conditions hold for the measurement or calculation of the individual noise figures and for the cascade connection, i.e. the match on both sides.

The determination of the noise figure F_t for the whole circuit is more difficult if arbitrarily mismatched two-ports are connected in cascade and if, additionally, the source and load impedances can also be chosen arbitrarily. Most of the difficulties are caused by the fact that, in general, the overall gain is not equal to the product of the individual gains.

The following consideration will be restricted to the cascade connection of only two two-ports. However, the extension to more than two stages will be obvious. Let the noise powers at the output of the single stages, generated by their internal sources, be denoted by ΔP_{n1} and ΔP_{n2} and let P_g be the available source power (Fig. 2.30).

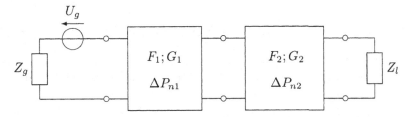

Fig. 2.30   Derivation of the cascade formula for mismatched two-ports.

Note that for the measurement or the definition of the noise figure, the load conditions for the single stages must be the same as for the cascade connection. If G_1 and G_2 denote the gain of the first and second stage, respectively, (Fig. 2.30), the following equations hold for the individual noise figures F_1 and F_2 of the first and second stage

$$F_1 = 1 + \frac{\Delta P_{n1}}{G_1 \cdot P_g}, \quad F_2 = 1 + \frac{\Delta P_{n2}}{G_2 \cdot P_g}, \tag{2.95}$$

provided that G_2 and F_2 are determined for the same source impedance which applies to the cascade connection.

In the following, a relation between the noise figure of the whole circuit and the noise figure of the individual stages will be derived. For the overall noise figure F_t the following equation is obtained where index av denotes the available power:

$$
\begin{aligned}
F_t &= \frac{G_2 \cdot (G_1 \cdot P_g)_{av} + G_2 \cdot (\Delta P_{n1})_{av} + \Delta P_{n2}}{G_2 \cdot (G_1 \cdot P_g)_{av}} \\
&= \underbrace{1 + \frac{(\Delta P_{n1})_{av}}{(G_1 \cdot P_g)_{av}}}_{F_1} + \frac{\Delta P_{n2}}{G_2 \cdot (G_1 \cdot P_g)_{av}}.
\end{aligned}
\tag{2.96}
$$

Note that the first part of Eq. (2.96) is equal to the noise figure F_1. Since the noise figure does not depend on the load impedance, the load can be chosen arbitrarily. For simplicity, a power match is assumed at the output. As a consequence, F_1 of Eq. (2.95) can be written as:

$$F_1 = 1 + \frac{\Delta P_{n1}}{G_1 \cdot P_g} = 1 + \frac{(\Delta P_{n1})_{av}}{(G_1 \cdot P_g)_{av}}. \tag{2.97}$$

The second term of Eq. (2.96) can be related to the available gain. According to the definition of the available gain we have:

$$(G_1 \cdot P_g)_{av} = G_{1av} \cdot P_g. \tag{2.98}$$

Thus Eq. (2.96) yields:

$$F_t = F_1 + \frac{\Delta P_{n2}}{G_{1av} \cdot G_2 \cdot P_g} = F_1 + \frac{F_2 - 1}{G_{1av}}. \tag{2.99}$$

This is the general relation between the overall noise figure F_t and the single noise figures F_1 and F_2. The influence of the noise figure of the second stage is reduced by the available gain of the first stage. However, if the first stage causes an attenuation, then G_{1av} is smaller than one and the influence of the noise of the second stage will be large.

The overall noise figure F_{t3} of a three-stage setup can be found by considering F_t of Eq. (2.99) as the first stage. The influence of F_3 can be taken into account by applying Eq. (2.99) once again. If the gain of the first two stages is denoted by G_{12av}, we get

$$F_{t3} = F_1 + \frac{F_2 - 1}{G_{1av}} + \frac{F_3 - 1}{G_{12av}}. \tag{2.100}$$

The available gain of two stages connected in cascade is equal to the product of the single available gains:

$$G_{12av} = G_{1av} \cdot G_{2av} \ . \tag{2.101}$$

This is comprehensible, if the available power P_{2av} at the output of the second stage is considered. For P_{2av} it is by definition

$$P_{2av} = G_{2av} \left(G_1 \cdot P_g \right)_{av} = G_{1av} \cdot G_{2av} \cdot P_g = G_{12av} \cdot P_g \ , \tag{2.102}$$

which leads to Eq. (2.101). The cascade formula of the noise figure is thus given by:

$$F_t = F_1 + \frac{F_2 - 1}{G_{1av}} + \frac{F_3 - 1}{G_{1av} \cdot G_{2av}} + \frac{F_4 - 1}{G_{1av} \cdot G_{2av} \cdot G_{3av}} + \dots \ . \tag{2.103}$$

For matched two-ports the relation (2.103) is equal to Eq. (2.94) due to $\kappa = G_{av}$.

Problem

2.17 Two amplifiers or, more general, linear two-ports have noise figures F_1 and F_2 and available gains G_{1av} and G_{2av}. In which order should the amplifiers be cascaded in order to achieve the lowest overall noise figure?

A passive network at the homogeneous temperature T_0 with the available gain G_{av} is connected to the input of an amplifier with noise figure F_2. According to the cascade formula, the following overall noise figure F_t results for this configuration:

$$F_t = \frac{1}{G_{av}} + \frac{F_2 - 1}{G_{av}} = \frac{F_2}{G_{av}} \ . \tag{2.104}$$

Expressing the noise figure in dB, then a lossy passive two-port, connected in front of a circuit, increases the noise figure by the dB-value of the available gain of the two-port. If the two-port represents a matched attenuator, then the noise figure in dB increases by the dB-value of the attenuation. In case of an unsymmetrical passive two-port, the available gain generally depends on the orientation of the device. The overall noise figure according to Eq. (2.104) may change, if the two-port is turned around. In addition, the noise figure F_2 of the amplifier generally depends on the output impedance of the two-port connected to its input.

2.3.5 Noise matching

For a given noisy two-port, the noise figure only depends on the source impedance Z_g. Setting the generator impedance Z_g to the value which minimizes the noise figure is called noise matching. If the source impedance cannot be changed directly, then an appropriate transformation circuit can be introduced between the generator and the device under test, which converts Z_g to its optimum value. The transformation can be performed with the help of transformers, reactances or transmission line elements. However, the loss of the transformation network should be as low as possible. In general, noise matching is not identical to power matching. For high frequencies and broadband amplifiers, systems are often designed for power matching. In the following, the noise matching will be discussed for an equivalent circuit with current and voltage sources at the input of the device under investigation, as shown in Fig. 2.31.

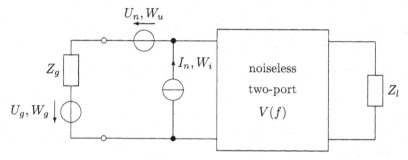

Fig. 2.31   For the discussion of noise matching.

Based on the discussion of section 2.3.2 and using the definitions

$$Z = Z_g = |Z|e^{j\varphi} \ , \quad W_{ui} = |W_{ui}|e^{j\vartheta} \ , \tag{2.105}$$

according to Eq. (2.84) the noise figure F is given by

$$F = 1 + \frac{W_u + |Z|^2 \cdot W_i + 2 \cdot |Z| \cdot |W_{ui}| \cdot \cos(\varphi + \vartheta)}{2kT_0 \cdot |Z| \cos \varphi} \ . \tag{2.106}$$

We are looking for the minimum noise figure as a function of $Z_g \equiv Z$. For this purpose, the partial derivatives with respect to $|Z|$ and φ are calculated. One condition for noise matching is

$$\frac{\partial F(|Z|, \varphi)}{\partial |Z|} = 0 \ . \tag{2.107}$$

From this the magnitude of the optimum source impedance Z_{opt} is obtained:

$$|Z_{opt}|^2 \cdot W_i = W_u \tag{2.108}$$

or

$$|Z_{opt}| = \sqrt{\frac{W_u}{W_i}} \ . \tag{2.109}$$

Thus the magnitude $|Z_{opt}|$ does not depend on the correlation or the cross-spectrum W_{ui}. From the second condition

$$\frac{\partial F(|Z|, \varphi)}{\partial \varphi} = 0 \qquad (2.110)$$

and with $|Z| = |Z_{opt}|$ according to Eq. (2.109) we get

$$\sin \varphi_{opt} = \frac{|W_{ui}|}{\sqrt{W_u \cdot W_i}} \sin \vartheta = |k_{ui}| \sin \vartheta \quad \text{with} \quad |\varphi_{opt}| \leq \frac{\pi}{2} \ . \qquad (2.111)$$

The phase of the optimum source impedance only depends on the correlation of the noise sources. The minimum noise figure F_{min} is obtained by inserting the optimum source impedance into Eq. (2.106):

$$F_{min} = 1 + \frac{\sqrt{W_u \cdot W_i}}{kT_0} \left(|k_{ui}| \cdot \cos \vartheta + \sqrt{1 - |k_{ui}|^2 \sin^2 \vartheta} \right) \ . \qquad (2.112)$$

For the case of a zero correlation, i.e. $k_{ui} = 0$ Eq. (2.112) yields

$$F_{min} = 1 + \frac{\sqrt{W_u \cdot W_i}}{kT_0} \ . \qquad (2.113)$$

For the case of maximum correlation, i.e. $|k_{ui}| = 1$, and with $|\vartheta| = \pi/2$ the result $F_{min} = 1$ is obtained from Eq. (2.112). This means that fully correlated noise sources may compensate each other.

Another conclusion is that for either $W_u = 0$ or $W_i = 0$ the optimum noise figure can also be $F_{min} = 1$. However, for these cases extreme impedance transformations are required, because due to Eq. (2.109), Z_{opt} will either converge towards zero or infinity.

Instead of considering the magnitude and phase of the source impedance, it is possible to perform a similar noise match derivation with respect to the real and imaginary part of the generator source admittance, G_g and B_g. The following equation for the noise figure is derived from Eq. (2.84) with $W_{ui} = C_r + jC_i$:

$$F = 1 + \frac{(G_g^2 + B_g^2) \cdot W_u + W_i + 2G_g C_r + 2B_g C_i}{4kT_0 G_g} \ . \qquad (2.114)$$

The derivative with respect to the conductance

$$\frac{\partial F}{\partial G_g} = \frac{W_u}{4kT_0} + \frac{B_g^2 W_u + W_i + 2B_g C_i}{4kT_0} \left(-\frac{1}{G_g} \right)^2 = 0 \qquad (2.115)$$

as well as the derivative with respect to the susceptance

$$\frac{\partial F}{\partial B_g} = \frac{2B_g W_u + 2C_i}{4kT_0 G_g} = 0 \qquad (2.116)$$

of the source admittance are set to zero in order to find the optimum source admittance. This leads straight to the following relations:

$$B_{opt} = -\frac{\mathsf{C}_i}{\mathsf{W}_u} \;, \tag{2.117}$$

$$G_{opt}^2 = \frac{\mathsf{W}_i}{\mathsf{W}_u} - B_{opt}^2 \;, \tag{2.118}$$

or, similar to Eq. (2.109), leads to the magnitude of the optimum source admittance:

$$|Y_{opt}|^2 = G_{opt}^2 + B_{opt}^2 = \frac{\mathsf{W}_i}{\mathsf{W}_u} \;. \tag{2.119}$$

Inserting these results into Eq. (2.114) yields a further expression for the minimum noise figure in dependence of the real and imaginary part of the source admittance

$$F_{min} = 1 + \frac{\mathsf{W}_i + G_{opt}\mathsf{C}_r + B_{opt}\mathsf{C}_i}{2kT_0 G_{opt}} \;. \tag{2.120}$$

Based on the so-called noise parameters of a linear two-port, namely the optimum source admittance, the minimum noise figure and further on the so-called equivalent noise resistance defined by

$$R_n = \frac{\mathsf{W}_u}{4kT_0} \;, \tag{2.121}$$

a further noise representation of a linear two-port is obtained. This description is equivalent to the representation on the basis of the noise spectra:

$$\mathsf{W}_u = 4kT_0 R_n \;, \tag{2.122}$$
$$\mathsf{W}_i = 4kT_0 R_n |Y_{opt}|^2 \;, \tag{2.123}$$
$$\mathsf{C}_r = 2kT_0 \cdot (F_{min} - 1 - 2R_n G_{opt}) \;, \tag{2.124}$$
$$\mathsf{C}_i = -4kT_0 R_n B_{opt} \;. \tag{2.125}$$

Replacing the noise spectra in Eq. (2.114) by noise parameters leads to a further equation for the noise figure with a parabolic characteristic:

$$F = F_{min} + \frac{R_n}{G_g} \cdot |Y_g - Y_{opt}|^2 \;. \tag{2.126}$$

Hence, this description will be called the parabolic noise figure relation in the following.

For passive, thermally noisy two-ports at a homogeneous temperature the noise figure is equal to the reciprocal value of the available gain (Eq. (2.92)). The noise figure thus becomes minimal, if the available gain is equal to the maximum available gain G_m, which can be achieved by power matching at both the input and the output. This is still valid if the homogeneous temperature differs from T_0 (Eq. (2.90)).

For passive two-ports with more than one temperature region, i.e. an inhomogeneous temperature distribution, the power match on both sides must not necessarily be identical to the noise match. However, for a passive, thermally noisy two-port at a homogeneous temperature the power match on both sides and the noise match are identical. Consequently, the optimum source impedance is $Z_g = Z_{opt}$. The minimum noise figure F_{min}, which is achievable by the choice of Z_{opt}, does not change, if the load impedance is varied, because the noise figure generally does not depend on the load impedance. For a load impedance deviating from the power match value at the output, normally the power match at the input will also be removed. Therefore, a power match on one side only of a passive two-port with a homogeneous temperature must not necessarily be equal to a minimum noise figure (noise matching).

As can be shown (problem 2.18), the contours of constant noise figure for a given noisy two-port, which does not need to be passive, are circles in the complex plane of the source impedance Z_g. These circles, which are not necessarily concentric, enclose the minimum noise figure F_{min}. The same applies for a constant available gain in dependance of the source impedance Z_g. These contours are also non concentric circles, which confine the point of maximum available gain.

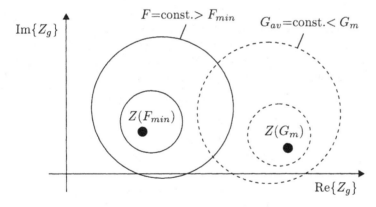

Fig. 2.32 The contours (circles) of constant noise figure and constant available gain in the complex plane of the source impedance Z_g.

For a low degradation of the overall noise figure by the second stage in a cascade circuit, the available gain of the first stage should be as high as possible. Thus, Z_g should be chosen such that the noise figure becomes as low as possible while the available gain should not become too small. As a compromise, Z_g should thus preferably be chosen in the vicinity of $Z(F_{min})$ as well as $Z(G_m)$.

Problem

2.18   Prove that the contours of constant noise figure are circles in the complex plane of the generator source impedance Z_g.

The noise match can also be discussed in terms of noise waves. The noise figure as a function of the source reflection coefficient Γ_g will be derived for a two-port where the noise is represented by the noise waves X_1 and X_2 (Fig. 2.33). The noisy two-port is described by the scattering matrix $[S]$. In

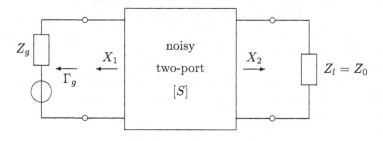

Fig. 2.33   Noise figure description with noise waves.

order to simplify the calculations, the load impedance Z_l is chosen equal to the reference impedance Z_0. The source reflection coefficient Γ_g is related to Z_g by:

$$\Gamma_g = \frac{Z_g - Z_0}{Z_g + Z_0} \ . \tag{2.127}$$

For the circuit in Fig. 2.33 the following expression can be found for the noise figure:

$$
\begin{aligned}
F &= 1 + \frac{|X_1 \cdot S_{21} \cdot \Gamma_g + (1 - S_{11} \cdot \Gamma_g) X_2|^2}{kT_0 \cdot (1 - |\Gamma_g|^2) \cdot |S_{21}|^2} \\
&= 1 + \frac{|X_1 \cdot \Gamma_g + (1 - S_{11} \cdot \Gamma_g) X_2 / S_{21}|^2}{kT_0 \cdot (1 - |\Gamma_g|^2)} \ .
\end{aligned}
\tag{2.128}
$$

With the abbreviations $W_1 = |X_1|^2$, $W_2 = |X_2/S_{21}|^2$, $W_{12} = X_1^* \cdot X_2/S_{21}$, $W_{12} = |W_{12}| \cdot e^{j\vartheta_1}$, $\Gamma_g = |\Gamma_g| \cdot e^{j\vartheta_2}$ the noise figure is given by:

$$
\begin{aligned}
F &= 1 + \frac{|\Gamma_g|^2 W_1 + |1 - S_{11}|\Gamma_g|e^{j\vartheta_2}|^2 W_2 + 2 \cdot |W_{12}| \cdot |\Gamma_g| \cdot \cos(\vartheta_1 - \vartheta_2)}{kT_0 (1 - |\Gamma_g|^2)} \\
&\quad - \frac{2\operatorname{Re}\{S_{11}\}|\Gamma_g|^2|W_{12}| \cos\vartheta_1}{kT_0 (1 - |\Gamma_g|^2)} \ .
\end{aligned}
\tag{2.129}
$$

From

$$\frac{\partial F(\Gamma_g, \vartheta_2)}{\partial |\Gamma_g|} = 0 \qquad (2.130)$$

an equation for the magnitude of the optimum source reflection coefficient is obtained:

$$|\Gamma_{g,opt}|^2 (-\text{Re}\{S_{11}\} \cos \vartheta_2 + |W_{12}| \cos(\vartheta_1 - \vartheta_2))$$
$$+ |\Gamma_{g,opt}|(W_1 + W_2(1 - |S_{11}|^2) - 2|W_{12}|\text{Re}\{S_{11}\} \cos \vartheta_1)$$
$$- \text{Re}\{S_{11}\} \cos \vartheta_2 + |W_{12}| \cos(\vartheta_1 - \vartheta_2) = 0 \ . \qquad (2.131)$$

Furthermore, from

$$\frac{\partial F(|\Gamma_g|, \vartheta_2)}{\partial \vartheta_2} = 0 \qquad (2.132)$$

the optimum phase of the reflection coefficient is given by:

$$\vartheta_{2,opt} = \arctan \left(\frac{-|W_{12}| \sin \vartheta_1}{\text{Re}\{S_{11}\} - |W_{12}| \cos \vartheta_1} \right) \ . \qquad (2.133)$$

If the two-port is matched on both sides with $S_{11} = S_{22} = 0$, then $\sin(\vartheta_1 - \vartheta_2) = 0$ and for the magnitude we obtain:

$$|\Gamma_{g,opt}| = \frac{W_1 + W_2 \pm \sqrt{(W_1 + W_2)^2 - 4 \cdot |W_{12}|^2}}{2 \cdot |W_{12}|} \ . \qquad (2.134)$$

For non-correlated two-port noise waves, i.e. $W_{12} = 0$, Eq. (2.133) for the optimum phase reduces to $\sin \vartheta_2 = 0$ and the magnitude of the optimum reflection coefficient is

$$|\Gamma_{g,opt}| = \frac{W_1 + W_2(1 + |S_{11}|^2)}{2\text{Re}\{S_{11}\}} \pm \sqrt{\left(\frac{W_1 + W_2(1 + |S_{11}|^2)}{2\text{Re}\{S_{11}\}} \right)^2 - 1} \ . \qquad (2.135)$$

It can be observed that, for an existing correlation, the noise figure can be minimized by a certain mismatch of the generator, that is $\Gamma_g \neq 0$. For thermally noisy two-ports at a homogeneous temperature being matched on both sides, it was noticed that the noise waves X_1 and X_2 are uncorrelated. For a minimum noise figure, the choice $\Gamma_g = 0$ and thus $Z_g = Z_0$, i.e. input matching, represents the best choice. This can also be derived from Eq. (2.131) with $S_{11} = 0$ and $W_{12} = 0$.

Similar to the parabolic relation for the noise figure as a function of the source admittance, the noise figure can also be represented by an equivalent relationship as a function of the generator source reflection coefficient:

$$F = F_{min} + 4\frac{R_n}{Z_0} \cdot \frac{|\Gamma_g - \Gamma_{opt}|^2}{|1 + \Gamma_{opt}|^2(1 - |\Gamma_g|^2)} \ . \qquad (2.136)$$

Problem

2.19   Prove that Eq. (2.126) and Eq. (2.136) are equivalent.

3

Measurement of Noise Parameters

The accurate determination of noise parameters is principally based on the precise measurement of noise powers. However, the precise measurement of noise powers is challenging and requires some experience. This is among others due to the fact that, in general, noise signals are very weak and can not be displayed without a sufficient preamplification. Furthermore, a band-pass filter is needed in order to identify the frequency dependence of the noise power. By applying, for example, a narrow band-pass filter with a variable center frequency, the noise power density as a function of frequency, i.e. the noise spectrum, can be measured. For a quantitative evaluation, obviously the pass-band shape of the filter has to be known. In order to calculate the available noise power of a two-port, in addition to the overall gain the matching properties between the device under test and the amplifier are required.

Concerning the necessary preamplification, one has to take into account that the first preamplifier also generates noise, in fact, often of the same order of magnitude as the device under test. As a consequence, it may be necessary to use special comparator circuits, so-called switching radiometers, for example, in order to eliminate the impact of the preamplifier noise as far as possible. In this situation, it is even more difficult to remove the noise contribution of the preamplifier, if the impedance of the device under test is unknown and frequency-dependent as well as complex. In this case, radiometers with a compensation circuit will be used. These radiometers require preamplifiers with not only uncorrelated input and output noise waves but also a well-defined input noise temperature. Then, the so-called available noise temperature of the device under test can be determined, which is a characteristic property of the device under test and which does not depend on the matching of the device.

An interesting variant of a radiometer is the so-called correlation radiometer, which, different from the previously mentioned switching radiometer, does not need a switch in front of the first preamplifier. Such a switch can influence the measurement results because of its non-ideal characteristics. For a correlation radiometer optional switches, preferably electronic switches, can be inserted advantageously after sufficient preamplification. Then the switches are less critical, because their noise contribution will be of minor importance as compared with the preamplified noise of the device under test. A section about the measurement of the cross spectrum and the cross correlation will help to improve the understanding of the correlation radiometer.

The determination of the noise factor or noise figure, respectively, belongs to the routine measurements tasks in high-frequency engineering. It is important that the measurement technique of the noise figure is closely linked to its definition. According to the definition of the noise figure the load resistance is assumed to be noise-free or noiseless. In practice, this can only be achieved by a sufficient amplification, so that the noise contribution of the load resistance can be neglected. However, the so-called post-amplifier influences the measured noise figure. With the help of the cascade formula the influence of the post-amplifier can be taken into account, so that the noise figure of the device under test can be determined without the contribution of the post-amplifier. A correction of the noise figure can be omitted, if the device under test itself has enough gain.

In addition to the determination of the noise figure, different methods for the measurement of the set of noise parameters, which completely describe the noise behavior of a thermally noisy linear two-port, will be presented. As already explained for the measurement of the noise figure, the noise parameters such as the minimum noise figure and the optimum generator admittance usually can not be measured directly. Low-noise amplification of the weak noise signals is required, so that an error correction of the measured data becomes necessary in order to determine the true parameters of the device under test. For this purpose, a so-called de-embedding method will be described.

All presented noise measurement procedures, such as the measurement of the noise temperature with a radiometer or the measurement of the noise figure and the noise parameters, have in common that an arbitrarily precise measurement is not possible, even if the pre-amplification, the mismatching and the noise contribution of the amplifier are exactly known. This is due to the fact that a principle error bound limits the precision of the measurements. This principle error bound results from the stochastic nature of the measurement signals. For a finite frequency bandwidth and a finite measurement time noise power can not be determined with arbitrary accuracy. The emerging error becomes smaller the larger the frequency bandwidth and the measurement time are. The high-frequency range is thus especially well suited for noise measurements, because large absolute bandwidths are in general available. This principle measurement error is the main difference between noise measurement techniques and measurements with coherent signals. A further important difference is that for the noise measurements the inevitable amplifier noise is generally of the same order of magnitude as the measurement quantity.

In comparison to this principle error bound, which cannot be avoided, all other uncertainties such as the post-amplifier noise, the mismatch of the device under test, the magnitude and drift of the amplifier gain can be kept arbitrarily small. It will thus be one aim to reduce these uncertainties down to the order of the principle measurement errors caused by the finite measurement time and bandwidth in order to achieve highly accurate noise measurements.

Furthermore, noise measurements can often become erroneous by cross talk and disturbing radiation. Such disturbing signals can influence the reproducibility of the measurements. It is thus necessary to pay attention to shielding and filtering in order to provide sufficient suppression of such interferences. Under favorable conditions, noise measurements are comparable to measurements with deterministic signals, with respect to reproducibility and precision.

In the following section, the measurement of the cross-correlation function and the cross-spectrum will be discussed first. The principle measurement error, which results from the finite measurement time and bandwidth, will be derived in section 3.1.8.

3.1 MEASUREMENT OF NOISE POWER

The accurate determination of noise parameters is principally based on the precise measurement of noise power. In this section, some of the most important methods for the measurement of the noise power will be presented.

3.1.1 Power measurement on the basis of a thermocouple

This method is based on the measurement of an increase of temperature with the help of a thermocouple. A thermocouple is heated by an RF signal. The increase of temperature causes a dc voltage, which can be measured and which is proportional to the absorbed RF power.

Initially, metals like a combination of bismuth and antimony (Bi, Sb) and gold were used as the contact material, as shown in Fig. 3.1. For the thermodynamic equilibrium, i.e. when all junctions are at the same temperature, all contact voltages will compensate. Then, no dc voltage can be measured between the outer connectors. However, if the Bi-Sb-junction is heated to the temperature T_1 while the Sb-Au-junctions remain at the ambient temperature T_0, then a thermoelectric voltage can be measured at the outer connectors, which is proportional to the temperature difference $T_1 - T_0$. The proportion-

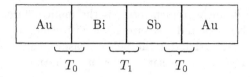

Fig. 3.1   Setup of a thermocouple on the basis of bismuth and antimony.

ality constant is called the thermoelectric coefficient. For the combination of Bi-Sb the thermoelectric coefficient is particularly high with a value of about $110\mu V/K$. Obviously, the junction area Bi-Sb must have a high thermal resistance towards its environment, in order to obtain a reasonable increase of temperature even for low power inputs.

On the other hand, the heat capacity must be sufficiently small, so that a temperature equilibrium can be reached quickly. For this reason, thin metal strips are used, which are arranged in such a way that the heated junction is not in contact with the substrate material (see Fig. 3.2). For instance, the commonly used sapphire substrate has a good heat conductivity. Thus one can achieve cold outer junctions that are almost at the same temperature T_0.

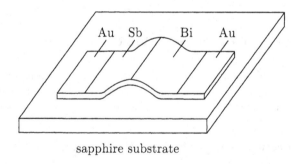

sapphire substrate

Fig. 3.2   Thermocouple of metal strips on a sapphire substrate.

The metal strips are designed with the aim to realize a constant high-frequency resistance of 100 Ω over a frequency range as broad as possible. The absorbed high-frequency power will be converted into heat, raising the temperature of the Bi-Sb-junction. The whole circuit can be realized as shown in Fig. 3.3. Preferably, two thermocouples are used, which are connected in series for the dc signals and in parallel for the RF-signals. In order to achieve an accurate match in a 50 Ω environment, a 100 Ω resistance is chosen.

By the use of two thermocouples the need for an inductor as a filter element can be avoided. This has the advantage that capacitors can be used instead of inductors. Capacitors can more easily be realized to cover a broad

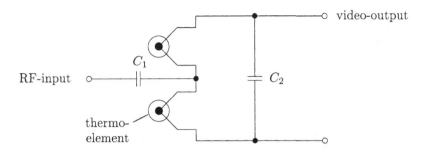

Fig. 3.3 Power measurement with two thermocouples.

frequency range. However, the Bi-Sb-thermocouple has a number of disadvantages, e.g. bad reproducibility, poor match and a small overload capacity.

State-of-the-art thermocouples with better performance are typically realized as metal-semiconductor devices in thin film technology. A standard structure is shown in Fig. 3.4. For this configuration a combination of highly

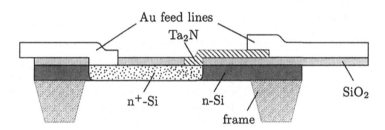

Fig. 3.4 Metal-semiconductor thermocouple.

doped silicon and tantalum nitride (n^+-Si/Ta_2N) is used as a thermocouple junction. The Ta_2N-film also serves as a 100 Ω high-frequency resistor.

The thermoelectric coefficient slightly depends on the doping. Hence, for a good reproducibility, a specific doping concentration has to be realized as precise as possible. The thermoelectric coefficient as designed for this application is about $250\mu V/K$. It is thus higher than for Bi-Sb. There is a certain dependence between the thermoelectric voltage and the absolute temperature, which must be compensated. The thermal time constant, determined by the thermal resistance and the thermal capacity, is of the order of about 0.1 ms. In order to measure the low dc voltages, the dc signals are amplified and modulated using, e.g. field effect transistors in a chopper amplifier with a clock frequency of e.g. 220 Hz. This chopper amplifier should be placed as close as possible to the thermocouple sensor, ideally on the same substrate. In order to eliminate further unwanted thermoelectrical voltages, all conductors are realized in gold. The ac signal of 220 Hz, which lies between the harmonics of 50 Hz, is connected to a power meter instrument, where it is further amplified, transformed back to a dc votage by a phase sensitive detector, A/D-converted and displayed.

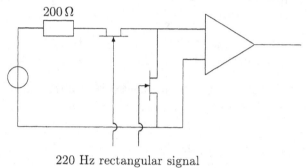

220 Hz rectangular signal

Fig. 3.5   Principle setup of a chopper amplifier.

3.1.2   Thermistor bridge

Thermistors are small samples of sintered metal oxide with contact wires on opposite sides of the sample. The resistance R_{th} of this material is strongly dependent on temperature, i.e. the resistance decreases with increasing temperature (NTC-resistors, Negative Temperature Coefficient). Two thermistors are placed in a bridge circuit, as shown in Fig. 3.6. A high voltage gain factor and strong feedback result in a nearly zero voltage E_1 at the input of the amplifier. This means that the bridge balances itself automatically, with $2R_{th} = R$. The resistors R are temperature independent. The balance requires a sufficiently high current through the thermistor, so that direct current heating increases the temperature and reduces the resistance R_{th}. The voltage E_2 at the bridge may be E_{20} for the state of the bridge without RF-power. Then the dc power supplied to the two thermistors is

$$P_{th0} = \frac{1}{4}\frac{E_{20}^2}{R} \ . \tag{3.1}$$

If RF-power is applied to the bridge, then the feedback again forces $2R_{th} = R$, but now the dc current through the thermistors is lower. The dc power absorbed in the thermistors is

$$P_{th1} = \frac{1}{4}\frac{E_{21}^2}{R} \ . \tag{3.2}$$

The reduction in dc power must be equal to the applied RF-power P_l.

$$P_l = P_{th0} - P_{th1} = \frac{1}{4R}\left(E_{20}^2 - E_{21}^2\right) \ . \tag{3.3}$$

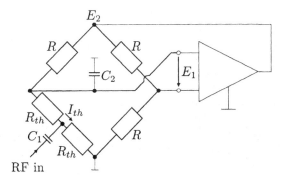

Fig. 3.6 Bridge circuit of a thermistor.

In order to account for variations of the ambient temperature, E_{20} may be measured by a second identical bridge, to which no RF-power is applied.

The thermistor bridge has a number of disadvantages compared with the thermocouple, e.g. a smaller dynamic range and a higher time constant. However, the thermistor bridge has the advantage, that due to the substitution principle, the RF-power measurement is an absolute power measurement. But this is not a real advantage for noise measurements, because most of the noise measurements, which are discussed in the next chapters, rely on relative power measurements only.

3.1.3 Power measurements with Schottky-diodes

A detector with Schottky-diodes can be used for highly sensitive power measurements, if the input signals are not too strong so that the detector is operated in the square-law region of the current-voltage characteristic of the diode.

The use of Schottky-diodes instead of pn-diodes in high-frequency detectors is motivated by the fact that Schottky-diodes are based on a majority carrier effect and are thus very fast. The nonlinear current-voltage characteristic of a Schottky-diode, which is principally based on a metal-semiconductor junction, can be described to a good approximation by the following formula:

$$i(t) = I_{SS} \left(\exp\left(\frac{u(t)}{U_T} \right) - 1 \right) \quad \text{and} \quad U_T = \frac{\tilde{n}\,kT}{q} \qquad (3.4)$$

with

$i(t), u(t)$ current or voltage of the Schottky-diode
I_{SS} reverse saturation current
U_T temperature voltage (25.9 mV for $\tilde{n} = 1$ and $T = 300$ K)
T temperature of the depletion layer
k Boltzmann constant
$\tilde{n}$ ideality factor, typically $1.05 \ldots 1.15$
q elementary charge

A detector diode in this application is generally operated without a bias voltage. The small signal conductance G_j for an unbiased operation is given by

$$G_j = \left. \frac{\mathrm{d}i}{\mathrm{d}u} \right|_{u=0} = \frac{I_{SS}}{U_T} \; . \tag{3.5}$$

The conductance G_j should preferably be high, so that as much high-frequency power as possible can reach the depletion layer. One possibility to achieve a high conductance is to increase the saturation current I_{SS}, which can be influenced, among others, by the contact potential. The contact potential basically depends on the metal type and the metallization conditions. In practice, the contact potential can be made small enough such that $1/G_j$ is of the order of some kΩ instead of the usual MΩ. Therefore, matching to a conventional 50 Ω-system is not possible without a narrow-band transformation circuit. Usually a brute-force match is achieved with the help of a resistor of approximately 50 Ω, as shown in the equivalent circuit in Fig. 3.7.

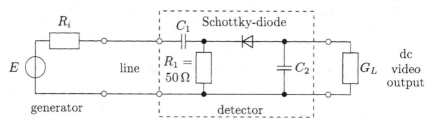

Fig. 3.7   Detector with resistor R_1 for a brute-force matching.

At the same time, the resistor R_1 for the forced matching closes the dc current path. The capacitor C_1 passes the high-frequency signal and blocks the dc voltage. The capacitor C_2 acts as a short-circuit for the high-frequency signal. Such detectors are available for the frequency range of approximately 0.1 MHz to more than 50 GHz. The lower frequency boundary is primarily determined by the capacitors C_1 and C_2, the upper frequency boundary is influenced by the parasitic elements of the Schottky diode. These parasitic elements consist of the bulk resistance, the depletion layer capacitance, the lead inductance and the package capacitance. In a practical detector circuit some compensation elements are introduced in order to obtain a flat frequency response up to the desired upper frequency.

In the case of an amplitude modulated measurement signal, the sensitivity and the inherent matching of the detector can be improved by biasing the detector diode, however, at the expense of the added complexity of a bias supply. In the case of an unmodulated signal, the rectified current cannot be distinguished from the impressed bias current.

For small input levels the video voltage is proportional to the incoming high-frequency power and the characteristic of the Schottky-diode can be approximated by a quadratic law. For higher input levels noticeable linearity deviations occur. Assuming the high-frequency signals to be sinusoidal, then the linearity deviations can be compensated by e.g. a look-up table.

For input signals with a sufficiently low signal level the Schottky diode is operated in the square-law region of its characteristic and the detector can be used for sensitive power measurements. Similar to the noise power measurements with a thermocouple, it is again recommended to transform the small dc voltages into ac signals with the help of a chopper circuit and to amplify the signals close to the Schottky diode. Undesired thermal voltages have again to be kept small. The further processing of the ac signals can be performed with a similar instrument to that used for the thermocouple sensor.

Quantitatively the dependence of the video signal on the high-frequency input power can be described as follows. The high-frequency signal is supposed to be sinusoidal. Then the voltage at the Schottky diode will also be nearly sinusoidal, because the feeding is realized with a low source resistance compared with the junction resistance. The time dependent voltage $u(t)$ at the depletion layer can be approximated by

$$u(t) = U_0 + \hat{U}_1 \cos \Omega t \; . \tag{3.6}$$

Here U_0 is the existing bias voltage. With the voltage from Eq. (3.6) one obtains for the current $i(t)$ through the Schottky diode by means of Eq. (3.4):

$$i(t) = I_{ss} \left[\exp \left(\frac{U_0}{U_T} \right) \exp \left(\frac{\hat{U}_1}{U_T} \cos \Omega t \right) - 1 \right] \; . \tag{3.7}$$

We are interested in the dc current I_0, i.e. the mean value versus time of $i(t)$. For this purpose, we must develop the expression $\exp[(\hat{U}_1/U_t) \cos \Omega t]$ into a Fourier series.

$$\exp \left[\frac{\hat{U}_1}{U_t} \cos \Omega t \right] = J_0 \left(\frac{\hat{U}_1}{U_T} \right) + 2 J_1 \left(\frac{\hat{U}_1}{U_T} \right) \cos \Omega t$$

$$+ 2 J_2 \left(\frac{\hat{U}_1}{U_T} \right) \cos 2\Omega t + \dots \; . \tag{3.8}$$

In the last equation the functions J_0, J_1, J_2 are the modified Bessel functions. Under the condition that $1/G_L \gg R_1$, one obtains for the dc components of the loop in Fig. 3.7:

$$I_0 = -G_L U_0 = I_{ss} \left[\exp \left(\frac{U_0}{U_T} \right) J_0 \left(\frac{\hat{U}_1}{U_T} \right) - 1 \right] . \qquad (3.9)$$

This equation describes, although in an implicit form, the relationship between the detector voltage and the RF-signal amplitude $\hat{U}_1$ as a function of the load admittance G_L and the reverse saturation current I_{ss}. For low amplitudes $\hat{U}_1$, i.e. $\hat{U}_1 \ll U_0$, this relation is of a quadratic characteristic, i.e. the detector output voltage is proportional to the RF input power. For higher RF input amplitudes the detector characteristic will deviate from a square law behavior.

3.1.4   Power measurements with field effect transistors

Field effect transistors on the basis of e.g. gallium arsenide with a metal semiconductor contact as the gate (GaAs MESFET, GaAs metal-semiconductor field effect transistor) can also be utilized as detectors or power meters. The basic characteristics and the noise behavior of field effect transistors are presented in some more detail in Section 4.6.1.

The non-linear current-voltage characteristics of the drain source channel under the assumption of a constant gate bias can serve as a detector for high-frequency signals. The drain source path is operated without a bias voltage. Thus the transistor operates in its ohmic region.

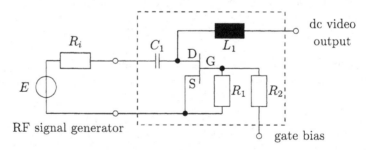

Fig. 3.8   Detector with a field effect transistor.

The gate bias can be chosen to achieve both a sensitive rectification and a good matching by a proper value of the channel resistance. A detector circuit as shown in Fig. 3.8 leads to a characteristic for the rectified voltage U_0 as shown schematically in Fig. 3.9 as a function of the high-frequency signal amplitude $\hat{U}_1$ and the gate bias voltage U_g as a parameter. In the graph, the voltages are normalized to the pinch-off voltage U_{pi}. The conductances are normalized to the channel conductance G_{ch}.

For low amplitudes $\hat{U}_1$, i.e. $\hat{U}_1 \ll U_{pi}$, the relation between $\hat{U}_1$ and the detector dc output voltage U_{l0} is of a quadratic behavior, i.e. the detector

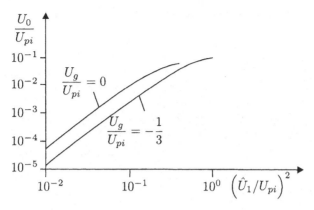

Fig. 3.9 Characteristics of the field effect transistor.

dc output voltage is proportional to the input RF-power. For higher RF input amplitudes the detector characteristic will deviate from a square law behavior.

The saturation of the detector typically occurs at higher power levels than for the Schottky-diode detector. This is basically due to the fact that the pinch-off voltage U_{pi} is typically of the order of some volts while the temperature voltage kT/q is much lower than 1 volt.

Furthermore, a detector with a field effect transistor can be operated in a phase sensitive detector or a controlled rectifier mode. For this purpose, a part of the high-frequency signal is directed to the gate via a 180° phase shifter. This phase shift is necessary in order to obtain a dc contribution with the same polarity as the part, which is directly rectified at the channel. The principle is depicted in Fig. 3.10.

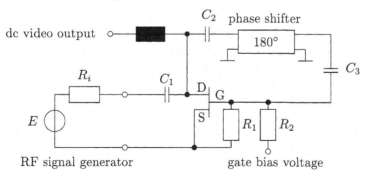

Fig. 3.10 FET-detector with a 180° phase shifter between the input and the gate.

3.1.5   Power measurements with analog multipliers

Four-quadrant analog multipliers are known from electronic measurement techniques. Often they are realized with bipolar transistors. The analog multiplier can be used as a power detector, if the RF-signal is applied to both input ports of the multiplier simultaneously and if the output signal is filtered by a low-pass filter. To a first-order approximation, for very high frequencies, the double balanced mixer may be used as a substitute for an analog multiplier. In this text, the ideal analog multiplier is sometimes employed as a component in an idealized system, that has to be analyzed.

3.1.6   Power measurements with a digital detector

The measurement of the noise power can also be performed digitally with the help of a digital detector. The block diagram of a possible setup is shown in Fig. 3.11. In this setup, the amplified noise signal of the device under test is bandpass-filtered and converted and digitized by an analog-to-digital converter (ADC). Behind the ADC, a digital signal processing algorithm can be performed, consisting of an additional digital band-pass filter (e.g. a FIR-BP, a finite impulse response bandpass filter), the calculation of the squared signal and its mean value. This digital signal processing can be performed by, e.g. a field programmable gate array (FPGA). Such a digital detector has the advan-

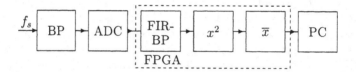

Fig. 3.11   Block diagram of the digital detector.

tage of a high linearity over a wide dynamic range. This is of great importance, e.g. for the very precise measurements of noise parameters. Furthermore, the measurements by such a digital detector are quite fast. However, one has to keep in mind that the standard deviation of the measurements is inversely proportional to the square root of the product of bandwidth and measurement time, as will be discussed in detail in the following section. For this reason, the measurements cannot be performed arbitrarily fast, if a high accuracy is required.

3.1.7   Power measurements with a spectrum analyzer

Measurements of noise power can be performed very conveniently with the help of a spectrum analyzer and a detector at the intermediate frequency of the spectrum analyzer (Fig. 3.12). The spectrum analyzer performs the func-

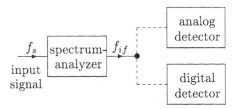

Fig. 3.12 Power measurements with a spectrum analyzer and a detector.

tion of a fixed narrow-band bandpass filter and, even more conveniently, the function of a wide-band tunable bandpass filter. The power detector at the intermediate frequency f_{if} of the spectrum analyzer may be realized according to an analog power measurement principle, as has been discussed before. The intermediate frequency f_{if} often lies around 20 MHz or even lower. Thus, it is quite feasible to employ a digital detector as a power detector, as has been discussed above (Fig. 3.11). For an intermediate frequency of e.g. 20 MHz, a digital power detector with excellent linearity and dynamic properties is state-of-the-art. Also digital bandpass filtering with a digitally adjustable bandwidth at the intermediate frequency f_{if} can be found in modern instruments but perhaps not in older ones. But practically all spectrum analyzers have an analog i.f. output port. Then it might be possible to use an external digital power detector, if an internal digital detector is not available.

In the following section, different measurement systems for the determination of noise parameters of one- and two-ports will be discussed. All described systems require the measurement of noise powers. The noise power measurements can be performed according to the methods described above. In the following, the method for the power measurement will not be specified and, therefore, the power detector will generally be symbolized simply by a detector diode.

However, before the theory of radiometers is discussed in some more detail, a quantitative derivation of the errors of noise power measurements with limited measurement time and restricted bandwidth will be presented in the next section. The principle error of noise power measurements under these conditions is non-zero. As will be seen, for a limited measurement time the frequency bandwidth should be as large as possible. However, a large bandwidth has the disadvantage of providing spectrum measurements with a lower spectrum resolution. For spectrum analyzers the maximum bandwidth is usually limited to a few MHz.

3.1.8 Errors in noise power measurements

The noise power or the power spectrum or generally the mean square value of a stochastic signal can not be determined exactly, because the measurement time and bandwidth are not unlimited. Therefore, a quantitative expression representing these uncertainties will be derived. A spectrometer, as used for the measurement of power spectra, is shown in a simplified version in Fig. 3.13. There, W_1, W_2, W_a denote the spectra and ρ_1, ρ_2 and ρ_a denote the corresponding auto correlation functions. A spectrometer filters a noise

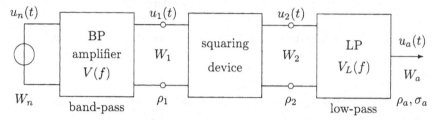

Fig. 3.13 Block diagram of a spectrometer.

signal in the spectral domain and calculates the mean square value of the noise signal for a certain time τ. We are interested in the mean-square value of the output voltage $u_a(t)$ or the mean-square value of samples of U_a, respectively:

$$
\begin{aligned}
\overline{u_a^2(t)} &= \langle U_a^2 \rangle = \mathrm{E}\{u_a^2(t)\} = \rho_a(0) \\
&= \int_{-\infty}^{+\infty} W_a(f)\, df = \int_{-\infty}^{+\infty} W_2(f) \cdot |V_L(f)|^2\, df \ .
\end{aligned}
\tag{3.10}
$$

In the last equation V_L is the transfer function of the low-pass filter.

The squaring device leads to the following equation for the instantaneous values of the voltages, where c is a constant:

$$
u_2(t) = c \cdot u_1^2(t) \ .
\tag{3.11}
$$

The expected value of the output voltage $u_a(t)$ is equal to

$$
\begin{aligned}
\overline{u_a(t)} &= \langle U_a \rangle = \mathrm{E}\{u_a(t)\} = V_L(0) \cdot \overline{u_2(t)} \\
&= c \cdot V_L(0) \cdot \int_{-\infty}^{+\infty} W_1(f)\, df = c \cdot V_L(0) \cdot \rho_1(0) \ .
\end{aligned}
\tag{3.12}
$$

Next, a relation between the spectrum W_2 and the spectrum W_1 has to be derived. Under the assumption of a Gaussian amplitude distribution the autocorrelation functions ρ_2 and ρ_1 are related as follows (cf. Eq. (1.64)):

$$
\begin{aligned}
\rho_2(\theta) &= \overline{u_2(t) \cdot u_2(t+\theta)} = \langle u_2(t) \cdot u_2(t+\theta) \rangle \\
&= c^2 \cdot \overline{u_1^2(t) \cdot u_1^2(t+\theta)} = c^2 \cdot \langle u_1^2(t) \cdot u_1^2(t+\theta) \rangle \\
&= c^2 \left(\rho_1^2(0) + 2\rho_1^2(\theta) \right) \ .
\end{aligned}
\tag{3.13}
$$

By means of a Fourier transformation, W_2 can be calculated from ρ_2. Here, the term $\rho_1^2(\theta)$ in Eq. (3.13), which is a product in the θ-range, is represented by a convolution in the frequency domain

$$W_2(f) = c^2 \cdot \rho_1^2(0) \cdot \delta(f) + 2c^2 \cdot \underbrace{\int_{-\infty}^{+\infty} W_1(f') \cdot W_1(f - f') \, df'}_{\text{convolution}} \ . \tag{3.14}$$

Based on Eq. (3.10) and Eq. (3.14) the variance σ_a^2 of the output voltage $u_a(t)$ is obtained as follows:

$$
\begin{aligned}
\sigma_a^2 \;=\; & \overline{u_a^2(t)} - \overline{u_a(t)}^2 = c^2 \cdot \rho_1^2(0) \int_{-\infty}^{+\infty} |V_L(f)|^2 \delta(f) \, df \\[2mm]
& + 2c^2 \int_{-\infty}^{+\infty}\!\!\!\int |V_L(f)|^2 W_1(f') \cdot W_1(f - f') \, df' \, df \\[2mm]
& - (c \cdot V_L(0) \cdot \rho_1(0))^2 \\[2mm]
=\; & 2c^2 \int_{-\infty}^{+\infty}\!\!\!\int |V_L(f)|^2 \cdot W_1(f') \cdot W_1(f - f') \, df' \, df \ .
\end{aligned}
\tag{3.15}
$$

The bandwidth f_L of the low-pass filter is much smaller than the bandwidth Δf of the high-frequency band-pass filter, i.e. $f_L \ll \Delta f$, so that in Eq. (3.15) the term $W_1(f - f') \approx W_1(-f') = W_1^*(f') = W_1(f')$. Furthermore, under the assumptions that the spectrum W_1 is constant within the pass-band Δf of the band-pass filter and that the band-pass filter has a rectangular shape around the center frequency f_0, Eq. (3.15) can be rewritten as

$$\overline{u_a^2(t)} - \overline{u_a(t)}^2 = 2c^2 \cdot 2 \cdot W_1^2(f_0) \cdot \Delta f \cdot \int_{-\infty}^{+\infty} |V_L(f)|^2 \, df \ . \tag{3.16}$$

With the relative variance $\tilde{\sigma}_a$ of the output signal, which is normalized to $\overline{u_a(t)}^2$, it follows with Eq. (3.12), using the one-sided spectra:

$$
\begin{aligned}
\tilde{\sigma}_a^2 \;=\; & \frac{\overline{u_a^2(t)} - \overline{u_a(t)}^2}{\overline{u_a(t)}^2} = \frac{2 \cdot c^2 \cdot W_1^2(f_0) \cdot \Delta f \cdot \int_0^{+\infty} |V_L(f)|^2 df}{c^2 \cdot W_1^2(f_0) \cdot (\Delta f)^2 \cdot V_L^2(0)} \\[2mm]
=\; & 2 \cdot \frac{1}{\Delta f} \cdot \frac{\int_0^{\infty} |V_L(f)|^2 df}{V_L^2(0)} \ .
\end{aligned}
\tag{3.17}
$$

An effective bandwidth Δf_L of the low-pass filter is defined as follows:

$$\Delta f_L = \frac{\int_0^{\infty} |V_L(f)|^2 df}{V_L^2(0)} \ . \tag{3.18}$$

Expressed by the fluctuation ΔT_m of the noise temperature T_m and taking into account the system noise temperature T_a of the spectrometer , Eq. (3.17) becomes

$$\tilde{\sigma}_a = \frac{\Delta T_m}{T_m + T_a} = \frac{\Delta T_m}{T_n} = \sqrt{2 \cdot \frac{\Delta f_L}{\Delta f}} \cdot \quad (3.19)$$

For a rectangular low-pass characteristic with the cut-off frequency f_L, we have $f_L = \Delta f_L$ and thus

$$\frac{\Delta T_m}{T_n} = \sqrt{2 \cdot \frac{f_L}{\Delta f}} \cdot \quad (3.20)$$

If the low-pass filter is realized as an ideal integrator with an integration time τ, then

$$|V_L(f)|^2 = c_i^2 \frac{\sin^2(\pi f \tau)}{(\pi f)^2} \quad , \quad (3.21)$$

with c_i being a constant and

$$\Delta f_L = \frac{1}{2\tau} \cdot \quad (3.22)$$

In this case, Eq. 3.20 yields

$$\frac{\Delta T_m}{T_n} = \frac{1}{\sqrt{\Delta f \cdot \tau}} \cdot \quad (3.23)$$

This important relation was first derived by Rice in 1946.

If the spectrum W_1 is not constant over the bandwidth Δf or if the high-frequency band-pass filter does not have a rectangular shape, then Eq. (3.19) is still valid, if instead of Δf an effective bandwidth Δf_{ef} is used, which is defined according to the following equation (cf. Fig. 3.13):

$$\Delta f_{ef} = \frac{\left[\int_0^\infty W_1(f)df\right]^2}{\int_0^\infty W_1^2(f)df} = \frac{\left[\int_0^\infty |V(f)|^2 \cdot W_n(f)df\right]^2}{\int_0^\infty |V(f)|^4 \cdot W_n^2(f)df} \cdot \quad (3.24)$$

Problem

3.1   Derive Eq. (3.22). Which high-frequency bandwidth is required, if for an integration time of 1s either 68% or 95% of all measurement values shall deviate less than 0.1K from a noise temperature of $T_m = 300$K?

3.2   MEASUREMENT OF THE CROSS-CORRELATION FUNCTION AND THE CROSS-SPECTRUM

The measurement technique for the cross-correlation function or the cross-spectrum results from the definition of these parameters. An analog method

for the measurement of the correlation function in the high-frequency range is
depicted in Fig. 3.14. One of the two input signals is delayed by an adjustable

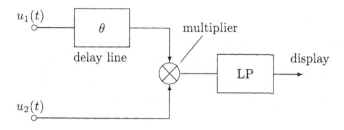

Fig. 3.14   Measurement of the cross-correlation function.

delay time θ. The product of both signals is generated by means of an analog
multiplier and the mean value is formed by a low-pass filter. In the high-
frequency range an analog multiplier can be realized by a double balanced
mixer.

In the frequency range up to perhaps 100 MHz a digital processing might
be preferred. Then, the amplitude characteristics of $u_1(t)$ and $u_2(t)$ are transformed by analog-to-digital converters and the following signal processing is
performed completely digitally.

The cross-spectrum can be measured with the circuit in Fig. 3.15. In
contrast to the measurement of the correlation function, here narrow band-
pass filters are used. The signals are described in the frequency domain. The

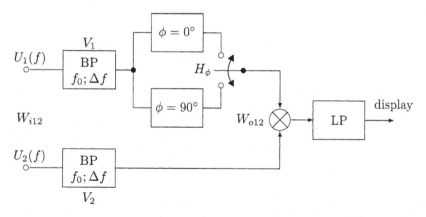

Fig. 3.15   Measurement of the cross-spectrum.

complex transfer functions of the band-pass filters including the amplifiers are
denoted by V_1 and V_2. The phase switch has the complex transfer function

H_ϕ. For the cross-spectrum W_{o12} at the output we have

$$W_{o12} = W_{i12} \cdot V_1^* \cdot H_\phi^* \cdot V_2 \ . \tag{3.25}$$

The cross-correlation function $\rho_{o12}(\theta = 0)$ is measured by the multiplier and the low-pass filter. Provided that the band-pass filters are sufficiently narrow, the output yields

$$\rho_{o12}(\theta = 0) = \int_{-\infty}^{+\infty} W_{o12}(f)\, df \quad \text{(because of } \theta = 0 \text{ and } \exp(j2\pi f\theta) = 1)$$

$$= \Big[V_1^*(f_0) \cdot H_\phi^*(f_0) \cdot V_2(f_0) \cdot W_{i12}(f_0)$$

$$+ V_1^*(-f_0) \cdot H_\phi^*(-f_0) \cdot V_2(-f_0) \cdot W_{i12}(-f_0) \Big] \Delta f \ . \tag{3.26}$$

Here, f_0 represents the center frequency of the band-pass filters. For equal band-pass filters with $V_1 = V_2 = V$ and for the phase switch in position $\phi = 0°$ or $H_\phi = 1$, it follows from Eq. (3.26) due to $V(-f_0) = V^*(f_0)$ and $H_\phi(-f_0) = H_\phi^*(f_0)$:

$$\rho_{o12} = |V(f_0)|^2 \, (W_{i12}(f_0) + W_{i12}^*(f_0)) \cdot \Delta f$$

$$= |V(f_0)|^2 \cdot 2 \cdot \text{Re}\{W_{i12}(f_0)\} \cdot \Delta f \ . \tag{3.27}$$

For the switch position $\phi = 0°$, the result is equal to the real part of the input cross-spectrum except for a proportionality constant. For the switch position $\phi = 90°$ it follows from Eq. (3.26) with $V_1 = V_2 = V$ and $H_\phi(f_0) = j$ or $H_\phi^*(-f_0) = H_\phi(f_0) = j$:

$$\rho_{o12} = \Big[-j|V(f_0)|^2 \cdot W_{i12}(f_0) + j|V(f_0)|^2 \cdot W_{i12}^*(f_0) \Big] \cdot \Delta f$$

$$= |V(f_0)|^2 \cdot 2 \cdot \text{Im}\{W_{i12}(f_0)\} \cdot \Delta f. \tag{3.28}$$

For the switch position $\phi = 90°$, a value results that is equal to the imaginary part of the input cross spectrum except for a proportionality constant which is the same as before. By varying the center frequency of the band-pass filter, both the real and the imaginary part of the cross spectrum can be measured as a function of frequency. For the calibration of the correlator the inputs can be supplied with two completely correlated noise signals. For $\phi = 90°$, a zero output signal should result. With completely uncorrelated signals it can be verified for both switch positions, that zero output signals result.

If the input signals of the measurement setup in Fig. 3.15 are identical, i.e. $U_1 = U_2 = U$, then the cross spectrum is equal to the power spectrum of U. For the measurements of power spectra only, the phase switch and one band-pass filter can be omitted and the simplified circuit in Fig. 3.16 results. Note that a multiplier with the input signals connected in parallel and a subsequent low-pass filter form a power meter as discussed before.

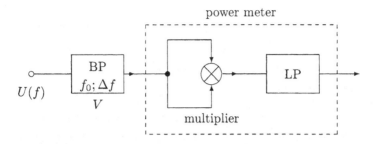

Fig. 3.16   Measurement of the spectrum with a multiplier or a power meter, respectively.

For the measurement of the cross-spectrum according to Fig. 3.15 it is often difficult to realize two band-pass filters, especially variable band-pass filters, with a good tracking behavior. The problem can be solved by converting both input signals with the help of **one** voltage controlled oscillator signal with a variable frequency to two identical fixed intermediate frequencies (heterodyne principle or double spectrum analyzer setup).

If only one variable band-pass filter is used, then the measurement can be performed sequentially in time on the basis of the setup in Fig. 3.17. At the

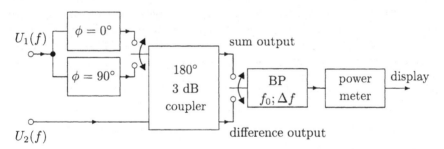

Fig. 3.17   Time serial measurement of the spectrum.

output of a 180°-coupler the sum and the difference of the input signals $U_1(f)$ and $U_2(f)$ are available. According to the mathematical identity $(a+b)^2 - (a-b)^2 = 4ab$, the spectra at the output of the 180°-coupler, representing the sum and the difference of the signals, are measured. Next, the difference of both measured values is calculated. In the symbolic notation, for the switch position $\phi = 0°$ we have:

$$\frac{1}{2}(U_1 + U_2)^* (U_1 + U_2)|V|^2 - \frac{1}{2}(U_1 - U_2)^* (U_1 - U_2) \cdot |V|^2$$
$$= (U_1^* \cdot U_2 + U_1 \cdot U_2^*) \cdot |V|^2$$
$$= 2|V|^2 \cdot \mathrm{Re}\{U_1^* U_2\} = 2|V|^2 \cdot \mathrm{Re}\{W_{i12}\} \ . \tag{3.29}$$

This result is proportional to the real part of the cross-spectrum at the input. In the same way, the imaginary part can be obtained by changing the position of the phase switch to 90°. If a 90°-3dB-coupler instead of the 180° type is used, then the real part results for the 90° position and the imaginary part for the 0° position of the switch.

Sometimes, however, the bandwidth of the band-pass filter is so wide, that the spectrum of the input signal is not constant within the pass-band. In this case, for the switch position $\phi = 0°$ and with $V_1 = V_2 = V$ according to Eq. (3.26) the measured value is proportional to the frequency average of the real part of the input cross spectrum. If f_1 and f_2 are the corner frequencies of the band-pass filter, then

$$\rho_{o12}(\theta = 0) = 2 \cdot \int_{f_1}^{f_2} |V(f)|^2 \cdot \mathrm{Re}\{W_{i12}\}\, df \ , \tag{3.30}$$

where $|V(f)|^2$ is a weighting function. A similar expression results for the switch position $\phi = 90°$.

3.3   ILLUSTRATIVE INTERPRETATION OF THE CORRELATION

The correlation between two noise signals can be described by two equivalent functions, namely the **cross-correlation function** ρ_{12} or the **cross-spectrum** W_{12}. For both representations a concise description of the correlation will be discussed in this section.

At first, the cross-correlation will be considered. Let $u_1(t)$ and $u_2(t)$ be continuous time signals, which are partly correlated. The signal $u_2(t)$ is separated into two components

$$u_2(t) = u_2'(t) + u_2''(t) \ , \tag{3.31}$$

in such a way that $u_2'(t)$ is identical in time to $u_1(t)$ except for a real proportionality factor γ

$$u_2'(t) = \gamma \cdot u_1(t) \ . \tag{3.32}$$

Such a partition is always possible and, at first, only represents a formal step. However, if γ is chosen properly, namely,

$$\gamma = \frac{\rho_{12}(0)}{\rho_{11}(0)} \ , \tag{3.33}$$

then $u_2''(t)$ is uncorrelated with $u_1(t)$. This can be shown directly by

$$\begin{aligned}
\overline{u_2''(t) \cdot u_1(t)} &= \overline{(u_2(t) - \gamma\, u_1(t)) \cdot u_1(t)} \\
&= \overline{u_2(t) \cdot u_1(t)} - \gamma \cdot \overline{u_1^2(t)} \\
&= \rho_{12}(\theta = 0) - \gamma \cdot \rho_{11}(\theta = 0) = 0 \ . \tag{3.34}
\end{aligned}$$

This consideration also works with two signals being shifted in time by θ relative to each other. The cross-correlation function thus represents, except for a normalization factor, that part of $u_2(t)$ which is identical in time to $u_1(t)$ and which could, therefore, be nulled by means of a balanced configuration, for example. As is generally known for an ac bridge, being supplied with sinusoidal signals, a complete balance to zero can be achieved due to the fact that two signals with a phase difference of 180° are interfering destructively. Uncorrelated signals cannot be balanced with a bridge at all, because for a superposition the signals have to be added according to the square root of the sum of the squares. For partly correlated signals, the completely correlated part $u_2'(t)$, which is described by ρ_{12}, can be nulled by a balance, whereas for the uncorrelated parts $u_2''(t)$ and $u_1(t)$, the square-law summation applies. Also for a finite time shift θ the signals can be separated into correlated and uncorrelated parts. The partition generally is a function of the time shift θ, similarly to the cross-correlation function, which generally depends on θ.

For a representation in the frequency domain, the correlation can be interpreted in a similar way. For this purpose, the phasor notation is used. Let $U_1(f)$ and $U_2(f)$ denote two partly correlated noise signals. Let k_{12} be a complex number, which represents that part of U_2 which is identical to U_1 after an appropriate phase- and amplitude change:

$$U_2 = U_2' + U_2'' = k_{12} \cdot U_1 + U_2'' \ . \tag{3.35}$$

Again, it can be shown that U_2'' is uncorrelated with U_1, if k_{12} is chosen properly:

$$U_1^* \cdot U_2'' = U_1^* \left(U_2 - k_{12} \cdot U_1 \right) = W_{12} - k_{12} \cdot W_1 \ . \tag{3.36}$$

Apparently, this relation is equal to zero, if k_{12} is chosen as

$$k_{12} = \frac{W_{12}}{W_1} \ . \tag{3.37}$$

3.4 MEASUREMENT OF THE EQUIVALENT NOISE TEMPERATURE OF A ONE-PORT

In this section it will be described, how the available noise power or equivalent noise temperature of a one-port at the center frequency f_0 in a given bandwidth Δf is measured as precisely as possible.

The fundamental measurement system, which is also called a radiometer or a spectrometer, is shown in Fig. 3.18. In principle, the measurement system consists of at least one cascade of amplifiers with an extremely low-noise preamplifier in the first stage, a band-pass filter, and a power meter as discussed in section 3.1. Such a noise measurement system poses a number of problems. First of all, the gain of the cascade of amplifiers has to be known exactly for the determination of the effective noise power of the device under

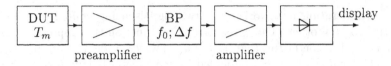

Fig. 3.18 Principle setup for noise temperature measurements.

test. Furthermore, the necessary high gain of the cascade of amplifiers might change with time, i.e. it drifts. Moreover, the first preamplifier, although being a low-noise type, also produces noise, which is often in the same order as the noise of the device under test. It is thus necessary to discriminate the preamplifier noise from the noise of the device under test. In addition, the measurement result should not depend on the gain. A possible setup to reduce these disturbing effects is illustrated in Fig. 3.19. Here, Z_g denotes the impedance of the device under test at the temperature T_m and Z_{in} is the input impedance of the amplifier. Z_g and Z_{in} may be complex. The noise

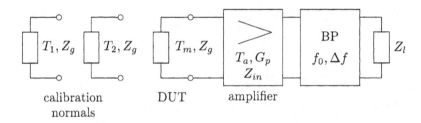

Fig. 3.19 Radiometer with different calibration standards.

contribution of the amplifier is also described by a temperature T_a (system or amplifier temperature). The amplifier can be represented by a thermally noisy generator resistance with the temperature T_a and the impedance Z_g, which is supposed to produce the same noise at the load impedance Z_l as the amplifier itself. For this model, the amplifier is assumed to be noise-free. In addition, two calibration standards with the known but different temperatures T_1 and T_2 are needed, both having the same impedance Z_g as the device under test. With the gain G_p of the amplifier, three noise powers can be measured at the output impedance Z_l, namely the noise power P_m for the measurement of the device under test and the noise powers P_1, P_2 for the measurements of the two calibration standards:

$$
\begin{aligned}
P_m &= G_p \left(T_m + T_a\right) \cdot k \cdot \Delta f \ , \\
P_1 &= G_p \left(T_1 + T_a\right) \cdot k \cdot \Delta f \ , \\
P_2 &= G_p \left(T_2 + T_a\right) \cdot k \cdot \Delta f \ .
\end{aligned}
\tag{3.38}
$$

Solving this set of equations for T_m yields

$$T_m = \frac{(p_{m2} - 1)\, p_{m1}}{p_{m2} - p_{m1}} \cdot T_1 - \frac{(p_{m1} - 1)\, p_{m2}}{p_{m2} - p_{m1}} \cdot T_2 \ , \tag{3.39}$$

which only depends on the known temperatures T_1 and T_2 and on the ratios of the noise powers $P_m/P_1 = p_{m1}$ and $P_m/P_2 = p_{m2}$. The result does not depend on the unknown gain G_p and the temperature of the amplifier T_a. It is advantageous to choose one of the temperatures equal to the ambient temperature T_0, e.g. $T_2 = T_0$, because the available noise power or available temperature is then known *a priori*.

One disadvantage of this method is that the three single measurements might take a relatively long time. Consequently, since the gain has to be constant during the measurements, the amplifier needs a good long-term stability. In Section 3.5, other radiometer circuits with reduced requirements for the long-term stability of the amplifier will be discussed.

Furthermore, the device under test might be mismatched and its impedance might be different from the impedance of the two calibration standards. Also for this case, radiometer circuits will be presented that may alleviate this problem.

3.5 SPECIAL RADIOMETER CIRCUITS

3.5.1 Dicke-Radiometer

With a Dicke-radiometer, also called switching radiometer, the noise of a one-port can be measured independently of the gain and the noise of the multistage amplifier. In Fig. 3.20 the block diagram is shown. By means of a

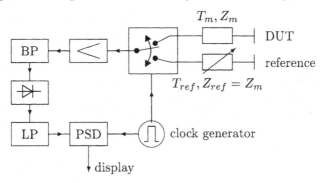

Fig. 3.20 Dicke-radiometer for a known impedance of the device under test.

switch, which should have low loss, the amplifier input is periodically switched between the one-port device under test with the unknown temperature T_m

and the impedance Z_m and a reference device with the variable temperature T_{ref} and the impedance Z_{ref}. The noise signal at the output of the band-pass filter versus time, as shown in Fig. 3.21, is composed of the contributions of T_m and T_{ref}, sequential in time, and of the contribution of the first preamplifier, which is constant in time. The noise signals of the device under test and

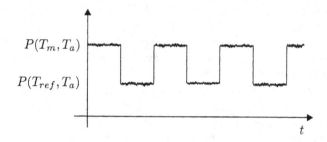

Fig. 3.21   Output power of the radiometer versus time.

the reference noise source are not correlated with the noise of the amplifier. However, the contribution of the amplifier is constant for both positions of the switch only, if the impedances of the device under test and the reference, Z_m and Z_{ref}, are identical. This is a precondition for the correct measurement with this radiometer. After having passed the amplifier and the band-pass filter, the noise signal is square-law rectified and filtered with the low-pass filter. Next, the a.c. signal caused by the switching procedure (at, for example, a frequency of 1 kHz) is filtered with e.g. a phase sensitive detector (PSD) and then displayed. For the measurement of T_m the reference temperature T_{ref} is varied until the ac-signal vanishes. In this balanced case

$$T_m = T_{ref} \ . \tag{3.40}$$

Since the additive noise contribution of the amplifier is equal for both switch positions, it does not influence the balance condition. Also drift effects of the gain do not affect the balance condition as long as they are slow by comparison with the period of the switching frequency. This is a noticeable advantage of the Dicke-radiometer by comparison with the fundamental radiometer of Fig. 3.18. The measurement time τ of the Dicke-radiometer can be arbitrarily larger than the switching period. The measurement time will be chosen large enough, so that the related measurement error for the noise power will be sufficiently small. However, it is absolutely essential that the power meter has a settling time much shorter than the switching period.

Often the switch of the Dicke-radiometer will not have the same reflection coefficient for both positions. In this case, the circuit of Fig. 3.20 can be modified as shown in Fig. 3.22. First, the device under test is connected to port 1. With the reference noise source at the switch position I a zero compensation is performed which leads to the value T_{ref}^I. Next, the reference

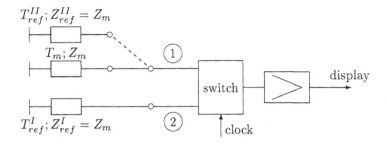

Fig. 3.22   Dicke-radiometer with two reference noise sources.

noise source II is connected to port 1 and by varying T_{ref}^{II} a further zero balance is performed. For equal impedances $Z_{ref}^{II} = Z_m$ the temperature of the device under test is given by

$$T_m = T_{ref}^{II} \ . \tag{3.41}$$

In this way, equal input reflection coefficients for both switch positions are not necessary. Only the long-term stability of the input reflection is needed. The reference noise source I at port 2 does not need to be calibrated, it just has to be variable and stable. The impedance Z_{ref}^{I} may be different from the impedance of the device under test Z_m. The impedance of the reference noise source II, however, must be identical to the impedance of the device under test, i.e. $Z_m = Z_{ref}^{II}$.

3.5.2   Problems with mismatched devices under test

In the previous sections, measurement methods for the determination of the noise temperature of a matched device or devices with a known impedance were presented. In this context, the word known can also have the meaning that the impedance of the device under test is equal to the impedance of the noise source. However, if the device under test is mismatched and, furthermore, its impedance is unknown, perhaps complex and frequency-dependent, then an exact measurement of the unknown noise temperature becomes more difficult because of two problems.

First problem: Because of the mismatch of the device under test with the temperature T_m, one does not measure the available noise power $P_{av} = k \cdot T_m \cdot \Delta f$ in the bandwidth Δf, but a smaller power P_l given by

$$P_l = (1 - |\rho|^2) \cdot k \cdot T_m \cdot \Delta f \ , \tag{3.42}$$

where $|\rho|$ denotes the magnitude of the reflection coefficient of the device under test. This relation will be derived next.

In Fig. 3.23 the load resistance Z_0, which also serves as the reference impedance, is assumed to be real. The noise power P_l at the load impedance

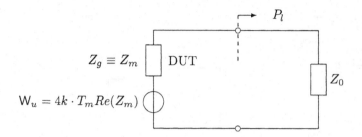

Fig. 3.23 Explanation of the measured noise power for a mismatched device.

Z_0, which is supposed to be noiseless, can be calculated as follows:

$$P_l = \frac{4k \cdot T_m \cdot \text{Re}\{Z_m\}}{|Z_0 + Z_m|^2} \cdot Z_0 \cdot \Delta f \ . \tag{3.43}$$

With

$$|\rho|^2 = \left| \frac{Z_m - Z_0}{Z_m + Z_0} \right|^2 \tag{3.44}$$

Eq. (3.43) yields:

$$
\begin{aligned}
P_l &= kT_m \cdot \Delta f \cdot \frac{2 \cdot (Z_m + Z_m^*) \cdot Z_0}{|Z_m + Z_0|^2} \\
&= kT_m \cdot \Delta f \frac{|Z_m + Z_0|^2 + 2(Z_m + Z_m^*) \cdot Z_0 - |Z_m + Z_0|^2}{|Z_m + Z_0|^2} \\
&= kT_m \cdot \Delta f \left[1 - \left| \frac{Z_m - Z_0}{Z_m + Z_0} \right|^2 \right] \\
&= kT_m \Delta f (1 - |\rho|^2) = P_{av}(1 - |\rho|^2) \ .
\end{aligned}
\tag{3.45}
$$

Here, $P_{av} = k \cdot T_m \cdot \Delta f$ is the available noise power of the device under test or the generator, respectively. The noise power P_l, which arrives at the load resistance Z_0, is reduced by the reflected part P_{re}

$$P_{re} = kT_m \cdot \Delta f \cdot |\rho|^2 \ . \tag{3.46}$$

The term $1 - |\rho|^2$ is equal to the gain G_p of the circuit:

$$G_p = \frac{P_l}{P_{av}} = 1 - |\rho|^2 \ . \tag{3.47}$$

If, as depicted in Fig. 3.24, the impedance of the device under test or the generator with $Z_m = Z_g$ as well as the load resistance Z_l are complex and not equal to the reference impedance Z_0, then the following well-known relation for the gain G_p results:

$$G_p = \frac{(1 - |\Gamma_g|^2)(1 - |\Gamma_l|^2)}{|1 - \Gamma_g \Gamma_l|^2} \ . \tag{3.48}$$

In Eq. (3.48) Γ_g and Γ_l are the reflection coefficients of the generator and the load impedance, with respect to the real reference impedance (Fig. 3.24)

$$\Gamma_g = \frac{Z_g - Z_0}{Z_g + Z_0} \ , \quad \Gamma_l = \frac{Z_l - Z_0}{Z_l + Z_0} \ . \tag{3.49}$$

Equation (3.48) is not well suited for an illustrative explanation. Such an

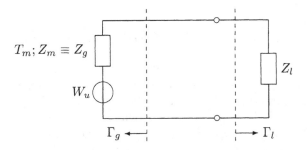

Fig. 3.24   Mismatched generator and load impedances.

explanation can be derived similarly to Eq. (3.45) and (3.46), if a modified reflection coefficient $\tilde{\rho}$ is introduced, which is defined with respect to the generally complex generator impedance $Z_g \equiv Z_m$. With

$$|\tilde{\rho}| = \left| \frac{Z_l - Z_m^*}{Z_l + Z_m} \right| = \left| \frac{Z_l - Z_g^*}{Z_l + Z_g} \right| \tag{3.50}$$

it follows for the power P_l at the load resistance:

$$P_l = P_{av}(1 - |\tilde{\rho}|^2) \ , \tag{3.51}$$

as will be shown in problem 3.2. Here, $P_{av} = kT_m \Delta f$ is again the available noise power (or the available power in general) of the device under test or the generator, respectively. The power P_{re} with

$$P_{re} = P_{av} - P_l = P_{av} \cdot |\tilde{\rho}|^2 \ , \tag{3.52}$$

can be interpreted as the reflected power, similar to Eq. (3.46), as will be seen more clearly later. The power P_l can be defined as the transmitted power. The available gain G_p is again given by

$$G_p = 1 - |\tilde{\rho}|^2 \ . \tag{3.53}$$

This term is identical to the term for G_p in Eq. (3.48).

Problem

3.2   Verify the validity of Eq. (3.51) as well as the identity of Eq. (3.53) and Eq. (3.48).

A second problem, which arises for a mismatched one-port device, is caused by a noise wave of the first preamplifier radiating toward the device under test. This noise wave can be reflected by the device under test and can thus return to the amplifier. In general, this reflected wave will be correlated with the noise wave at the output of the amplifier. For a measurement setup as shown in Fig. 3.20, the noise contributions of the amplifier are no longer equal for both positions of the switch and, consequently, it has an impact on the noise balance. This problem can be solved by connecting a matched ferrite isolator between the switch and the amplifier (Fig. 3.25). Such a passive non-reciprocal device of a homogenous temperature shows uncorrelated noise waves at its input and output, as has already been discussed. Additionally, its noise contribution at its input is known quantitatively, if the physical temperature of the isolator is known. Furthermore, the noise wave of the

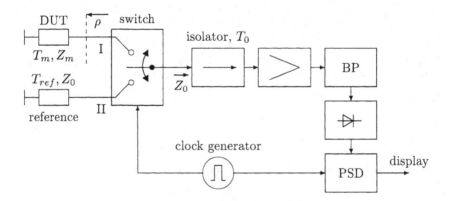

Fig. 3.25   Radiometer setup with an isolator in front of the first preamplifier.

preamplifier cannot reach the device under test, due to the backward isolation of the isolator.

If the radiometer is balanced, i.e. if the temperature of the matched reference is tuned until the noise powers at the output of the cascade of amplifiers are equal for both positions of the switch corresponding to a zero output signal of the phase-sensitive detector, then the noise powers at the input of the isolator are given by:

position I $P_I = k \cdot T_m \cdot \Delta f \cdot (1 - |\rho|^2) + k \cdot T_0 \cdot |\rho|^2 \cdot \Delta f$

position II $P_{II} = k \cdot T_{ref} \cdot \Delta f$.

For position I the second term represents the contribution of the isolator. For a balance, that means $P_I = P_{II}$, we have

$$T_m \cdot (1 - |\rho|^2) + T_0 \cdot |\rho|^2 = T_{ref} \ . \tag{3.54}$$

If $|\rho|^2$ is known from measurements, then T_m can be calculated for a known T_{ref} and a known temperature T_0 of the isolator by

$$T_m = \frac{T_{ref} - T_0 \cdot |\rho|^2}{1 - |\rho|^2} \ . \tag{3.55}$$

If a passive ferrite isolator is not available, as e.g. for frequencies below 500 MHz, then the uncorrelated amplifier of Fig. 2.23 can be used. However, its temperature at its input has to be determined. If a high degree of decorrelation is needed, then it is also possible to combine an isolator and an uncorrelated amplifier.

An evaluation of Eq. (3.55) becomes difficult, if the reflection coefficient ρ depends on the frequency within the measurement bandwidth. Then, $|\rho(f)|^2$ in Eq. (3.55) has to be replaced by its mean value.

A more elegant way for the measurement of the noise temperature of mismatched objects is to use so-called compensation radiometers, where the reflection coefficient of the device under test does not need to be known, because it is not part of the balance condition. Such compensation radiometers will be discussed in the next section.

3.5.3   Compensation radiometers

As can be seen in Fig. 3.26, the device under test and the reference noise source are connected alternately to the input of the amplifier by a switchable circulator. The direction of circulation can be reversed by changing the direction of the magnetizing field of the circulator. This, however, is not a very practical solution and therefore a more suitable circuit will be proposed later. For operation in position I, the noise power P_I enters the input of the first preamplifier. The noise power P_I consists of a noise contribution from the reference device, which is reflected by the device under test, and a second contribution from the device under test. Both contributions are uncorrelated. Because of the possible mismatch of the device under test, described by the reflection coefficient ρ, the noise power of the device under test is reduced by a factor of $1 - |\rho|^2$.

$$P_I = k \cdot T_{ref} \cdot |\rho|^2 \cdot \Delta f + k \cdot T_m \cdot (1 - |\rho|^2) \cdot \Delta f \ . \tag{3.56}$$

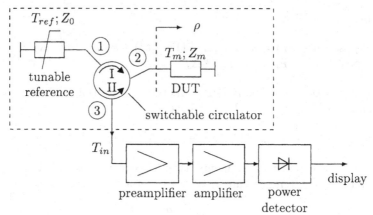

Fig. 3.26   Compensation radiometer with a switchable circulator.

For the orientation II of the circulator and under the assumption of an ideal circulator, the noise power of the reference source is measured:

$$P_{II} = k \cdot T_{ref} \cdot \Delta f \ . \tag{3.57}$$

By varying the reference temperature T_{ref} the noise powers P_I and P_{II} are balanced for the two directions of circulation:

$$
\begin{aligned}
P_I = P_{II} &= k \cdot T_{ref} |\rho|^2 \cdot \Delta f + k \cdot T_m (1 - |\rho|^2) \cdot \Delta f \\
&= k \cdot T_{ref} \Delta f \ .
\end{aligned}
\tag{3.58}
$$

From this relation it follows that

$$T_m = T_{ref} \ , \tag{3.59}$$

provided that the magnitude of the reflection coefficient ρ of the device under test is not equal to one. The result of Eq. (3.59) does not depend on the value of the reflection coefficient ρ. Hence, ρ may be frequency-dependent. This means that $\rho(f)$ may vary within the bandwidth of the measurements. The described method is called a compensation method, because the reduced noise contribution of the device under test, due to its mismatch, is compensated by an equivalent contribution of the reference.

One can also argue in another way: the part of the circuit within the dashed box in Fig. 3.26 has an input noise temperature T_{in}. If this part is at a homogeneous temperature $T_{ref} = T_m$ and if, furthermore, the circulator is lossless, then $T_{in} = T_{ref} = T_m$.

The compensation principle will be explained once more with the help of Fig. 3.27. In this circuit, an isolator at the temperature T_0 is placed between the device under test and the amplifier. For clarity reasons, this isolator is realized on the basis of an ideal circulator. One of the three ports of the

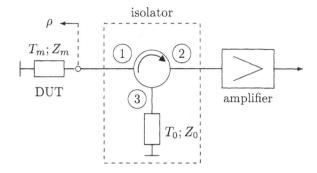

Fig. 3.27   Principle of the compensation radiometer.

circulator is terminated by the real reference impedance Z_0, which is at the temperature T_0. One can recognize that a wave radiated from the input of the amplifier is absorbed at port 3 of the circulator. The wave at the output of the amplifier is constant and does not depend on the impedance of the device, Z_m. This output wave can be eliminated for the measurements. Another noise wave originates from the termination Z_0 of port 3 of the circulator. It propagates towards the device under test where it is reflected. Then, it passes the circulator resulting in the power P_{re} at the input of the amplifier

$$P_{re} = k \cdot T_0 \cdot \Delta f \cdot |\rho|^2 = P_{av} \cdot |\rho|^2 \ . \tag{3.60}$$

The device under test emits the power P_l to the input of the amplifier. With $T_m = T_0$ the power P_l is given by:

$$\begin{aligned} P_l &= k \cdot T_m \cdot \Delta f \cdot (1 - |\rho|^2) = k \cdot T_0 \cdot \Delta f \cdot (1 - |\rho|^2) \\ &= P_{av}(1 - |\rho|^2) \ . \end{aligned} \tag{3.61}$$

This power is lower than the available power P_{av}, due to the mismatch of the device under test. The missing part is compensated by P_{re}. Thus, the input circuit is described by the equivalent circuit of Fig. 3.23.

Even if the load resistance Z_l is mismatched, i.e. complex and not equal to a real reference impedance Z_0 but still at the temperature T_0, an exact compensation can be achieved. For this purpose, the circuit of Fig. 3.24 is considered. Extracting the lossless imaginary part $j\mathrm{Im}\{Z_l\}$ from the load resistance Z_l, Fig. 3.24 can be modified according to Fig. 3.28. Shifting the reference plane from 1-1' to 2-2' in Fig. 3.28, the magnitude of the reflection coefficient $\tilde{\rho}$ remains unchanged. Adopting the expression for $\tilde{\rho}$ from Eq. (3.50) leads to

$$|\tilde{\rho}| = \left| \frac{Z_l - Z_g^*}{Z_l + Z_g} \right| = \left| \frac{\mathrm{Re}\{Z_l\} - (Z_g + j\mathrm{Im}\{Z_l\})^*}{\mathrm{Re}\{Z_l\} + (Z_g + j\mathrm{Im}\{Z_l\})} \right| \ . \tag{3.62}$$

At the reference plane 2-2' the impedance is equal to $Z_g + j\mathrm{Im}\{Z_l\}$. The impedance $\mathrm{Re}\{Z_l\}$ can be chosen as a new real reference impedance with

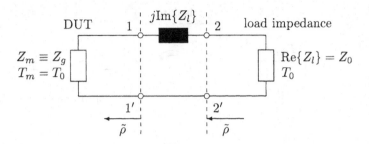

Fig. 3.28 Explanation of the compensation effect for a complex load resistance.

$Re\{Z_l\} \equiv Z_0$. Then, Eq. (3.60) and Eq. (3.61) can be applied and the case of a complex load resistance Z_l is traced back to the case of a real load resistance Z_0, to which the following considerations will be restricted.

Instead of the switchable circulator, which will probably cause switching spikes and which might be limited to low switching rates, it is also possible to use a fixed circulator, a signal divider and an ordinary mechanical or electronic switch, as depicted in Fig. 3.29.

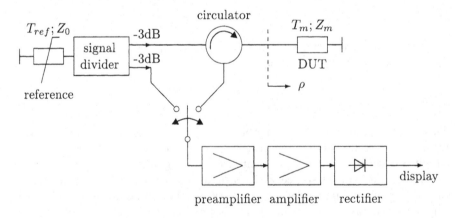

Fig. 3.29 Compensation radiometer with a fixed circulator, a signal divider and a switch.

A compensation radiometer can also be realized without a circulator, of which the real properties often differ noticeably from its ideal characteristics. The compensation radiometer can also be built by means of couplers, wave terminations, an attenuator, a switch and a non-reciprocal passive isolator, as shown in Fig. 3.30.

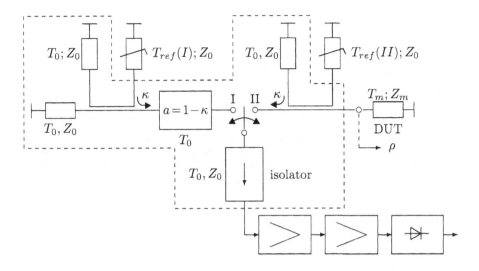

Fig. 3.30 Compensating radiometer with couplers, a switch, an isolator and two reference sources.

Problem

3.3 Derive the balance condition for the radiometer of Fig. 3.30. The couplers are assumed to be lossless with an attenuation κ for the power coupling. The matched attenuator has the power attenuation $a = 1 - \kappa$. All passive components are at the ambient temperature T_0. Except for the device under test, which has the reflection coefficient ρ, all other components are matched. The variable reference noise source with the temperature T_{ref} is available twice, with exact tracking of the two noise temperatures, i.e. $T_{ref}^I = T_{ref}^{II}$.

Provided that all passive components, i.e. all components within the dashed box in Fig. 3.30, including the isolator, are at the same homogenous temperature, e.g. T_0, then an arbitrarily mismatched device under test ($\rho \neq 0$) at the temperature $T_m = T_0$ will necessarily be measured exactly with this temperature. This means that for $T_m = T_0$ there is no offset error, even if e.g. the couplers and the connecting lines have losses. Such an offset test can be performed if, for example, a wave termination with the temperature T_0 is used as the device under test combined with a series or parallel reactance to cause a mismatch. The reactance may have losses, if these are also at the temperature T_0. An offset error generally occurs, if the isolator or the input of the preamplifier are not at the temperature T_0. If the input temperature is lower than T_0, then it can be raised artificially to the value T_0 by additive noise.

An offset error can also be caused by correlation effects of the preamplifier. The isolator/preamplifier combination itself radiates some noise power towards the device under test, i.e. a noise wave is emitted by the input of the isolator/preamplifier, reflected by the mismatched device under test and then returned to the isolator/preamplifier. In addition, the isolator/preamplifier generates a noise wave at the output. For all radiometer setups discussed so far, it was assumed for the balance condition that the noise waves at the input and output are uncorrelated, and that the noise wave at the output thus only adds a constant contribution for both switch positions, which disappears in the final balance condition. This is true for a matched isolator at a homogenous temperature with sufficient reverse isolation. A low-noise preamplifier does generally not possess this characteristic except for the case of de-correlation, as for example the de-correlated amplifier of Fig. 2.23. In general, the isolator has a finite backward isolation. Therefore, the combination of an isolator and a preamplifier can again possess a finite de-correlation only. In fact, a high degree of de-correlation is needed for the described compensation radiometers, in order to keep the measurement errors small, as will also be shown in problem 3.4.

Problem

3.4   Consider the radiometer circuits of Figs. 3.26, 3.29 and 3.30. Which de-correlation of the preamplifier with isolator is required for a measurement error below 1K? The reflection coefficient of the device under test is assumed to be −6 dB.

It may be necessary to utilize both isolators as well as de-correlated amplifiers and there might even be the need for further provisions to reduce the effective correlation. For the correlators of Fig. 3.26 and 3.29, the circulator also reduces a possible correlation. The two variable reference noise sources in Fig. 3.30 must have a good tracking behavior. These two noise sources can also be realized by means of **one** noise source and a matched and decoupling signal divider or by means of **one** noise source and a switch.

The compensating radiometer can be modified by adding a further variable noise source so that mismatched devices under test with a temperature below the ambient temperature T_0 can be measured without using a cold noise source (cf. problem 3.5).

Problem

3.5   With one further hot noise source with $T_{aux} > T_0$ the compensating radiometer according to Fig. 3.30 or Fig. 3.26, 3.29, respectively, can be extended such that mismatched cold devices under test with $T_m < T_0$ can be measured. The variable noise source with the temperature T_{aux} can be inserted into the circuit of Fig. 3.30 via a further directional coupler in the measurement path, so that the overall noise in this branch is increased and a zero compensation by the reference temperature T_{ref}^I becomes possible. Furthermore, the compensating branch is extended by an attenuator with the attenuation $a_2 = 1 - \kappa_2$. After having increased the temperature T_{aux} in a defined way, e.g. by a factor η, a second balance is performed by increasing the reference temperature T_{ref}^{II}. Derive an equation to determine the temperature T_m of the device from T_{ref}^{II}.

3.5.4 Correlation radiometer

The operation of a correlation radiometer will first be explained for a matched device under test. The noise signal of the device under test with the temperature T_m, described by the noise wave A_m, is superimposed to the noise wave A_{ref} of the variable and matched reference noise source. The superposition is realized with a directional coupler, e.g. a 3 dB 180°-coupler, such that the sum of the signals A_m and A_{ref} is obtained at one output port and the difference of the signals A_m and A_{ref} at the other output port (Fig. 3.31). The

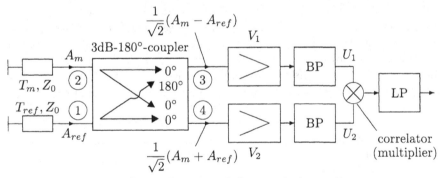

Fig. 3.31 Principle setup of the correlation radiometer.

noise waves of the 180°-coupler are amplified separately by amplifiers with the complex voltage amplification factors V_1 and V_2. The signals are correlated in the correlator, which can be described in the time domain as a multiplier with a subsequent low-pass filter, thus producing the function: $u_1(t) \cdot u_2(t)$. A description of the correlator in the frequency domain leads to the expression:

$$\text{Re}\{W_{12}\} = \text{Re}\{U_1^*(f) \cdot U_2(f)\} \ . \tag{3.63}$$

With

$$U_1 = \frac{1}{\sqrt{2}} \cdot (A_m - A_{ref}) \cdot V_1$$

$$U_2 = \frac{1}{\sqrt{2}} \cdot (A_m + A_{ref}) \cdot V_2 \qquad (3.64)$$

and

$$|A_m|^2 = k \cdot T_m \Delta f$$
$$|A_{ref}|^2 = k \cdot T_{ref} \Delta f \qquad (3.65)$$

Eq. (3.63) becomes

$$\mathrm{Re}\{W_{12}\} = \frac{1}{2} \cdot \mathrm{Re}\{(A_m - A_{ref})^* \cdot (A_m + A_{ref}) \cdot V_1^* \cdot V_2\}$$
$$= \frac{1}{2} \cdot k(T_m - T_{ref}) \cdot \Delta f \cdot \mathrm{Re}\{V_1^* \cdot V_2\} \;, \qquad (3.66)$$

if A_m and A_{ref} are not correlated. For a circuit as in Fig. 3.31, one can expect that A_m and A_{ref} are not correlated, because they originate from different sources. A vanishing correlation, i.e. $\mathrm{Re}\{W_{12}\} = 0$ or a zero display at the output of the correlator, respectively, corresponds to $T_m = T_{ref}$, independently of V_1 or V_2. Thus, the balance of the correlation radiometer is, in principle, similar to the balance of the switching radiometer: the reference temperature is varied until a zero output signal is obtained. The significant advantage of the correlation radiometer by comparison with the switching radiometer is that no switch in front of the preamplifier is needed. Such a switch can produce measurement errors especially due to its switching spikes, its non-zero and variable attenuation and its noise contribution. Nevertheless, the additional use of a switch might be advantageous even for a correlation radiometer, as shown in Fig. 3.32. In general, an analog multiplier shows a finite dc voltage as an offset at its output, even if the correlation is zero. For this reason, it is useful to introduce a 180°-phase shifter behind **one** of the two amplifiers of the correlation radiometer (Fig. 3.32), which periodically changes the polarity at a low frequency of, e.g. $f_{if} = 10$ kHz. At the output of the multiplier (correlator) the resulting 10 kHz alternating signal is amplified and displayed. By this provision, the influence of an offset voltage of the multiplier can be eliminated.

For the zero balance it does not matter, if the 0°/180°-phase shifter shows a slightly different attenuation in both switching positions as expressed by the amplification factor $V_1' = V_1 \cdot (1 - \Delta V)$, where ΔV is real. The amplitude of the 10 kHz signal at the output of the multiplier is proportional to the difference of $\mathrm{Re}\{W_{12}\}$ for both of the phase states , i.e. 0° (I) and 180° (II) of the 180°-phase shifter (compare with Eq. (3.27)).

$$\mathrm{Re}\{W_{12}^{\mathrm{I}}\} - \mathrm{Re}\{W_{12}^{\mathrm{II}}\}$$
$$= \frac{1}{2} \cdot k\Delta f \cdot (T_m - T_{ref})\mathrm{Re}\{V_1^* V_2 - V_1'^* V_2 \cdot e^{-j180°}\}$$
$$= \frac{1}{2} \cdot k\Delta f \cdot (T_m - T_{ref})\mathrm{Re}\{V_1^* V_2 - V_1^*(1 - \Delta V)V_2 \cdot e^{-j180°}\}$$
$$= \frac{1}{2} \cdot k\Delta f \cdot (T_m - T_{ref})\mathrm{Re}\{V_1^* V_2(2 - \Delta V)\} \qquad (3.67)$$

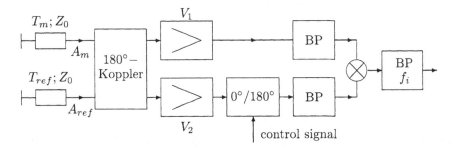

Fig. 3.32   Correlation radiometer with a 180°-switch.

Obviously, a balance again leads to $T_m = T_{ref}$. However, if the correlator has a finite offset error, then a different attenuation for the two switching states of the 180°-phase shifter leads to an error at the output of the correlator. In other words, a parasitic amplitude modulation of the phase shifter leads to a measurement error, if the correlator is not exactly balanced.

It should be mentioned that the 180° phase difference of the phase shifter is not critical and that for a deviation from the 180° phase shift the sensitivity is only marginally reduced.

Instead of a 180°-coupler in the receiver stage of the correlation radiometer, a 90°-3dB-coupler can be employed as well. This will be discussed in more detail in problem 3.6.

Problem

3.6   The correlation radiometer will be realized on the basis of a 90°-3dB-coupler at the input. What further changes of the circuit are necessary in order to measure noise temperatures with such a setup?

For mismatched devices under test the compensating methods derived for the switching radiometer can also be transferred to the correlation radiometer. For the correlation radiometer of Fig. 3.26, a modified version is depicted in Fig. 3.33 as an example.

The relation for the balance of the radiometer circuit is equal to Eq. (3.58). The circulator can also be replaced by a directional coupler as depicted in Fig. 3.30. Two variable reference noise sources should have tracking properties as good as possible. However, the two noise sources should not be correlated. Therefore, it is not advisable to derive them from a single noise source with the help of a signal divider. For the radiometers discussed here, it is not necessary that the device under test and the reference noise source have the same amplitude statistics.

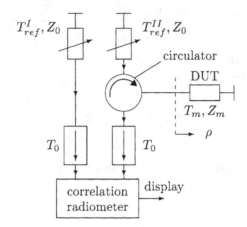

Fig. 3.33   Compensation-correlation radiometer for a mismatched device under test.

3.5.5   Fundamental errors of noise power or noise temperature measurements

As already discussed in Section 3.1.8, the power of a stationary noise signal cannot be measured without errors, because the available measurement time and the measurement bandwidth is limited. The standard deviation $\sigma_m = \Delta T_m$ for a temperature measurement according to Eq. (3.23) is inversely proportional to the square root of the measurement time and the bandwidth $B = \Delta f$. If T_a denotes the system temperature of the amplifier, then

$$\sigma_m = \Delta T_m = \frac{1}{\sqrt{\tau \cdot \Delta f}} \cdot (T_m + T_a) \ . \tag{3.68}$$

A switching radiometer compares the noise temperature T_m (standard deviation ΔT_m) of a one-port with the noise temperature T_{ref} (standard deviation ΔT_{ref}) of a reference one-port. The balance of the radiometer yields the result $T_m = T_{ref}$. However, this balance can be achieved only with an error ΔT_{bal} (standard deviation), because the temperature of the DUT as well as the temperature of the reference are measured with an error. With Eq. (1.52) for uncorrelated noise signals, the variance of the error for the balance of the temperature results as the sum of the variances of the measured temperatures of the device under test and the reference temperature.

$$\Delta T_{bal}^2 = \Delta T_m^2 + \Delta T_{ref}^2 \ . \tag{3.69}$$

For equal measurement times τ' for the reference and the measurement device and for a balance of the setup, we have $\Delta T_m = \Delta T_{ref}$ and

$$\Delta T_{bal} = \sqrt{2} \cdot \Delta T_m = \frac{\sqrt{2}}{\sqrt{\tau' \cdot B}} (T_m + T_a) \quad \text{with} \quad B = \Delta f \ . \tag{3.70}$$

Concerning the switching radiometer the measurement times τ' for the measurement device and the reference device will normally be identical. With a total measurement time $\tau = 2\tau'$ the temperature error ΔT_{sw} increases by a factor of $\sqrt{2}$ to

$$\Delta T_{sw} = \sqrt{2} \cdot \sqrt{2} \cdot \Delta T_m = \frac{2}{\sqrt{\tau \cdot B}}(T_m + T_a) \ . \qquad (3.71)$$

The total measurement time in general is independent of the switching frequency. The increase of the measurement error by a factor of $\sqrt{2}$ or 3 dB for an equal distribution of the measurement time between the object and the reference can be avoided, if a double or transfer switch and two amplifying channels are used as depicted in Fig. 3.34. The transfer switch consecutively

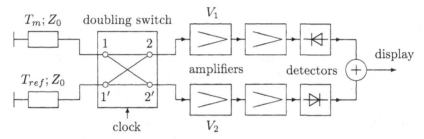

Fig. 3.34   Two-channel radiometer with a transfer switch.

contacts 1 with 2 and $1'$ with $2'$ or 1 with $2'$ and $1'$ with 2, respectively. The output signals of the two detectors or power meters are in phase and can be subtracted or, as shown in Fig. 3.34, can be added, if the polarity of one detector diode is reversed. This radiometer with two channels makes maximum use of the measurement time. The radiometer thus has a temperature error according to Eq. (3.70). This means an improvement of 3 dB by comparison with the simple switching radiometer.

The temperature error increases, if the DUT with a reflection coefficient ρ is mismatched. This will be derived quantitatively in problem 3.7.

Problem

3.7   Calculate the temperature error for a compensation radiometer as depicted in Fig. 3.26 as a function of the reflection coefficient ρ.

The previous error considerations are independent of the switching frequency. Alternatively, the necessary measurements can also be performed subsequently, the results can be stored and the differences can then be computed.

In addition to these principle errors, which can be kept small by choosing a large bandwidth, further errors arise due to, for example, drift effects, quantization errors, non-uniform heating, etc. The reasons for these errors can be as manifold as known from other measurement procedures. For mismatched devices under test further measurement errors might appear, if the input temperature of the preamplifier and the isolator are not precisely equal to T_0 and if the isolator-amplifier combination has partly correlated noise waves at its input and output.

For a switching radiometer with a zero balance, the characteristic of the detector has no influence, i.e. the detector must not necessarily have a quadratic characteristic, as long as the device under test and the reference have the same statistical amplitude behavior, for example, if they are both normally distributed.

3.5.6   Principle errors of a correlation radiometer or correlator

One would expect that a correlation radiometer has the same temperature error (Eq. 3.70) as a two-channel switching radiometer, because of their similarity. In fact, this is true as will be shown in problem 3.8.

Problem

3.8   Show for a correlation radiometer (Figs. 3.31 or 3.32) that it has the same principle measurement error as the radiometer with the double switch of Fig. 3.34.

Furthermore, one can prove that also for a correlation radiometer the multiplier within the correlator must not necessarily be ideal, if a zero balance is performed. In the GHz frequency range a broadband multiplier is realized e.g. as a double balanced mixer which shows some deviations from an ideal multiplier characteristic. Under the balance condition and for an equal amplitude distribution of the device under test and the reference, the correlator must not necessarily have a perfect multiplier characteristic. This will be demonstrated in problem 3.9 by a direct calculation.

Problem

3.9   Show by using the characteristic function that vanishing correlation is measured correctly with a correlator, even if its characteristic deviates from an ideal multiplier. The correlator contains a 180°-phase shifter according to Fig. 3.32.

3.6 MEASUREMENT OF THE NOISE FIGURE

According to its definition the measurement of the noise figure is based on a variation of the generator's noise power and the observation of a corresponding change of the noise power at the output. For the **3dB-method** an adjustable calibrated noise generator is needed, the temperature T_g of which should be known (Fig. 3.35). For the quantitative determination of the noise figure

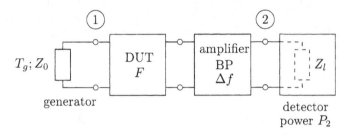

Fig. 3.35 Principle setup for the measurement of the noise figure.

F, the noise temperature T_g of the generator has to be increased until the noise power P_2 at the output of the device under test or the post-amplifier, respectively, has doubled. The doubled noise power is denoted by P_2'. The noise contribution of the two-port to the output noise is given by ΔP_2 and the noise contribution of the generator with the temperature T_0 is represented by P_{20} with:

$$
\begin{aligned}
P_2 &= \Delta P_2 + P_{20} \\
P_2' &= \Delta P_2 + P_{20} \cdot \frac{T_g}{T_0} \\
&= 2\left(\Delta P_2 + P_{20}\right) .
\end{aligned}
\tag{3.72}
$$

Solving Eq. (3.72) for ΔP_2,

$$
\Delta P_2 = P_{20}\left[\frac{T_g}{T_0} - 2\right] ,
\tag{3.73}
$$

and inserting the result into Eq. (2.70) for the noise figure yields

$$
F = 1 + \frac{\Delta P_2}{P_{20}} = \frac{T_g}{T_0} - 1 = \frac{T_{ex}}{T_0} .
\tag{3.74}
$$

The expression $T_g - T_0$ is also called excess noise temperature T_{ex}. The noise figure is thus proportional to the excess temperature of the noise generator normalized to T_0, which is necessary for doubling the noise power at the output. Instead of an increase by 3 dB of the output power it is also possible to choose any other arbitrary value. The noise figure is measured at the center frequency f_0 of the band-pass filter of the amplifier cascade, strictly speaking as a mean noise figure within the bandwidth Δf of the filters. The noise figure is generally measured for the cascade of the device under test and the first amplifier. For the determination of the noise figure of the device under test itself, the contribution of the first amplifier has to be eliminated by using the cascade formula. This is not necessary, if the device under test is an amplifier with sufficient gain. Then, the noise contribution from the following stage can be neglected. The noise powers at the output of the device under test can also be measured directly by means of a radiometer. In this case, the noise figure of the following post-amplification stages will not influence the measurement result.

For the **Y-factor-method** a noise generator with a fixed excess temperature $T_{g0} - T_0$ is used, which can be switched on and off periodically. Such a noise source can be realized, for example, by an avalanche diode and a matched attenuator. A common value for T_{g0}/T_0 is 16 dB. For the on-state of the noise generator with the temperature T_{g0} the amplified noise power shall be denoted by P_2'. For the off-state of the noise generator with the temperature T_0 the noise power at the output is assumed to be P_2. The ratio of P_2' and P_2 is called the Y-factor:

$$Y = \frac{P_2'}{P_2} \ . \tag{3.75}$$

With Eq. (3.72) and Eq. (3.74) and a known Y, the following relation results for the noise figure

$$F = \frac{T_{g0}/T_0 - 1}{Y - 1} = \frac{T_{ex}}{T_0(Y - 1)} \ . \tag{3.76}$$

Similar to other noise parameters, the noise figure cannot be measured arbitrarily precisely due to the finite measurement time and the restricted bandwidth. In problem 3.10 it will be shown that the error caused by the calculation of the mean square value of the noise signals is of minor importance in comparison to the influence of the other measurement errors.

Problems

3.10 How large is the error due to the stochastic nature of the noise signal for a noise figure measurement in a bandwidth of 5MHz and for a measurement time of 0.1s? The noise figure is assumed to be 6dB and the excess temperature has a value of 16dB.

3.11 How can the gain of a device under test be determined with a noise figure meter?

In addition to the measurement of the noise figure, the knowledge of the minimum noise figure and the optimum generator admittance of a linear two-port is of great interest for the design of low-noise circuits e.g low-noise amplifiers. Based on the measured noise parameters, the noise behavior of a linear circuit can be simulated completely. In the following section, different methods for the measurement of the noise parameters of linear two-ports will be presented.

3.7   MEASUREMENT OF THE MINIMUM NOISE FIGURE AND OPTIMUM SOURCE IMPEDANCE

A variety of methods for the measurement of the noise parameters of linear two-ports, i.e. the minimum noise figure F_{min}, the optimum generator admittance $Y_{opt} = G_{opt} + jB_{opt}$ and the equivalent noise resistance R_n will now be discussed. Alternatively, in a spectral representation, the spectra W_u, W_i, $W_{ui} = C_r + jC_i$ for the equivalent circuit will represent the noise parameters and have to be determined.

In principle, all methods are based on noise power measurements. The simplified block diagram of the measurement setup is depicted in Fig. 3.36.

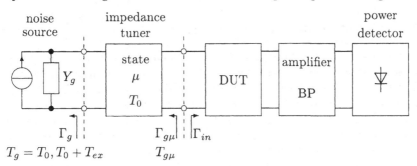

Fig. 3.36   Block diagram of the noise parameter measurement setup.

Similar to the measurement of the noise figure, the setup consists of a noise source with a cold and a hot state, described by the noise temperatures T_0 and $T_{ex} + T_0$ for the noise generator temperature T_g. At the output of the DUT the noise signals are amplified and band-pass-filtered and the noise power can be measured by a power meter. As already seen in Eq. (2.114) or Eq. (2.126), the relation between the noise figure and the noise parameters only depends on the generator admittance. The determination of the noise parameters can thus be performed with the help of an impedance tuner with different impedance states denoted by μ, which transforms the generator admittance of the noise source into a variety of different admittances $Y_{g\mu}$. Such an impedance tuner can be inserted between the noise source and the device under test (DUT).

Besides changing the generator admittance, the tuner at the temperature T_0 also causes a change of the generator temperature for the hot state of the source leading to an effective temperature $T_{g\mu}$ for the μ-th state of the impedance tuner at the input of the DUT.

In order to determine the noise parameters of a linear two-port, it is possible to measure the noise figure of the DUT according to the Y-factor-method, as described above, and to vary the generator admittance by means of the tuner until a minimum of the noise figure is obtained. For this optimum state, the generator admittance can be measured with a network analyzer. This is quite a tedious and unprecise method for the noise parameter measurement, because on the one hand it might take quite some time to find the minimum for the noise figure and on the other hand the impedance range of the tuner may be restricted, so that the tuner may not necessarily be able to realize the optimum generator admittance. Furthermore, the noise temperature for the hot state of the noise generator may change with the position of the tuner, thus changing the measurement value of the noise figure. For these reasons, other methods for the measurement of the noise parameters will be presented in the following section, e.g. the hot-cold or paired method which is based on noise figure measurements and the cold or unpaired method as well as the 7-state-method, which are based on noise power measurements.

3.7.1   Hot-cold method or paired method

The hot-cold or paired method for the determination of the noise parameters is based on noise figure measurements F_μ for a variety of different generator admittances $Y_{g\mu} = G_{g\mu} + jB_{g\mu}$. As already introduced in Eq. (3.75), the measurement of the noise figure requires two noise power measurements according to the Y-factor method. For the μ-th impedance position of the tuner with the corresponding generator admittance $Y_{g\mu}$, the noise figure F_μ can be calculated from the noise power measurements $P_{c\mu}$ and $P_{h\mu}$ for the temperatures $T_{c\mu} = T_0$ and $T_{h\mu} = T_0 + T_{ex\,\mu}$. The noise figure of the μ-th tuner position can be written as a function of the spectral noise parameters (Eq. (2.114)):

$$F_\mu = 1 + \frac{|Y_{g\mu}|^2 \mathsf{W}_u + \mathsf{W}_i + 2G_{g\mu}\mathsf{C}_r + 2B_{g\mu}\mathsf{C}_i}{4kT_0 G_{g\mu}} \qquad (3.77)$$

Rearranging this equation

$$(F_\mu - 1)4kT_0 G_{g\mu} = |Y_{g\mu}|^2 \mathsf{W}_u + \mathsf{W}_i + 2G_{g\mu}\mathsf{C}_r + 2B_{g\mu}\mathsf{C}_i \ , \qquad (3.78)$$

one obtains a linear equation for the spectral noise parameters W_u, W_i, C_r and C_i. Thus, the measurement of the noise figure for different tuner positions leads to a system of linear equations for the determination of the unknown spectral noise parameters. For the calculation of the unknowns, at least four noise figure measurements for different generator admittances $Y_{g\mu}$ have to be performed. In order to minimize the measurement error, it is advantageous to perform more than the necessary four measurements, so that the parameters can be determined on the basis of the least-squares method for an over-determined system of linear equations. Moreover, for the enhancement of the measurement accuracy, weighting factors can be introduced in order to reduce the influence of data which seem to be less accurate.

The spectral noise parameters are thus known. The parabolic noise parameters can directly be calculated with (cf. Section 2.3.5):

$$B_{opt} = -\frac{C_i}{W_u} \tag{3.79}$$

$$G_{opt} = \sqrt{\frac{W_i}{W_u} - B_{opt}^2} \tag{3.80}$$

$$R_n = \frac{W_u}{4kT_0} \tag{3.81}$$

$$F_{min} = 1 + 2R_nG_{opt} + \frac{C_r}{2kT_0} . \tag{3.82}$$

For this hot-cold or paired method measurement errors arise, because the effective generator noise temperature $T_{g\mu}$ for the hot state of the noise source has to be known exactly for the different source admittances $Y_{g\mu}$. Furthermore, the source reflection coefficient $\Gamma_{g\mu}$ has to be the same for the hot and the cold state of the noise source.

3.7.2 Cold method or unpaired method

The problems discussed above can be minimized by a determination of the noise parameters mainly from unpaired noise power measurements with a so-called cold noise source at the ambient temperature T_0. For the noise equivalent circuit in Fig. 3.37, the noise power at the output of the two-port can be calculated in the frequency band Δf as follows:

$$\begin{aligned} P_{L\mu} &= |U_2|^2 \mathrm{Re}\{Y_L\}\Delta f = \kappa|U_1|^2\mathrm{Re}\{Y_L\}\Delta f \\ &= \underbrace{\kappa\mathrm{Re}\{Y_L\}\Delta f}_{m} \\ &\qquad \cdot \frac{[4kT_{g\mu}G_{g\mu} + |Y_{g\mu}|^2W_u + W_i + 2G_{g\mu}C_r + 2B_{g\mu}C_i]}{|Y_{g\mu} + Y_{in}|^2} . \end{aligned} \tag{3.83}$$

Here, κ represents the power gain and Y_{in} is the input admittance of the device under test.

With Eq. (3.83) it follows that, in contrast to the noise figure, the noise power also depends on the product of the power gain κ, the real part of the load admittance Y_L and the bandwidth Δf, all together abbreviated by a factor m. Furthermore, it depends on the unknown input admittance of the two-port Y_{in}.

A partly unpaired method still needs one noise figure measurement, i.e. one paired noise power measurement. Based on this paired measurement with the

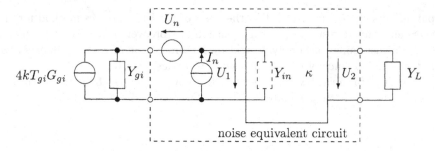

Fig. 3.37 Noise equivalent circuit for the calculation of the noise power at the load admittance Y_L.

hot and cold noise power P_h and P_c for the generator temperature T_h and T_c and for the same generator admittance Y_g

$$P_h = \frac{m}{|Y_{in} + Y_g|^2}(4kT_hG_g + |Y_g|^2W_u + W_i + 2G_gC_r + 2B_gC_i) \quad (3.84)$$

$$P_c = \frac{m}{|Y_{in} + Y_g|^2}(4kT_cG_g + |Y_g|^2W_u + W_i + 2G_gC_r + 2B_gC_i) \quad (3.85)$$

the factor m can be determined, if the input admittance Y_{in} is known:

$$m = \frac{(P_h - P_c)|Y_{in} + Y_g|^2}{4k(T_h - T_c)G_g} \quad . \quad (3.86)$$

For the determination of all noise parameters, further noise power measurements, with e.g. the noise source operated at the ambient temperature T_0, have to be considered. On the basis of cold noise power measurements for different tuner positions μ, one obtains with Eq. (3.85)

$$P_{c\mu}\frac{|Y_{in} + Y_{g\mu}|^2}{m} - 4kT_cG_{g\mu} = |Y_{g\mu}|^2W_u + W_i + 2G_{g\mu}C_r + 2B_{g\mu}C_i \quad (3.87)$$

a system of linear equations for the unknown parameters W_u, W_i, C_r and C_i, if Y_{in} is known. With at least one noise figure and four noise power measurements the noise parameters are calculable.

Another algorithm makes it possible to determine the noise parameters on the basis of unpaired noise power measurements only. For this purpose, at least five noise power measurements have to be performed for different source admittances. For at least one of these measurements, the noise source has to be operated in its hot state. For this method identical values of the source admittance for the hot and cold state of the noise source are not necessary.

For the determination of the noise parameters according to the unpaired method different ratios of the measured noise powers denoted by p_n, $n = 1 \ldots 4$, for the μ-th and ν-th tuner position are considered. In this way, the dependence on the load admittance Y_L and the power gain κ or the factor m can be eliminated:

$$p_n = \frac{P_{L\mu}|Y_{in} + Y_{g\mu}|^2}{P_{L\nu}|Y_{in} + Y_{g\nu}|^2} \ . \qquad (3.88)$$

With the help of Eq. (3.83) the following linear relation as a function of the unknown spectral noise parameters W_u, W_i, C_r and C_i results:

$$
\begin{aligned}
4k(p_n T_{g\nu} G_{g\nu} - T_{g\mu} G_{g\mu}) &= 2C_r(G_{g\mu} - p_n G_{g\nu}) + 2C_i(B_{g\mu} - p_n B_{g\nu}) \\
&\quad + W_u(|Y_{g\mu}|^2 - p_n|Y_{g\nu}|^2) + W_i(1 - p_n) \ .
\end{aligned}
$$
$$(3.89)$$

At least one noise power measurement has to be performed with the noise source operated at a hot temperature different from the ambient temperature T_0, so that a system of four linearly independent equations results. As will be shown later, the optimum generator admittance Y_{opt} can already be determined on the basis of cold noise power measurements only.

For this unpaired or cold method, the input admittance Y_{in} of the device under test has to be known. It can be determined with the help of a network analyzer. However, this additional measurement can lead to additional errors. For example, reproducibility errors might arise because of the necessary multiple connections and disconnections of the connectors.

Also for these methods it is advantageous to perform more than just the minimum number of power measurements and to calculate the unknown noise parameters on the basis of the least-squares method for over-determined linear equations.

3.7.3   The 7-state-method

The 7-state-method allows to determine the noise parameters on the basis of unpaired noise power measurements. The optimum generator admittance Y_{opt} and the input admittance Y_{in} of the DUT can be calculated from cold noise power measurements only. The measurement of the DUT's input admittance with a network analyzer is not necessary. The minimum noise figure F_{min} and the equivalent noise resistance R_n can be determined on the basis of cold noise power measurements only, except for the previously defined factor m. In order to determine the unknown factor m, one further noise power measurement with a hot noise source has to be performed.

The theory of the 7-state method is based on noise power measurements according to Eq. (3.83). This relation can be rewritten as follows:

$$P_\mu \cdot |Y_{g\mu} + Y_{in}|^2 = m[4kT_{g\mu}G_{g\mu} + W_i + |Y_{g\mu}|^2 W_u + 2G_{g\mu}C_r + 2B_{g\mu}C_i] \ , \quad (3.90)$$

leading to

$$S_\mu = P_\mu \cdot |Y_{g\mu} + Y_{in}|^2 - m[4kT_{g\mu}G_{g\mu} + W_i + |Y_{g\mu}|^2 W_u + 2G_{g\mu}C_r + 2B_{g\mu}C_i]$$

$$
\begin{aligned}
= \underbrace{P_\mu |Y_{g\mu}|^2}_{a_{\mu 1}} + \underbrace{2P_\mu G_{g\mu}}_{a_{\mu 2}} G_{in} + \underbrace{2P_\mu B_{g\mu}}_{a_{\mu 3}} B_{in} + \underbrace{P_\mu}_{a_{\mu 4}} \left(G_{in}^2 + B_{in}^2 \right) \\
- \underbrace{4kT_{ex,\mu} G_{g\mu}}_{a_{\mu 5}} m - \underbrace{\frac{1}{}}_{a_{\mu 6}} m \underbrace{W_i}_{\tilde{W}_i} - \underbrace{|Y_{g\mu}|^2}_{a_{\mu 7}} m \underbrace{W_u}_{\tilde{W}_u} \\
- \underbrace{2G_{g\mu}}_{a_{\mu 8}} m \underbrace{(2kT_0 + C_r)}_{\tilde{C}_r} - \underbrace{2B_{g\mu}}_{a_{\mu 9}} m \underbrace{C_i}_{\tilde{C}_i}
\end{aligned} \tag{3.91}
$$

with the source temperature

$$
T_{g\mu} = T_0 + T_{ex,\mu} \ . \tag{3.92}
$$

For the different tuner positions μ, a system of equations results, which depends non-linearly on the input admittance $Y_{in} = G_{in} + jB_{in}$ at the input of the DUT and linearly on the noise parameters and the factor m. The products of the noise parameters and the factor m are substituted by normalized noise parameters $\tilde{W}_i$, $\tilde{W}_u$ and $\tilde{C}_i$, also called modified noise parameters. As the real part of the cross spectrum C_r and the part of the noise temperature at ambient temperature, $4kT_0$, both depend on the real part of the source input admittance, they are combined as $\tilde{C}_r$. Thus, only the term $a_{\mu 5}$ in front of the factor m depends on the excess noise temperature $T_{ex,\mu}$.

The noise parameters, the input admittance and the factor m can be determined on the basis of eight noise power measurements, where at least one measurement has to be performed with a hot noise source temperature. In order to enhance the measurement accuracy with the help of the least-squares method a higher number of measurements is recommended. For the elimination of the non-linear dependence on the DUT's input admittance, the sum S of the squares of S_μ of Eq. (3.91)

$$
S = \sum_{\mu=1}^{n} S_\mu^2 \tag{3.93}
$$

is differentiated with respect to the seven unknown terms consisting of the four normalized noise parameters, $\tilde{W}_i$, $\tilde{W}_u$, $\tilde{C}_r$ and $\tilde{C}_i$, the real and the imaginary part of the input admittance, G_{in} and B_{in}, and the factor m. This leads to a set of seven equations:

$$
\frac{\partial S}{\partial G_{in}} = 2 \sum_{\mu=1}^{n} S_\mu \cdot (a_{\mu 2} + 2a_{\mu 4} G_{in}) \overset{!}{=} 0 \tag{3.94}
$$

$$
\frac{\partial S}{\partial B_{in}} = 2 \sum_{\mu=1}^{n} S_\mu \cdot (a_{\mu 3} + 2a_{\mu 4} B_{in}) \overset{!}{=} 0 \tag{3.95}
$$

$$
\frac{\partial S}{\partial m} = -2 \sum_{\mu=1}^{n} S_\mu \cdot a_{\mu 5} \overset{!}{=} 0 \tag{3.96}
$$

$$\frac{\partial S}{\partial \tilde{\mathsf{W}}_i} \;=\; -2\sum_{\mu=1}^{n} S_\mu \cdot a_{\mu 6} \overset{!}{=} 0 \tag{3.97}$$

$$\frac{\partial S}{\partial \tilde{\mathsf{W}}_u} \;=\; -2\sum_{\mu=1}^{n} S_\mu \cdot a_{\mu 7} \overset{!}{=} 0 \tag{3.98}$$

$$\frac{\partial S}{\partial \tilde{\mathsf{C}}_r} \;=\; -2\sum_{\mu=1}^{n} S_\mu \cdot a_{\mu 8} \overset{!}{=} 0 \tag{3.99}$$

$$\frac{\partial S}{\partial \tilde{\mathsf{C}}_i} \;=\; -2\sum_{\mu=1}^{n} S_\mu \cdot a_{\mu 9} \overset{!}{=} 0 \;, \tag{3.100}$$

which can be solved for the unknown parameters. This is demonstrated in problem 3.12.

Problem

3.12 Derive a solution for the unknown parameter G_{in} with the help of the least square error minimization of Eqs. (3.94) to (3.100). Discuss the possibility to determine G_{in} with the help of cold measurements only, i.e. with the generator temperature $T_{ex} = 0$.

As described in problem 3.12, eliminating B_{in} and solving for G_{in} leads to a polynomial of 8th degree with the coefficients h_i, $i = 1, \ldots, 8$,

$$h_8 G_{in}^8 + h_7 G_{in}^7 + h_6 G_{in}^6 + h_5 G_{in}^5 + h_4 G_{in}^4 + h_3 G_{in}^3 + h_2 G_{in}^2 + h_1 G_{in} + h_0 = 0 \;, \tag{3.101}$$

which is unambiguously solvable numerically. The determination of G_{in} or B_{in} can either be performed on the basis of only cold, only hot or a combination of both measurements. Using only cold noise power measurements with $T_{ex,\mu} = 0$ leads to $a_{\mu 5} = 0$ in Eq. (3.91) and the unknown factor m can thus be eliminated. Consequently, on the basis of cold noise power measurements, it is possible to calculate G_{in} and B_{in} and the normalized noise parameters $\tilde{\mathsf{W}}_i, \tilde{\mathsf{W}}_u, \tilde{\mathsf{C}}_r$ and $\tilde{\mathsf{C}}_i$. The spectral noise parameters are thus known, except for the factor m. As the optimum generator admittance Y_{opt} only depends on quotients of the spectral noise parameters, Y_{opt} is calculable using the results of the cold noise measurements only:

$$G_{opt} \;=\; \sqrt{\frac{\mathsf{W}_i}{\mathsf{W}_u} - B_{opt}^2} = \sqrt{\frac{\tilde{\mathsf{W}}_i}{\tilde{\mathsf{W}}_u} - B_{opt}^2} \tag{3.102}$$

$$B_{opt} \;=\; -\frac{\mathsf{C}_i}{\mathsf{W}_u} = -\frac{\tilde{\mathsf{C}}_i}{\tilde{\mathsf{W}}_u} \;. \tag{3.103}$$

For the determination of the factor m at least one hot noise power measurement is necessary. For $T_{ex\,\mu} \neq 0$ it follows $a_{\mu 5} \neq 0$ in Eq. (3.91) and the factor m can be calculated:

$$m = \frac{1}{a_{\mu 5}}\left[P_\mu \cdot |Y_{g\mu} + Y_{in}|^2 - a_{\mu 6}\tilde{W}_i - a_{\mu 7}\tilde{W}_u - a_{\mu 8}\tilde{C}_r - a_{\mu 9}\tilde{C}_i\right] \quad . \quad (3.104)$$

The 7-state-method has the advantage that an additional measurement of the DUT's input admittance with a network analyzer is not necessary. Reproducibility errors due to multiple connections can thus be avoided. Furthermore, nearly all unknowns, as e.g. the DUT input admittance and the optimum generator admittance, can be determined on the basis of cold noise power measurements. The so-called cold measurements, which are performed at ambient temperature, help to reduce measurement uncertainties because their noise temperature is well known by definition and is not influenced by the state of the tuner. Figure 3.38 shows experimental results for the noise parameters of a low noise high electron mobility transistor (HEMT) as an active element.

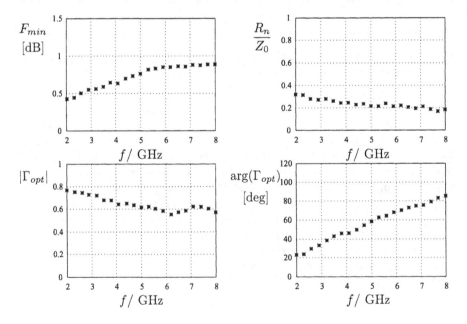

Fig. 3.38   Noise parameters of a high electron mobility transistor.

3.8   DE-EMBEDDING OF THE NOISE PARAMETERS

In general, the weak noise signals of the device under test have to be amplified sufficiently, so that a noise power measurement becomes possible. As a consequence, for the noise parameter measurement system in Fig. 3.36 only the noise parameters of the cascade connection of the device under test and the amplifier can be determined. The direct measurement of the DUT noise parameters is not possible. For the correction of the measurement data, in order to eliminate the influence of the preamplifier, a so-called de-embedding has to be performed. The following correction procedure is based on a noise correlation matrix description with chain matrices. For the equivalent circuit of a two-port a correlation matrix representation is derived. For the input and output currents and voltages we have:

$$\left[\begin{array}{c} U_1 \\ I_1 \end{array} \right] = [A_d] \left[\begin{array}{c} U_2 \\ -I_2 \end{array} \right] + \left[\begin{array}{c} U_d \\ I_d \end{array} \right] , \tag{3.105}$$

where $[A_d]$ represents the chain matrix of the two-port. The noise parameters are defined by a noise correlation matrix $[C_A]$:

$$[C_A] = \left[\begin{array}{c} U_d \\ I_d \end{array} \right] \left[\begin{array}{c} U_d \\ I_d \end{array} \right]^\dagger = \left[\begin{array}{c} U_d \\ I_d \end{array} \right] [\, U_d^* \quad I_d^* \,] = \left[\begin{array}{cc} |U_d|^2 & U_d I_d^* \\ U_d^* I_d & |I_d|^2 \end{array} \right] . \tag{3.106}$$

where the dagger-sign ($\dagger$) denotes the hermitian conjugate. The hermitian conjugate is equal to the transposed complex conjugate matrix.

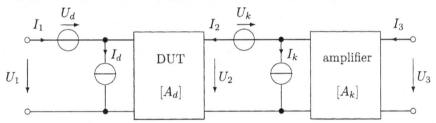

Fig. 3.39   Noise equivalent circuit of the cascade connection of the device under test and the preamplifier in the chain matrix representation.

Based on the equivalent circuits in Fig. 3.39 and Fig. 3.40 for the cascade connection of the device under test and the preamplifier, the following relations can be derived with a chain matrix description. With Eq. (3.105) and

$$\left[\begin{array}{c} U_2 \\ -I_2 \end{array} \right] = [A_k] \left[\begin{array}{c} U_3 \\ -I_3 \end{array} \right] + \left[\begin{array}{c} U_k \\ I_k \end{array} \right] , \tag{3.107}$$

we get the following relation:

$$\left[\begin{array}{c} U_1 \\ I_1 \end{array} \right] = [A_d] [A_k] \left[\begin{array}{c} U_3 \\ -I_3 \end{array} \right] + [A_d] \left[\begin{array}{c} U_k \\ I_k \end{array} \right] + \left[\begin{array}{c} U_d \\ I_d \end{array} \right] \tag{3.108}$$

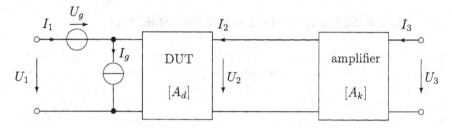

Fig. 3.40   Noise equivalent circuit of the device under test and the preamplifier in cascade connection with noise equivalent sources at the input of the cascade connection.

for the noise equivalent circuit of Fig. 3.39. Further on, the noise equivalent circuit in Fig. 3.40 is described by:

$$\begin{bmatrix} U_1 \\ I_1 \end{bmatrix} = [A_d] [A_k] \begin{bmatrix} U_3 \\ -I_3 \end{bmatrix} + \begin{bmatrix} U_g \\ I_g \end{bmatrix} . \tag{3.109}$$

A comparison of the coefficients of Eq. (3.108) and Eq. (3.109) leads to

$$\begin{bmatrix} U_g \\ I_g \end{bmatrix} = [A_d] \begin{bmatrix} U_k \\ I_k \end{bmatrix} + \begin{bmatrix} U_d \\ I_d \end{bmatrix} . \tag{3.110}$$

For the determination of the spectral noise parameters the noise correlation matrix $[C_g]$ is calculated for the setup in Fig. 3.40:

$$[C_g] = \begin{bmatrix} U_g \\ I_g \end{bmatrix} \begin{bmatrix} U_g \\ I_g \end{bmatrix}^\dagger = \begin{bmatrix} U_g \\ I_g \end{bmatrix} \begin{bmatrix} U_g^* & I_g^* \end{bmatrix} = \begin{bmatrix} |U_g|^2 & U_g I_g^* \\ U_g^* I_g & |I_g|^2 \end{bmatrix} . \tag{3.111}$$

Finally, from Eq. (3.110) and Eq. (3.111), a relation for the correlation matrices $[C_d]$ of the DUT and $[C_k]$ of the preamplifier can be derived. We can benefit from the fact that U_d, I_d and U_k, I_k are uncorrelated.

$$
\begin{aligned}
[C_g] &= \begin{bmatrix} U_g \\ I_g \end{bmatrix} \begin{bmatrix} U_g \\ I_g \end{bmatrix}^\dagger = \left([A_d] \begin{bmatrix} U_k \\ I_k \end{bmatrix} + \begin{bmatrix} U_d \\ I_d \end{bmatrix} \right) \left([A_d] \begin{bmatrix} U_k \\ I_k \end{bmatrix} + \begin{bmatrix} U_d \\ I_d \end{bmatrix} \right)^\dagger \\
&= [A_d] \underbrace{\begin{bmatrix} U_k \\ I_k \end{bmatrix} \begin{bmatrix} U_k \\ I_k \end{bmatrix}^\dagger}_{[C_k]} [A_d]^\dagger + \underbrace{\begin{bmatrix} U_d \\ I_d \end{bmatrix} \begin{bmatrix} U_d \\ I_d \end{bmatrix}^\dagger}_{[C_d]} \\
&= [A_d] [C_k] [A_d]^\dagger + [C_d] .
\end{aligned}
\tag{3.112}
$$

This equation can be solved for $[C_d]$ so that the noise parameters of the device under test are calculated as follows:

$$[C_d] = [C_g] - [A_d] [C_k] [A_d]^\dagger , \tag{3.113}$$

if the noise parameters of the whole setup as well as the noise parameters of the preamplifier and additionally the chain matrix of the DUT are known. In order to perform a de-embedding of the DUT noise parameters, it is thus necessary to perform noise measurements both with and without the DUT.

The noise parameters of a device under test are thus calculable. As derived for this de-embedding procedure, a noise characterization of the setup itself has to be performed first, leading to the noise parameters of the pre- and postamplifier cascade of the measurement setup. Then, the DUT's noise parameters can be determined. The noise behavior of the measurement setup will thus completely be characterized as part of the noise parameter determination.

3.9 ALTERNATIVE METHOD FOR THE DETERMINATION OF THE NOISE TEMPERATURE OF A ONE-PORT

In addition to the calculation of the noise parameters of two-port networks, the noise parameter measurement system can also be utilized for the measurement of noise temperatures of one-ports as has already been discussed previously on the basis of radiometers.

With the knowledge of the noise parameters W_u, W_i, C_r, C_i, the input impedance Y_{in} and the factor m of the measurement setup, the noise temperature of a one-port device can be calculated, for example, via (Eq. 3.114). On the basis of a noise power measurement

$$P_L = m \cdot \frac{4kT_gG_g + |Y_g|^2\mathsf{W}_u + \mathsf{W}_i + 2G_g\mathsf{C}_r + 2B_g\mathsf{C}_i}{|Y_g + Y_{in}|^2} \tag{3.114}$$

the noise temperature can be determined by

$$T_g = \frac{1}{4kG_g} \cdot \left(P_L \frac{|Y_g + Y_{in}|^2}{m} - |Y_g|^2\mathsf{W}_u - \mathsf{W}_i - 2G_g\mathsf{C}_r - 2B_g\mathsf{C}_i \right) \ . \tag{3.115}$$

Contrary to the previously described radiometers, a decorrelation of the preamplifier is not necessary for this method, because the noise behavior of the amplifier is completely characterized and thus known, so that its contribution is accounted for correctly in this solution.

4
Noise of Diodes and Transistors

Apart from thermal noise, shot noise is one of the fundamental noise phenomena of electronic devices. Shot noise is closely related to the fact that the current does not flow continuously but in small portions, due to the discrete charge of the electrons. Furthermore, the transition of the electrons takes place irregularly in time. A direct current is thus constant only as a time average but not for short periods of time. Consequently, the direct current is superimposed by a fluctuation current. The spectrum of the current fluctuations of high-frequency devices can be constant up to high frequencies, similar to the thermal noise. As described by the so-called Schottky-relation, the spectrum depends on the value of the direct current. For example, pn-diodes and Schottky-diodes show shot noise. Relating the shot noise for a given direct current to the impedance of the diode allows one to introduce an effective noise temperature for the pn- and Schottky-diodes, similar to the temperature description for thermally noisy resistors. On the following pages it will be shown that in most cases the thus defined noise temperature of Schottky diodes is lower than that of thermally noisy resistors at the same physical temperature.

With the PIN-diode a device will be presented, which has a noise temperature nearly equal to the physical temperature.

Noise equivalent circuits of bipolar and field effect transistors can mostly be based on thermal noise equivalent sources and on shot noise equivalent sources. In addition, passive elements and controlled sources are needed for the equivalent circuits. With the help of equivalent circuits, the devices can be implemented in the circuitry of e.g. amplifiers, so that complete equivalent circuits for small signal amplifiers can be found. On the basis of such noise equivalent circuits, the noise figure of amplifiers can be calculated and compared with measurements.

4.1 SHOT NOISE

The current of electron tubes shows noise, because the transport of the electrons is not a continuous process, but rather it relies on discrete charges. Noise arises because the flow of the electrons, i.e. their number per unit time, is not constant but is subject to statistical fluctuations. A fairly concise model is a vacuum tube or, more specifically, a vacuum diode in the saturation region. The entire current $i(t)$ can be separated into a direct current I_0 and an alternating current $i_{sh}(t)$. Let z denote the mean number of electrons per unit time and q the elementary charge. Then, we can write

$$i(t) = I_0 + i_{sh}(t) \tag{4.1}$$

with

$$I_0 = z \cdot q \ . \tag{4.2}$$

The time average of the alternating part of the current $i_{sh}(t)$ is assumed to be zero

$$\overline{i_{sh}(t)} = 0 \ . \tag{4.3}$$

The alternating or noisy part $i_{sh}(t)$ of the current $i(t)$ is assumed to be an ergodic fluctuation phenomenon. This means that the time average and the ensemble average are assumed to be equal.

A vacuum diode with a metal cathode generates a saturation current, if it is operated at a high anode voltage. The saturated diode is also a fairly concise model for a number of semiconductor devices, which show a similar noise behavior. For the model of the saturated vacuum diode, the following assumptions are valid to a large extent:

a) The electrons are emitted from the hot cathode statistically independent from each other.

b) The path-time characteristic of the single electron in the region between the cathode and the anode is independent of the presence of other electrons, i.e. the influence of a space charge is neglected.

Furthermore, the following assumptions are made in order to simplify the model:

c) The electrons have no thermal initial speed at the cathode.

d) All electrons follow the same path versus time.

e) No secondary electrons are emitted at the anode.

Each electron, which passes through the vacuum diode, induces a current pulse of length τ in the outer circuit. Here, τ is the travel time of an electron from the cathode to the anode.

The shape of the induced current pulse is equal for each electron according to the assumptions made and it is described by a function $\xi(t)$. In principle, $\xi(t)$ can be calculated if the voltage of the anode and the geometry of the vacuum diode are known. As will be seen, however, a detailed knowledge of $\xi(t)$ is not necessary. It is assumed that $\xi(t)$ is equal to zero at the starting time of the electron. The electron starts its travel at $\vartheta = 0$ and ends its travel at $\vartheta = \tau$:

$$\xi(\vartheta) = 0 \quad \text{for} \quad \vartheta \leq 0 \quad \text{and} \quad \vartheta \geq \tau \ . \tag{4.4}$$

Thus, the ν-th current pulse, which starts at t_ν, has the following time characteristic:

$$i_\nu(t) = q \cdot \xi(t - t_\nu) \ . \tag{4.5}$$

For some current pulses the time characteristic is shown in Fig. 4.1. Each

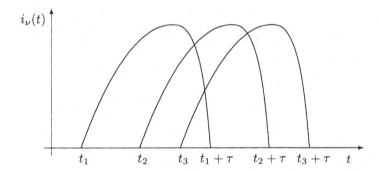

Fig. 4.1 Time characteristic of the current pulse in the outer circuit.

electron transports the elementary charge q, so that a normalization condition can be formulated for the single current pulse:

$$\int_{t_\nu}^{t_\nu + \tau} i_\nu(t) \, dt = \int_{t_\nu}^{t_\nu + \tau} q \cdot \xi(t - t_\nu) \, dt = q \tag{4.6}$$

or

$$\int_{0}^{\tau} \xi(\vartheta) \, d\vartheta = 1 \ . \tag{4.7}$$

The total current $i(t)$ results from the superposition of the single current pulses:

$$i(t) = I_0 + i_{sh}(t) = q \cdot \sum_{\nu} \xi(t - t_\nu) \ . \tag{4.8}$$

Thus the direct current is given by

$$I_0 = z \cdot \int\limits_0^\tau q \cdot \xi(\vartheta)\, d\vartheta = q \cdot z \ . \tag{4.9}$$

Because of $\overline{i_{sh}(t)} = 0$, we get for the mean square values:

$$\overline{i^2(t)} = I_0^2 + \overline{i_{sh}^2(t)} \ . \tag{4.10}$$

In order to determine $\overline{i_{sh}^2(t)}$, the variance of a statistically independent sequence of single pulses has to be calculated (the so-called Campbell theorem).

It will be the aim to determine the autocorrelation function of $i(t)$. An irregular pulse sequence $p(t)$ (a so-called Poisson process), consisting of arbitrarily starting but equal and normalized single pulses $\xi(t - t_\nu)$, is considered. Let $z \cdot dt$ be the probability that an impulse lies within an infinitesimal time interval dt. Here, z is constant and independent of the position of the time interval. Furthermore, the pulses are assumed to be independent from each other, i.e. the appearance of a pulse at a time t_ν has no influence on the appearance of further pulses. Arbitrary pulse overlaps are allowed.

With the help of the Dirac or δ-function, the single pulse can also be represented by an integral form:

$$\xi(t - t_\nu) = \int\limits_{-\infty}^{+\infty} \xi(t') \cdot \delta(t - t_\nu - t')\, dt' \ . \tag{4.11}$$

For the pulse sequence $p(t)$, the following expression can be obtained if the order of summation and integration is changed:

$$\begin{aligned}
p(t) &= \int\limits_{-\infty}^{+\infty} \xi(t') \cdot \sum_\nu \delta(t - t_\nu - t')\, dt' \\
&= \int\limits_{-\infty}^{+\infty} \xi(t') \cdot x(t - t')\, dt' \ .
\end{aligned} \tag{4.12}$$

Here, $x(t)$ stands for an irregular sequence of δ-pulses. For the autocorrelation function $\rho_p(\theta)$ of $p(t)$ the following expression results, where $P(\theta)$ denotes the autocorrelation function of a sequence of δ-pulses:

$$\begin{aligned}
\rho_p(\theta) &= \int\limits_{-\infty}^{+\infty}\!\!\int \xi(t') \cdot \xi(t'')\, \mathrm{E}\{x(t - t')x(t - t'' + \theta)\}\, dt'\, dt'' \\
&= \int\limits_{-\infty}^{+\infty}\!\!\int \xi(t') \cdot \xi(t'') \cdot P(t' - t'' + \theta)\, dt'\, dt'' \ .
\end{aligned} \tag{4.13}$$

The autocorrelation function of an irregular pulse sequence is thus expressed by the autocorrelation function of an irregular sequence of Dirac pulses. As can be anticipated and as will be shown in problem 4.1, the autocorrelation function $P(\theta)$ of an irregular sequence of Dirac impulses is again a Dirac function.

$$P(\theta) = z \cdot \delta(\theta) + z^2 \ . \tag{4.14}$$

The term z^2 on the right-hand side of Eq. (4.14) results from the constant term of $\mathrm{E}\{x(t)\}$. For the term $\mathrm{E}\{p(t)\}$ we have:

$$
\begin{aligned}
\mathrm{E}\{p(t)\} &= \int\limits_{-\infty}^{+\infty} \xi(t') \cdot \mathrm{E}\{x(t - t')\}\, dt' \\
&= z \cdot \int\limits_{-\infty}^{+\infty} \xi(t)\, dt = z \ .
\end{aligned}
\tag{4.15}
$$

Considering only the fluctuating part $P_{sh}(\theta)$ of the autocorrelation function, we observe that this part consists of a pure Dirac function without constant terms:

$$
\begin{aligned}
P_{sh}(\theta) &= \mathrm{E}\{(x(t) - \mathrm{E}\{x(t)\})(x(t + \theta) - \mathrm{E}\{x(t)\})\} \\
&= P(\theta) - z^2 = z \cdot \delta(\theta) \ .
\end{aligned}
\tag{4.16}
$$

Problem

4.1   Derive Eqs. (4.14) and (4.16).

With the result of Eq. (4.14), i.e. the autocorrelation function for an irregular sequence of Dirac pulses, Eq. (4.13) can be evaluated. The result is

$$\rho_p(\theta) = z \cdot \int\limits_{-\infty}^{+\infty} \xi(t) \cdot \xi(t + \theta)\, dt + z^2 \ . \tag{4.17}$$

This relation is called Campbell's theorem. The power spectrum W_p of the irregular pulse sequence is obtained as the Fourier transform of the autocorrelation function $\rho_p(\theta)$. The Fourier transform of the single pulse $\xi(t)$ is

$$S(f) = \int\limits_{-\infty}^{+\infty} \xi(t) e^{-j2\pi ft}\, dt \ . \tag{4.18}$$

The power spectrum W_p of this so-called Poisson process can thus be determined immediately, because the Fourier transform of the convolution integral in Eq. (4.17) is just $|S(f)|^2$. Hence we have:

$$W_p(f) = \int_{-\infty}^{+\infty} \rho_p(\theta)e^{-j2\pi f\theta}\, d\theta = z \cdot |S(f)|^2 + z^2 \cdot \delta(f) \ . \qquad (4.19)$$

The frequency characteristic is solely determined by the shape of the single pulse. The spectrum becomes broader with shorter pulses. The term $z^2 \cdot \delta(f)$ in Eq. (4.19) describes the constant part and the term $z \cdot |S(f)|^2$ represents the fluctuation part or the noise spectrum of the irregular pulse sequence, respectively.

For a Poisson process the shape of the autocorrelation function or the power spectrum is entirely determined by the shape of the single pulse. The pulse density enters as a multiplying factor only. It is of no relevance whether there are overlapping pulses or not. In contrast, the amplitude distribution of the stochastic signals depends on the probability of pulse overlapping. Assuming that the single pulses have a rectangular shape with a rate that overlapping appears very rarely, then the amplitude distribution will have two values only, namely the two distinct values of the rectangular signals. For strong pulse overlapping it follows from the central limit theorem that the amplitude distribution can be approximated by a Gaussian distribution, independently of the shape of the single pulses. For the shot noise the case of an intense overlapping is practically always given. As a consequence, it can be assumed that the amplitude distribution is a Gaussian distribution. However, for the shape of the power spectrum the assumption of strong overlapping of the single pulses is without relevance, as has already been discussed. With the fluctuation part

$$p_{sh}(t) = p(t) - \mathrm{E}\{p(t)\}$$

and

$$\overline{i_{sh}^2(t)} = q^2 \cdot \overline{p_{sh}^2(t)} \ . \qquad (4.20)$$

Equation (4.19) yields the one-sided power spectrum W_{sh} of the current fluctuations, i.e. the spectrum of the shot noise:

$$W_{sh} = 2 \cdot q^2 \cdot z \cdot |S(f)|^2 = 2 \cdot q \cdot I_0 \cdot |S(f)|^2 \ . \qquad (4.21)$$

The frequency dependence of the noise is determined by the form of the single pulse and its spectrum $|S(f)|^2$.

For the region of low frequencies, i.e. frequencies f that are small compared to the reciprocal time length τ of the pulses, i.e.

$$f \ll \tau^{-1} \ , \qquad (4.22)$$

the noise can be calculated without knowledge of the particular pulse shape. For this case we have:

$$|S(f)|^2 = \left| \int_0^\tau \xi(\theta)e^{-j\,2\pi f\,\theta}\, d\theta \right|^2$$

$$\approx \left| \int_0^\tau \xi(\theta)\, d\theta \right|^2 = 1 \quad \text{for} \quad f \cdot \theta \approx 0 \ . \tag{4.23}$$

The low-frequency part of the noise spectrum W_{sh} is thus independent of the pulse form and entirely determined by the direct current I_0:

$$W_{sh} = 2q \cdot I_0 \ . \tag{4.24}$$

This so-called Schottky relation is of great practical relevance, because it can be applied to a multitude of devices. Especially for very fast semiconductor devices, the Schottky relation is valid up to very high frequencies. The spectrum $|S(f)|^2$ only weakly depends on the pulse shape. Typically, it shows a characteristic similar to a si^2 function as shown in Fig. 4.2.

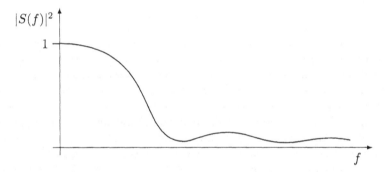

Fig. 4.2 Typical characteristic of the pulse spectrum.

4.2 SHOT NOISE OF SCHOTTKY DIODES

In the following section, metal-semiconductor junctions, so-called Schottky-diodes, will be considered. Because these diodes are based on an effect of majority carriers, they are fast enough that the Schottky relation in its frequency independent form is valid up to very high frequencies (e.g. ≈ 100 GHz). First, a Schottky diode without bulk resistance is considered. The following relation holds between the current I and the voltage U:

$$I = I_{ss} \cdot \left(\exp\left(\frac{q \cdot U}{\tilde{n} \cdot kT} \right) - 1 \right) \ . \tag{4.25}$$

In Eq. (4.25) I_{ss} is the so-called saturation current, T is the physical temperature of the depletion layer and $\tilde{n}$ is an empirical so-called ideality factor, which describes a deviation from the ideal exponential behavior. Typical values are

$\tilde{n} = 1.02\ldots1.3$. The current I_0 for the bias voltage U_0 can be considered to be composed of a current in the forward direction I_f and a current I_r in the reverse direction. The current in the forward direction consists of an electron current from the conduction band of the semiconductor to the metal. With increasing bias voltage this electron current finds a decreasing potential barrier from the semiconductor to the metal. The current in the reverse direction consists of an electron current from the metal to the semiconductor. In this case, the electrons have to surmount a nearly constant potential barrier from the metal to the semiconductor. For reverse voltages, the barrier from the semiconductor to the metal increases rapidly, so that the current in the forward direction becomes so small that it can be neglected. The current in the reverse direction, that is the electron current from the metal to the semiconductor, remains nearly constant. Therefore, the saturation current I_{ss} is approximately equal to the reverse current I_r, i.e.

$$I_r = I_{ss} \quad \text{and} \quad I_f = I_0 + I_{ss} \ . \tag{4.26}$$

Without a bias voltage the magnitudes of the currents in the forward and reverse directions are equal. For the current in the forward direction as well as for the current in the reverse direction similar assumptions apply as for the vacuum diode, which led to the derivation of the Schottky relation. As long as the charge carriers move in regions with relatively large numbers of fixed charges, the interaction of the mobile charge carriers will remain small. Therefore, both the current in the forward direction as well as the current in the reverse direction show shot noise, which is uncorrelated from each other. For the corresponding noise spectra of the forward current, W_f, the reverse current, W_r, and the total current, W_{is}, the following relations apply:

$$\begin{aligned} W_f &= 2q(I_0 + I_{ss}) \\ W_r &= 2qI_{ss} \\ W_{is} &= 2q(I_0 + 2I_{ss}) \end{aligned} \tag{4.27}$$

As long as the transit time and the life time of the carriers can be neglected, that is up to frequencies in the mm-wave region, the small signal conductance G_s of the Schottky diode is given by:

$$\begin{aligned} G_s(U_0) = \left.\frac{dI}{dU}\right|_{U=U_0} &= \frac{q}{\tilde{n} \cdot kT} I_{ss} \cdot \exp\left(\frac{q \cdot U_0}{\tilde{n} \cdot kT}\right) \\ &= \frac{q}{\tilde{n} \cdot kT}(I_0 + I_{ss}) \ . \end{aligned} \tag{4.28}$$

With the small signal conductance G_s as the source conductance and the noise spectrum W_{is} of an ideal current source a noise equivalent circuit of a Schottky diode can be specified (Fig. 4.3). The noise equivalent circuit according to Fig. 4.3 can be extended by a thermally noisy bulk resistance R_b with the temperature T (Fig. 4.4). With the source conductance G_s the

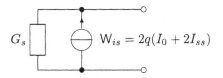

Fig. 4.3   Noise equivalent circuit of a Schottky diode without bulk resistance.

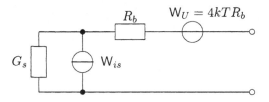

Fig. 4.4   Noise equivalent circuit of a Schottky diode with bulk resistance R_b.

shot noise of a Schottky diode can be described by an effective temperature, similar to the thermal noise. This temperature will be named the effective temperature T_{ef}. According to Eq. (4.28) the spectrum of the shot noise W_{is} can be represented by the small signal conductance G_s. Thus

$$W_{is} = 2\tilde{n}kT \cdot G_s + 2 \cdot q \cdot I_{ss} \ . \tag{4.29}$$

As for the thermal noise, the effective temperature is defined by the following equation:

$$W_{is} = 4k \cdot T_{ef} \cdot G_s \ , \tag{4.30}$$

from which T_{ef} can be determined:

$$T_{ef} = \frac{1}{2}\tilde{n} \cdot T + \frac{1}{2}\tilde{n} \cdot T \cdot \frac{I_{ss}}{I_0 + I_{ss}} \ . \tag{4.31}$$

In practice, the second term on the right side of Eq. (4.31) is negligible for an operating point in the conduction region with $I_0 \gg I_{ss}$. Consequently, the interesting result can be obtained that a Schottky diode with a negligible bulk or series resistance behaves like a thermally noisy impedance with the temperature $\tilde{n} \cdot T/2$. The available noise power P_{av} in the bandwidth Δf is independent of the frequency and equal to

$$P_{av} = \frac{1}{2}\tilde{n} \cdot kT \cdot \Delta f \ . \tag{4.32}$$

Here, T is the thermodynamic or physical temperature of the junction of the Schottky diode.

For thermodynamic equilibrium, i.e. $I_0 = 0$, the effective temperature is expected to be equal to the ambient temperature with $T_{ef} = T$. In Eq. (4.31) this is true for $\tilde{n} = 1$, however, it does not apply for $\tilde{n} > 1$. This is a hint of the heuristic approach when introducing the ideality factor $\tilde{n}$.

Figure 4.5 shows the effective noise temperature of a Schottky diode as a function of the bias current in forward direction. In problem 4.2 it is shown that this characteristic can be described quite well by a model as depicted in Fig. 4.4 with one noise source for the shot noise and another for the thermal noise of the bulk resistor R_b.

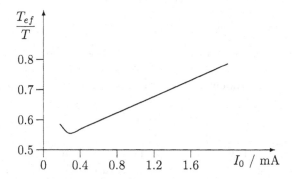

Fig. 4.5   Effective noise temperature of a Schottky diode as a function of the bias current.

Problem

4.2   Calculate the effective temperature T_{ef} of a commercial Schottky diode with $R_b = 9.8\,\Omega$, $\tilde{n}{=}1.2$ and $I_{ss} = 8\,\text{nA}$ according to the model in Fig. 4.4.

As can be deduced from Fig. 4.5, with Schottky diodes it is possible to realize un-cooled noise generators with noise temperatures below room temperature.

At low frequencies, for example below 1 MHz, additional noise mechanisms might appear for instance as flicker or $1/f$-noise and recombination noise. As a consequence, the effective temperature can increase even above T_0. Generally, the flicker noise is more distinct for gallium arsenide (GaAs) diodes than for silicon (Si) devices.

A considerably increased noise level can be observed if the reverse voltage is increased into the region of the breakdown. This so-called avalanche noise is virtually independent of frequency, similar to the shot noise and the thermal noise.

4.3   SHOT NOISE OF PN-DIODES

The current I of a pn-junction can be assumed to be composed of four parts:

First, a current of the majority carriers, which pass through the depletion region and become minority carriers in the adjacent diffusion region, where they recombine. The current of the majority carriers can consist of a current of holes, I_{pf}, as well as a current of electrons, I_{nf}. For an unsymmetrical pn-junction one carrier type will dominate.

Second, a current of minority carriers, which pass through the depletion region and recombine in the adjacent diffusion regions. The currents are I_{pr} and I_{nr}, respectively.

For each of these four partial currents a situation known from the vacuum diode applies: The transition of the carriers through the depletion region is subject to statistical fluctuations. An interaction of the carriers rarely occurs, because in the depletion region the density of the mobile charge carriers generally is much smaller than the density of the fixed ionized acceptors or donors. The adjacent $p-$ and $n-$regions are electrically neutral. Consequently, the four listed currents generate shot noise contributions, which are uncorrelated. The corresponding shot noise spectra are:

$$
\begin{aligned}
W_{if} &= 2q \cdot (|I_{pf}| + |I_{nf}|) = 2q(I + I_{ss}) \\
W_{ir} &= 2q \cdot (|I_{pr}| + |I_{nr}|) = 2q \cdot I_{ss} \\
W_{is} &= W_{if} + W_{ir} = 2q(I + 2I_{ss}) \ .
\end{aligned}
\tag{4.33}
$$

The saturation I_{ss} current in the reverse direction usually is very small. The shot noise spectra thus possess the same spectral form as observed for the Schottky diode. The equations (4.25) and (4.29) are also valid. However, pn-junctions show, even at moderate frequencies, for example a few 100 MHz, a noticeable deviation of the small signal conductance $G(f)$ from the low frequency value.

Under certain restricting conditions, the shot noise spectrum can be described empirically but in accordance with the experiment as follows:

$$
W_{is} = 4kT \cdot \mathrm{Re}\{G(f) - G_{ch}\} + 2q(I_0 + 2I_{ss}) \ .
\tag{4.34}
$$

It is thus assumed that the part of the conductance $G(f)$, which differs from the low-frequency conductance G_{ch}, i.e. $G(f) - G_{ch}$, generates thermal noise. Consequently, the effective noise temperature increases.

4.4   NOISE OF PIN DIODES

A PIN diode made of silicon consists of the series connection of a p-zone, an intrinsic i-zone and a n-zone (Fig. 4.6). If the PIN-diode is biased in forward direction, then holes and electrons from the p- and n-regions are injected into

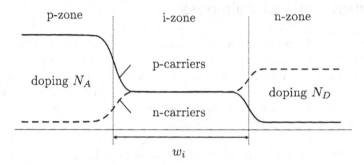

Fig. 4.6   PIN-structure in the forward conduction state.

the i-zone. Temporarily, they are stored in the i-zone until they contribute to a current due to recombination. However, part of the recombination also takes place in the boundary regions and the contacts. Neglecting the recombination in the bulk regions, then in principle, the current at the pi-junction consists of holes, which are injected into the i-zone. This current I_0 is equal to the total charge Q_p of all holes divided by their life time τ_p.

$$I_0 = \frac{Q_p}{\tau_p} = \frac{q \cdot p \cdot A_i \cdot w_i}{\tau_p} \ . \tag{4.35}$$

Here, A_i is the area, q is the elementary charge, p is the mean density of the charge carriers in the i-zone and w_i is the width of the i-zone. A corresponding term can be written for the in-junction.

$$I_0 = \frac{Q_n}{\tau_n} = \frac{q \cdot n \cdot A_i \cdot w_i}{\tau_n} \ . \tag{4.36}$$

Both types of charge carriers in the i-zone are in a charge equilibrium, i.e. the resulting space charge is zero. Furthermore, it is assumed that the i-zone is free of fixed space charges. Because the holes and the electrons mainly recombine with one another in the i-zone, we have

$$n \approx p \ , \quad \tau_p \approx \tau_n = \tau_i \ . \tag{4.37}$$

The high-frequency resistance R_i of the i-zone can approximately be calculated via the specific conductivity σ_i. If μ_p is the mobility of the holes, μ_n the mobility of the electrons in the i-zone and μ_i their mean value, $(\mu_p + \mu_n)/2$, we get

$$\sigma_i = q \cdot (\mu_p \cdot p + \mu_n \cdot n) = 2 q \mu_i \cdot p \ . \tag{4.38}$$

The high-frequency resistance R_i of the i-zone is thus given by

$$R_i = \frac{w_i}{\sigma_i \cdot A_i} = \frac{w_i^2}{2 q \mu_i p \cdot A_i \cdot w_i} = \frac{w_i^2}{2 \mu_i \tau_i I_0} \ . \tag{4.39}$$

The higher the current, the more mobile charge carriers will be in the i-zone and the lower the high frequency resistance will be. In order to realize small high-frequency resistances, the width w_i should be small and the life time τ_i should be large. The total resistance of the adjacent bulk regions also should be as small as possible.

The minority carriers stored in the bulk regions cause diffusion capacitances. For alternating signals these capacitances can be assumed to be connected in series to the resistance of the i-zone. Beyond some kHz the impedance of the diffusion capacitances becomes so small that it can be neglected compared with the bulk resistance R_b and the resistance of the i-zone R_i.

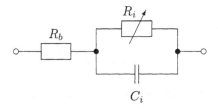

Fig. 4.7   Equivalent circuit of a PIN-diode for alternating signals.

For a reverse biased PIN-junction, a small and virtually voltage independent depletion capacitance C_i arises, which is determined by the width w_i of the i-zone, its area A_i, and its dielectric constant ϵ_i. Fig. 4.7 shows an equivalent circuit which applies for high-frequency signals. The PIN-diode is highly resistive under reverse bias and of low resistance in forward direction. The resistance can be controlled by the current and is proportional to $1/I_0$ (Eq. 4.39). For a sufficiently high current I_0 the resistance R_i becomes small compared with the bulk resistance R_b.

PIN-diodes can thus be used as electronically controlled high-frequency switches or as continuously controlled high-frequency resistances or as attenuators.

Due to the equilibrium of the space charges in the i-zone it is expected that the high-frequency resistance R_i of the i-zone shows thermal noise. This means that it is thermally noisy with the physical temperature of the i-zone. This can be confirmed fairly well by the experiment. Figure 4.8 shows the effective noise temperature of a PIN-diode as a function of the control current. Within the measurement accuracy the effective noise temperature is equal to the physical temperature T. This is caused by the fact that the bulk resistance and the i-zone show the same noise temperature. Thus, the temperature partition between the bulk resistance and the PIN junction does not change as a function of the bias current I_0. This is a different situation compared with the Schottky diode in Fig. 4.5.

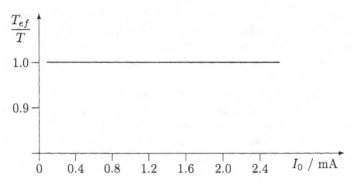

Fig. 4.8   Effective noise temperature of a PIN-diode as a function of the bias current I_0.

4.5 NOISE EQUIVALENT CIRCUITS OF BIPOLAR TRANSISTORS

In the following, a *pnp*-transistor will be considered, the basic structure of which is shown in Fig. 4.9. It is assumed that the operating frequency is so high that low-frequency noise effects as flicker noise or recombination noise are negligible. The basic remaining noise mechanisms are shot noise and thermal noise. Some of the symbols used in the following section are given first:

I_e, I_b, I_c direct current of the emitter, base or collector node
I_{ee} emitter saturation current
I_{cc} collector saturation current
W_i^e noise spectrum of the emitter-base junction
W_i^c noise spectrum of the collector-base junction
G_{e0} small signal conductance of the emitter-base junction
U_{eb} emitter-base direct voltage
α_0 direct current gain
W_i^{ec} cross spectrum of the emitter and collector noise currents
R_b base bulk resistance
I^e, I^b, I^c emitter, base, collector noise currents in a symbolic oriented arrow notation
$\tilde{I}^e, \tilde{I}^b, \tilde{I}^c$ signal currents of emitter, base and collector

The emitter-base junction is a forward biased *pn*-junction, which shows shot noise according to the Schottky-relation. Furthermore, I_e is the emitter current and I_{ee} is the emitter saturation current. For the shot noise spectrum W_i^e, Eq. (4.27) yields

$$W_i^e = 2q(I_e + I_{ee}) + 2q \cdot I_{ee} = 2q(I_e + 2 \cdot I_{ee}) = |I^e|^2 \ . \tag{4.40}$$

In the usual way, the noise spectrum W_i^e is set equivalent to the magnitude squared of the corresponding oriented arrow, $|I^e|^2$. Note that the dimension of

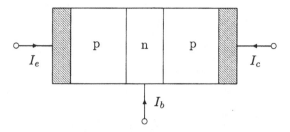

Fig. 4.9   Schematic setup of a *pnp*-transistor.

the symbolic oriented arrow I^e - or corresponding arrows - does not necessarily belong to a current. However, for the case of a current the dimension of W_i^e is equal to $A^2 \cdot s$ and thus the dimension of I^e is equal to $A \cdot \sqrt{s}$.

The index i indicates that $W_i^e = |I^e|^2$ represents a noise current source, which is connected in parallel with the conductance G_{e0} in the equivalent circuit of Fig. 4.10. The spectrum W_i^e can as well be expressed with the help of the small signal conductance G_{e0} of the emitter base junction. With

$$I_e = I_{ee} \cdot \left(\exp \left(\frac{q \, U_{eb}}{kT} \right) - 1 \right)$$

and

$$G_{e0} = \frac{dI_e}{dU_{eb}} = \frac{q}{kT} \cdot I_{ee} \cdot \exp \left(\frac{q \, U_{eb}}{kT} \right) = \frac{q}{kT}(I_e + I_{ee}) \qquad (4.41)$$

we have

$$\begin{aligned} W_i^e &= 2q(I_e + 2 \cdot I_{ee}) = 4kT \cdot G_{e0} - 2q \cdot I_e \\ &\approx 2kT \cdot G_{e0} \quad \text{for} \quad I_{ee} \ll I_e \ . \end{aligned} \qquad (4.42)$$

The forward current of the emitter-base junction proceeds to the base-collector junction except for some minor recombination losses. This is described by the current gain α_0. Thus, the term $\alpha_0(I_e + I_{ee})$ is the main part of the collector current. A small additional part is given by the collector reverse current (saturation current) I_{cc}. The collector current is thus given by

$$-I_c = \alpha_0(I_e + I_{ee}) + I_{cc} \ . \qquad (4.43)$$

The minus sign for I_c results from the current direction defined in Fig. 4.9. Both parts of the collector current consist of minority carriers (here holes). For this reason, the spectrum is calculated as

$$W_i^c = 2q|I_c| = 2q\alpha_0(I_e + I_{ee}) + 2qI_{cc} = |I^c|^2 \ . \qquad (4.44)$$

The noise currents I^e and I^c are strongly correlated. It can be anticipated that the cross-spectrum

$$W_i^{ec} = (I^e)^* \cdot I^c \qquad (4.45)$$

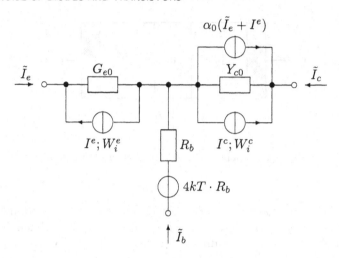

Fig. 4.10   Noise equivalent circuit of a bipolar transistor with correlated noise sources.

is proportional to the power spectrum of the noise of the common forward current

$$|I_f^e|^2 = 2q(I_e + I_{ee}) \ . \tag{4.46}$$

Thus the cross-spectrum is given by

$$\begin{aligned} W_i{}^{ec} &= -(I_f^e)^* \cdot \alpha_0 \cdot I_f^e = -\alpha_0 |I_f^e|^2 \\ &= -\alpha_0 2q(I_e + I_{ee}) = -2kT \cdot \alpha_0 \cdot G_{e0} \ . \end{aligned} \tag{4.47}$$

The minus sign follows from the definition of the direction of I_c. In Fig. 4.10 the thermal noise of the bulk resistance R_b of the base zone is taken into account by a voltage source. The bulk resistances of the emitter and collector zone are neglected.

A drawback of the equivalent circuit in Fig. 4.10 is that the noise sources are largely correlated. In this respect, the equivalent circuit in Fig. 4.11 is more convenient, where two noise current sources with $\alpha_0 \cdot I^e$ and I^c are combined into one, i.e. I^a. Furthermore, the noise current source I^e is converted into a noise voltage source U^e. The following relation results for the spectrum:

$$W_u^e = |U^e|^2 = 2kT \cdot \frac{1}{G_{e0}} = 2kT \cdot R_{e0} \ . \tag{4.48}$$

Furthermore, the spectrum W_i^a can be derived by means of the equations (4.42), (4.44) and (4.47):

$$\begin{aligned} W_i^a &= |I^a|^2 = (\alpha_0 I^e + I^c)^* \cdot (\alpha_0 I^e + I^c) \\ &= \alpha_0^2 |I^e|^2 + |I^c|^2 + 2 \cdot \alpha_0 \cdot \mathrm{Re}\{(I^e)^* \cdot I^c\} \end{aligned}$$

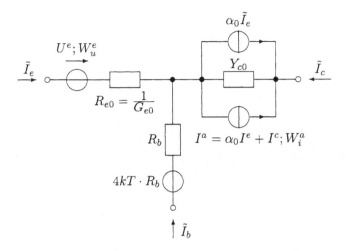

Fig. 4.11 Noise equivalent circuit of a bipolar transistor with uncorrelated noise sources.

$$
\begin{aligned}
&= \; \alpha_0^2(4kT \cdot G_{e0} - 2qI_e) + 2q|I_c| - 4\alpha_0^2 \cdot kT \cdot G_{e0} \\
&= \; 2q(|I_c| - \alpha_0^2 \cdot I_e) = 2q \cdot \alpha_0(I_e + I_{ee}) + 2qI_{cc} - 2q \cdot \alpha_0^2 \cdot I_e \\
&= \; 2q\alpha_0 \cdot (1 - \alpha_0)(I_e + I_{ee}) + 2q(\alpha_0^2 I_{ee} + I_{cc}) \\
&= \; 2kT \cdot G_{e0} \cdot \alpha_0(1 - \alpha_0) + 2q(\alpha_0^2 \cdot I_{ee} + I_{cc}) \; .
\end{aligned}
\tag{4.49}
$$

Finally, the cross-spectrum W_{ui}^{ea} becomes:

$$
\begin{aligned}
W_{ui}^{ea} &= \; \left(\frac{1}{G_{e0}} \cdot I^e\right)^* \cdot (\alpha_0 \cdot I^e + I^c) \\
&= \; \alpha_0 \cdot \frac{1}{G_{e0}} \cdot W_i^e + \frac{1}{G_{e0}} \cdot W_i^{ec} \\
&= \; \alpha_0 \cdot \frac{1}{G_{e0}}(4kT \cdot G_{e0} - 2q \cdot I_e - 2kT \cdot G_{e0}) \; .
\end{aligned}
\tag{4.50}
$$

With the approximation of Eq. (4.42) one can find that W_{ui}^{ea} is approximately zero:

$$
W_{ui}^{ea} \approx 0 \; .
\tag{4.51}
$$

The equivalent circuit of Fig. 4.11 thus has the convenient characteristic, that the three equivalent noise sources are uncorrelated. With the help of noise equivalent circuits such as the one of Fig. 4.11, for example, the noise figure of a small signal amplifier can be calculated. Figure 4.12 shows a common emitter circuit. For simplification, the small conductance Y_{c0} has been neglected.

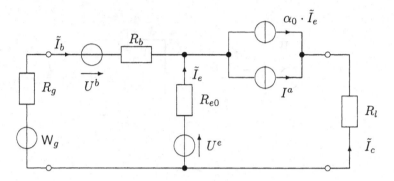

Fig. 4.12   Small signal amplifier as a common emitter circuit with equivalent noise sources.

As will be shown in problem 4.3, the following solution results from a small signal noise calculation:

$$F = 1 + \frac{R_b}{R_g} + \frac{R_{e0}}{2R_g} + \frac{(R_g + R_b + R_{e0})^2}{2 \cdot R_{e0} \cdot R_g \cdot \alpha_0^2}$$
$$\cdot \left(\alpha_0(1 - \alpha_0) + \frac{\alpha_0^2 I_{ee} + I_{cc}}{I_e + I_{ee}} \right) \tag{4.52}$$

Problems

4.3   Determine for the equivalent circuit in Fig. 4.12 the noise figure as well as the optimum noise figure. Calculate the minimum noise figure for $R_{e0} = 15\,\Omega$, $R_b = 40\,\Omega$, $\alpha_0 = 0.98$ and $(\alpha_0^2 I_{ee} + I_{cc})/(I_e + I_{ee}) = 10^{-2}$.

4.4   Calculate with the same data as in problem 4.3 and with the noise equivalent circuit of Fig. 4.11 the noise figure and the optimum noise figure for a small signal amplifier in common base configuration.

For high frequencies, the transit time of the carriers in the *pn*-junction as well as in the base zone have to be taken into account. All parameters will, in general, be frequency dependent and complex. The conductance of the emitter-base junction, $Y_e(f)$, split into a real and an imaginary part, G_e and B_e, can be written as

$$Y_e(f) = G_e(f) + jB_e(f) = \frac{1}{Z_e(f)}$$

and

$$Y_c = Y_c(f) \ . \tag{4.53}$$

According to Eq. (4.34), it is assumed that the part deviating from G_{e0}, that is $G_e(f) - G_{e0}$, shows thermal noise. For the spectra of the noise equivalent circuit in Fig. 4.10 we get:

$$
\begin{aligned}
\mathsf{W}_i^e(f) &= (I^e)^* \cdot I^e = 4kT \cdot G_e(f) - 2q \cdot I_e \\
\mathsf{W}_i^c(f) &= (I^c)^* \cdot I^c = 2q \cdot |I_c| \\
\mathsf{W}_i^{ec}(f) &= (I^e)^* \cdot I^c = -2kT \cdot \alpha(f) \cdot Y_e(f) \ .
\end{aligned}
\tag{4.54}
$$

For the current gain with the cut-off frequency f_α we have:

$$
\alpha(f) = \frac{\alpha_0}{1 + jf/f_\alpha} \ .
\tag{4.55}
$$

Finally, the frequency dependent spectra for the noise equivalent circuit in Fig. 4.11 can be determined.

The spectrum for the noise voltage U^e can be written as

$$
\mathsf{W}_u^e(f) = (4kT \cdot G_e(f) - 2q \cdot I_e) \cdot |Z_e(f)|^2 \ .
\tag{4.56}
$$

The spectrum of I^a, that is W_i^a, can be derived from the equivalent circuit in Fig. 4.12 with the parallel connection of $\alpha \cdot I^e$ and I^c and with W_i^e, W_i^c and W_i^{ec} from Eq. (4.54).

$$
\begin{aligned}
\mathsf{W}_i^a(f) &= (\alpha I^e + I^c)^* \cdot (\alpha I^e + I^c) \\
&= |\alpha|^2 \mathsf{W}_i^e(f) + \mathsf{W}_i^c + \alpha(\mathsf{W}_i^{ec}(f))^* + \alpha^* \mathsf{W}_i^{ec}(f) \\
&= |\alpha|^2 (4kT \cdot G_e(f) - 2q \cdot I_e) + 2q \cdot |I_c| - |\alpha|^2 2kT \cdot 2 \cdot \mathrm{Re}\{Y_e(f)\} \\
&= 2q(|I_c| - |\alpha|^2 \cdot I_e) \ .
\end{aligned}
\tag{4.57}
$$

With the result for $|I_c|$ of Eq. (4.43) we get:

$$
\begin{aligned}
\mathsf{W}_i^a(f) &= 2q(\alpha_0(I_e + I_{ee}) + I_{cc} - |\alpha|^2 I_e) \\
&= 2qI_e(\alpha_0 - |\alpha(f)|^2) + 2q(\alpha_0 I_{ee} + I_{cc}) \ .
\end{aligned}
\tag{4.58}
$$

The last equation still missing is the cross-spectrum W_{ui}^{ea} between the noise voltage U^e and the noise current I^a:

$$
\begin{aligned}
\mathsf{W}_{ui}^{ea}(f) &= (I^e \cdot Z_e)^* \cdot (\alpha I^e + I^c) = Z_e^* \cdot \alpha \cdot \mathsf{W}_i^e + Z_e^* \cdot \mathsf{W}_i^{ec} \\
&= \alpha(f) \cdot Z_e^*(4kT \cdot G_e(f) - 2q \cdot I_e - 2kT \cdot G_e(f) \\
&\quad - 2kT \cdot jB_e(f)) \\
&= \alpha(f) \cdot Z_e^*(f) \cdot (2kT \cdot Y_e^*(f) - 2q \cdot I_e) \ .
\end{aligned}
\tag{4.59}
$$

Thus frequency-dependent spectra for the noise equivalent circuit according to Fig. 4.11 have also been derived.

However, it should be noted that such equivalent circuits can offer an empirical approximation only. A refinement of the model should as well consider parasitic circuit elements of the transistor case and the connection of the transistor to the case.

However, the linear parasitic circuit elements can also be combined with the external linear circuitry of the transistor. A further approximation of the models discussed is the neglected feedback of the output to the input.

4.6 NOISE OF FIELD EFFECT TRANSISTORS

4.6.1 Static characteristics and small signal behavior

Field effect transistors (FETs) of different types are utilized in numerous high frequency circuits. Although amplifiers are the main field of application, FETs are also suited for the realization of oscillators, mixers, switches and other circuits. Field effect transistors made of gallium arsenide with Schottky-type gates (GaAs-MEtal Semiconductor FET, GaAs-MESFET) can be used at frequencies up to the mm-wave range ($>$ 100 GHz).

The analysis of the noise behavior of FETs is closely related to the principle of operation of the device. Therefore, we will start with a description of the static and small signal behavior of the FET. Due to its dominant role in RF-applications, only the junction FET will be considered.

Figure 4.13 shows the basic structure of a junction FET. On a substrate

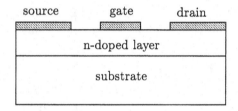

Fig. 4.13   Principal setup of an n-channel-junction field effect transistor.

there is an active semiconductor layer, which in most high frequency transistors is n-doped. In this layer a current flows from the drain to the source contact. The gate contact and the n-conducting channel form a junction, i.e. either a pn-junction (Junction Field Effect Transistor, JFET) or a Schottky junction (MESFET). Via the gate potential the current flow between the drain and source can be controlled.

A detailed analysis will be restricted to the inner FET, i.e. the region below the gate contact. Figure 4.14 shows a cross-section view of the inner FET, the dimensions of which are specified by the length l and width a of the gate and the thickness d of the active layer.

A quantitative relation between the drain current I_d as a function of the gate-source voltage U_g and the drain-source voltage U_d was first given by Shockley in 1952. The theory of Shockley is based on two assumptions. With the so called *gradual channel approximation* it is assumed that the width $w(x)$ of the space charge region smoothly changes along the channel and, therefore, the electrical field E approximately has a y-component only in the space charge region and a x-component only within the conducting channel (see Fig. 4.14). Furthermore, an ohmic behavior is assumed in the channel, i.e. proportionality between the current density and the electric field. This

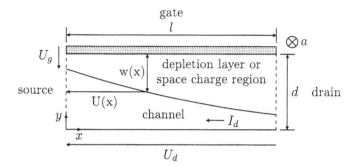

Fig. 4.14 Cross-section of the inner FET.

means that the mobility μ of the electrons is assumed to be constant. In particular, this latter assumption is not really valid for gallium arsenide and related compound FETs. A number of models have been developed to take into account the dependence of the carrier mobility on the electric field. However, these models will not be discussed here, because a simple analytical noise theory for the FET so far only exists for the Shockley model.

Using the assumptions of the Shockley model, the drain current is given by the expression

$$I_d = -q\mu N_D\, a(d - w(x))E_x(x) \ , \tag{4.60}$$

with q as the elementary charge and N_D as the doping density. The electrical field E_x in x-direction and the depletion layer width w depend on the voltage $U(x)$ in the channel:

$$E_x(x) = -\frac{dU(x)}{dx} \ , \tag{4.61}$$

$$w(x) = \sqrt{\frac{2\epsilon_0\epsilon_r}{qN_D}(U_{Df} - U_g + U(x))} \ , \tag{4.62}$$

with ϵ_0 as the vacuum permittivity and ϵ_r as the relative dielectric constant of the semiconductor material. The quantity U_{Df} is the so-called diffusion voltage or built-in barrier potential of the gate contact.

With the definitions of the pinch-off voltage

$$U_{pi} = \frac{qN_D d^2}{2\epsilon_0\epsilon_r} \tag{4.63}$$

and the normalized voltage $V(x)$ according to

$$V(x) = \frac{U_{Df} - U_g + U(x)}{U_{pi}} \tag{4.64}$$

the depletion layer width and the drain current may be expressed as

$$w(x) = d\sqrt{V(x)} \ , \tag{4.65}$$

$$I_d = q\mu N_D\, a\, d\, U_{pi}(1 - \sqrt{V(x)})\frac{dV(x)}{dx} \; . \tag{4.66}$$

The solution of the differential equation Eq. (4.66) is

$$I_d = G_{ch}\, U_{pi}\, F_1(V_d) \; . \tag{4.67}$$

Here,

$$G_{ch} = q\mu N_D\frac{a\, d}{l} \tag{4.68}$$

is the conductance of the channel for vanishing depletion layer ($w(x) \equiv 0$) and the function F_1 is defined by

$$F_1(V) = V\left(1 - \frac{2}{3}\sqrt{V}\right) - V_g\left(1 - \frac{2}{3}\sqrt{V_g}\right) \; . \tag{4.69}$$

The quantities V_g and V_d are the normalized voltages at both ends of the conducting channel:

$$V_g = V(x=0) = \frac{U_{Df} - U_g}{U_{pi}} \; , \tag{4.70}$$

$$V_d = V(x=l) = \frac{U_{Df} - U_g + U_d}{U_{pi}} \; . \tag{4.71}$$

For a constant gate voltage U_g the drain current I_d increases with an increasing drain voltage U_d until U_d is equal to the saturation value

$$U_{dsat} = U_{pi} - U_{Df} + U_g \; . \tag{4.72}$$

For $U_d = U_{dsat}$ we have $V(l) = 1$ and therefore $w(l) = d$. At the drain side end of the channel the depletion layer extends across the total thickness d of the active layer. In the simple Shockley model, it is assumed that a further increase of the drain current is not possible and that for $U_d > U_{dsat}$ the current I_d remains at the saturation value:

$$I_{dsat} = I_d(U_{dsat}) = G_{ch}U_{pi}\, F_1(1) \; . \tag{4.73}$$

The drain current I_d can be controlled by the gate voltage U_g. The drain current decreases if U_g becomes more negative. For $U_g = U_{Df} - U_{pi}$ the drain current is zero, i.e. $I_d = 0$, independent of the drain voltage. In this case, the transistor is completely switched off. For $U_g = U_{Df}$ the maximum drain current $I_d = G_{ch}U_{pi}/3$ is obtained. Then, $w(0) = 0$ and a further increase of the gate voltage will cause significant current flow across the gate. Usually, this state is avoided by a proper choice of the bias voltages and the signal amplitudes.

As an example, the Fig. 4.15 shows the typical I-V characteristics of a field effect transistor. The range with $U_d < U_{dsat}$ is called the linear or ohmic region. The range of the characteristics with $U_d > U_{dsat}$ is called the saturation region. For an amplifier, the bias points are always chosen within the saturation region.

For silicon devices the characteristics calculated with the Shockley model show quite good agreement with measurements. However, a slight deviation from the model may be observed in the saturation region, because the drain current still increases somewhat with increasing drain-source voltage. Although the characteristics of GaAs devices are qualitatively similar to the curves in Fig. 4.15, a precise quantitative calculation is not possible with the Shockley model.

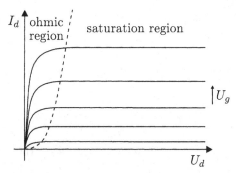

Fig. 4.15   I-V characteristics of a field effect transistor.

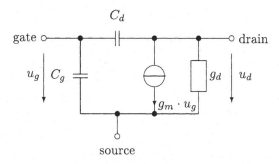

Fig. 4.16   Small signal equivalent circuit of the inner FET.

If we superimpose small signal voltages u_g and u_d on the bias voltages U_g and U_d, then the behavior of the inner FET for the small signals can be described by the equivalent circuit of Fig. 4.16. The transconductance g_m and the output conductance g_d are obtained as partial derivatives of Eq. (4.67):

$$g_m = \frac{\partial I_d}{\partial U_g} = G_{ch} \left(\sqrt{V_d} - \sqrt{V_g} \right) , \tag{4.74}$$

$$g_d = \frac{\partial I_d}{\partial U_d} = G_{ch} \left(1 - \sqrt{V_d} \right) . \tag{4.75}$$

In a similar way, the gate-source capacitance C_g and the drain-gate capacitance C_d are derived from the charge Q_g on the gate contact:

$$Q_g = -Q_0 \frac{F_2(V_d)}{F_1(V_d)} \qquad (4.76)$$

with

$$Q_0 = qN_D \, a \, d \, l \qquad (4.77)$$

and the function

$$F_2(V) = V \left(\frac{2}{3}\sqrt{V} - \frac{1}{2}V \right) - V_g \left(\frac{2}{3}\sqrt{V_g} - \frac{1}{2}V_g \right) \, . \qquad (4.78)$$

Via the partial derivatives of Eq. (4.76) with respect to U_g and U_d we get the gate and drain capacitances:

$$C_g = 2C_0 \frac{1 - \sqrt{V_g}}{F_1(V_d)} \left[\frac{F_2(V_d)}{F_1(V_d)} - \sqrt{V_g} \right] \, , \qquad (4.79)$$

$$C_d = 2C_0 \frac{1 - \sqrt{V_d}}{F_1(V_d)} \left[\sqrt{V_d} - \frac{F_2(V_d)}{F_1(V_d)} \right] \, , \qquad (4.80)$$

with

$$C_0 = \epsilon_0 \epsilon_r \frac{a \, l}{d} = \frac{Q_0}{2 \, U_{pi}} \, . \qquad (4.81)$$

All elements of the equivalent circuit are functions of the voltages U_g and U_d. Of particular interest are the values for $U_d = U_{dsat}$, i.e. along the boundary of the saturation region. As for the drain current, one can also assume for the quantities g_m, g_d, C_g and C_d, that their values are not changed essentially by a further increase of the drain voltage. Setting $V_d = 1$ in equations (4.74), (4.75) and (4.79), (4.80) yields:

$$g_m(U_{dsat}) = G_{ch}(1 - \sqrt{V_g}) \, , \qquad (4.82)$$

$$g_d(U_{dsat}) = 0 \, , \qquad (4.83)$$

$$C_g(U_{dsat}) = 2C_0 \frac{1 - \sqrt{V_g}}{F_1(1)} \left[\frac{F_2(1)}{F_1(1)} - \sqrt{V_g} \right]$$

$$= 3C_0 \frac{1 + \sqrt{V_g}}{(1 + 2 \cdot \sqrt{V_g})^2} \, , \qquad (4.84)$$

$$C_d(U_{dsat}) = 0 \, . \qquad (4.85)$$

Thus, for an operating point in the saturation region, the small signal equivalent circuit of the inner FET is approximately reduced to the capacitance C_g at the input and a voltage controlled current source with the transconductance

g_m. The ratio of both values is defined as the transit or cut-off frequency ω_{tr} of the transistor:

$$\omega_{tr} = \frac{g_m}{C_g} \; . \tag{4.86}$$

At the angular frequency ω_{tr} the current gain of the inner FET has decreased to one.

4.6.2 Thermal noise of the inner FET

According to the Shockley model, within the channel the current density and the electric field are proportional to each other, i.e. they follow Ohm's law. Therefore, thermal noise will be generated in the channel. But since the channel does not have constant cross section dimensions due to the active behavior of the FET, the relationship between the inherent physical noise sources and the resulting equivalent noise sources referred to the outer terminals is rather complicated.

With respect to its small signal properties, the transistor can be regarded as a linear two-port. Therefore, the usual two-port equivalent circuits with two noise sources can also be applied to the field effect transistor. For a fixed operating point the noise properties are completely described by the spectra and cross-spectra of these two noise sources. Often a model with two current noise sources as shown in Fig. 4.17 is chosen for the FET.

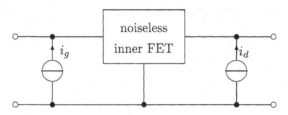

Fig. 4.17   Noise equivalent circuit of the inner FET.

We will proceed with the description of a general method to calculate the equivalent short circuit noise currents i_g and i_d with the help of the Shockley relations. This calculation is based on the so-called two-transistor model. In this approach an infinitesimal channel section of length dx at a distance x from the source is considered, as depicted in Fig. 4.18a. The thermal noise of this channel section induces noise currents di_g and di_d in the external short circuit. For their calculation, the transistor is separated into two noise-free transistors with gate lengths x and $l - x$, respectively. As shown in Fig. 4.18b, the noise of the infinitesimal channel section is described by a noise voltage source du at the interface between the two sub-transistors. If for each sub-transistor an equivalent circuit as shown in Fig. 4.16 is assumed, then the spectra dW_g, dW_d and dW_{gd} of the currents di_g and di_d can be calculated from the spectrum

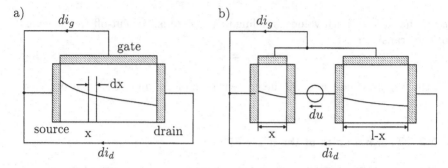

Fig. 4.18   Calculation of the FET noise with the two-transistor model.

dW_u of the noise voltage source. The spectra W_g, W_d and the cross-spectrum W_{gd} of the complete noise currents i_g and i_d are obtained by an integration over the noise contributions of all infinitesimal channel sections. For this integration, the assumption is made that the noise voltages du from different positions x in the channel are uncorrelated.

The values of the equivalent circuit elements of both sub-transistors depend on the position x of the interface. Eqs. (4.74), (4.75) and (4.79), (4.80) can be applied, if the different geometries and voltages of the sub-transistors are correctly accounted for. With V as the normalized voltage at the position x, we obtain for the source-sided sub-transistor:

$$g_{m1} \;=\; G_{ch}\frac{l}{x}\left(\sqrt{V(x)}-\sqrt{V_g}\right)\;, \tag{4.87}$$

$$g_{d1} \;=\; G_{ch}\frac{l}{x}\left(1-\sqrt{V(x)}\right)\;, \tag{4.88}$$

$$C_{g1} \;=\; 2C_0\,\frac{x}{l}\,\frac{1-\sqrt{V_g}}{F_1(V(x))}\left(\frac{F_2(V(x))}{F_1(V(x))}-\sqrt{V_g}\right)\;, \tag{4.89}$$

$$C_{d1} \;=\; 2C_0\,\frac{x}{l}\,\frac{1-\sqrt{V(x)}}{F_1(V(x))}\left(\sqrt{V(x)}-\frac{F_2(V(x))}{F_1(V(x))}\right)\;. \tag{4.90}$$

For the drain-sided sub-transistor we get:

$$g_{m2} \;=\; G_{ch}\frac{l}{l-x}\left(\sqrt{V_d}-\sqrt{V(x)}\right)\;, \tag{4.91}$$

$$g_{d2} \;=\; G_{ch}\frac{l}{l-x}\left(1-\sqrt{V_d}\right)\;, \tag{4.92}$$

$$C_{g2} \;=\; 2C_0\,\frac{l-x}{l}\,\frac{1-\sqrt{V(x)}}{F_1(V_d)-F_1(V)}\left(\frac{F_2(V_d)-F_2(V)}{F_1(V_d)-F_1(V)}-\sqrt{V(x)}\right) \tag{4.93}$$

$$C_{d2} \;=\; 2C_0\,\frac{l-x}{l}\,\frac{1-\sqrt{V_d}}{F_1(V_d)-F_1(V)}\left(\sqrt{V_d}-\frac{F_2(V_d)-F_2(V)}{F_1(V_d)-F_1(V)}\right)\;. \tag{4.94}$$

Since the ohmic region usually is not used for amplification, we can restrict the further discussion to the case $V_d = 1$. This yields

$$g_{d2} = C_{d2} = 0 \ . \tag{4.95}$$

Due to the external short circuit between gate and source of the source-sided sub-transistor, the elements C_{g1} and g_{m1} become ineffective. For the calculation of the short circuit noise currents we can use the equivalent circuit of Fig. 4.19. Because the circuit is linear, a noise analysis can be performed with

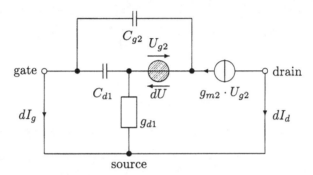

Fig. 4.19 Noise equivalent circuit of the FET for the calculation of the short circuit noise currents.

the usual complex phasors dU, dI_g and dI_d. The result is

$$dI_g \ = \ \frac{j\omega(C_{g2}g_{d1} - C_{d1}g_{m2})}{g_{m2} + g_{d1} + j\omega(C_{d1} + C_{g2})} dU \ , \tag{4.96}$$

$$dI_d \ = \ \frac{g_{m2}(g_{d1} + j\omega C_{d1})}{g_{m2} + g_{d1} + j\omega(C_{d1} + C_{g2})} dU \ . \tag{4.97}$$

If the above relations for the equivalent circuit elements are inserted into these equations, the resulting expressions are very complicated. So far, a closed form analytical solution on the basis of the Eqs. (4.96) and (4.97) is not known. The expressions are simplified by low frequency approximations. With

$$\omega \ll \frac{g_{m2} + g_{d1}}{C_{d1} + C_{g2}} \tag{4.98}$$

and

$$\omega \ll \frac{g_{d1}}{C_{d1}} \tag{4.99}$$

Eqs. (4.96) and (4.97) are reduced to

$$dI_g \ = \ \frac{j\omega(C_{g2}\, g_{d1} - C_{d1}\, g_{m2})}{g_{m2} + g_{d1}} dU \ , \tag{4.100}$$

$$dI_d \ = \ \frac{g_{m2}\, g_{d1}}{g_{m2} + g_{d1}} dU \ . \tag{4.101}$$

The conditions (4.98) and (4.99) are only met if frequencies below the transit frequency ω_{tr} according to Eq. (4.86) are considered. With the Eqs. (4.82), (4.84), (4.88), (4.90), (4.91), (4.93) and

$$\frac{x}{l} = \frac{F_1(V)}{F_1(1)} \qquad (4.102)$$

one finally obtains

$$dI_g = j\omega C_g \frac{1 - \sqrt{V}}{1 - \sqrt{V_g}} \frac{\dfrac{F_2(1)}{F_1(1)} - \sqrt{V}}{\dfrac{F_2(1)}{F_1(1)} - \sqrt{V_g}} dU , \qquad (4.103)$$

$$dI_d = g_m \frac{1 - \sqrt{V}}{1 - \sqrt{V_g}} dU . \qquad (4.104)$$

For the corresponding power spectra dW_g, dW_d, and the cross-spectrum dW_{gd} we get:

$$dW_g = (\omega C_g)^2 \left(\frac{\left(1 - \sqrt{V}\right)\left(\gamma - \sqrt{V}\right)}{\left(1 - \sqrt{V_g}\right)\left(\gamma - \sqrt{V_g}\right)} \right)^2 dW_u , \qquad (4.105)$$

$$dW_d = g_m^2 \left(\frac{1 - \sqrt{V}}{1 - \sqrt{V_g}} \right)^2 dW_u , \qquad (4.106)$$

$$dW_{gd} = -j\omega C_g g_m \left(\frac{1 - \sqrt{V}}{1 - \sqrt{V_g}} \right)^2 \frac{\gamma - \sqrt{V}}{\gamma - \sqrt{V_g}} dW_u . \qquad (4.107)$$

Here, the abbreviation

$$\gamma = \frac{F_2(1)}{F_1(1)} = \frac{\dfrac{1}{6} - V_g \left(\dfrac{2}{3}\sqrt{V_g} - \dfrac{1}{2}V_g \right)}{\dfrac{1}{3} - V_g \left(1 - \dfrac{2}{3}\sqrt{V_g} \right)}$$

$$= \frac{1}{2} \frac{1 + 2 \cdot \sqrt{V_g} + 3V_g}{1 + 2 \cdot \sqrt{V_g}} \qquad (4.108)$$

was used. The quantity γ has a simple physical meaning. According to Eq. (4.76) γ is the ratio of the space charge $-Q_g$ of the depletion layer to the maximum charge Q_0, or the averaged normalized depletion layer charge width $\overline{w}/d$ of the inner FET. The noise dW_u of the noise voltage source is the thermal noise of the infinitesimal channel section of length dx with the resistance dR:

$$dW_u = 2kTdR , \qquad (4.109)$$

with

$$dR = \frac{dx}{q\mu N_D a(d-w)} = \frac{dx}{G_{ch}l\left(1-\sqrt{V}\right)} \quad . \tag{4.110}$$

From Eqs. (4.66) to (4.68) a relationship between dx and dV is obtained:

$$dx = \frac{l}{F_1(1)}\left(1-\sqrt{V}\right)dV \quad . \tag{4.111}$$

This leads to an expression for the resistance of the channel section:

$$dR = \frac{dV}{G_{ch}\,F_1(1)} \quad . \tag{4.112}$$

Of the three spectra W_g, W_d and W_{gd} the power spectrum of the noise current i_d is the easiest to calculate. By insertion of the Eqs. (4.109) and (4.112) into the Eq. (4.106) we get

$$dW_d = 2kTg_m\frac{\left(1-\sqrt{V}\right)^2}{F_1(1)\left(1-\sqrt{V_g}\right)}dV \quad . \tag{4.113}$$

The integration results in

$$
\begin{aligned}
W_d &= \int dW_d \\
&= 2kTg_m\frac{1}{F_1(1)\left(1-\sqrt{V_g}\right)}\int_{V_g}^{1}\left(1-2\sqrt{V}+V\right)dV \\
&= 2kTg_m\frac{1}{F_1(1)\left(1-\sqrt{V_g}\right)}\left(\frac{1}{6}-V_g\left(1-\frac{4}{3}\sqrt{V_g}+\frac{1}{2}V_g\right)\right) \quad .
\end{aligned}
\tag{4.114}
$$

After some algebraic manipulations the result can be expressed as

$$W_d = 2kTg_m\tilde{P}(V_g) \tag{4.115}$$

with the function

$$\tilde{P}(V_g) = \frac{\left(1+3\sqrt{V_g}\right)\left(1-\sqrt{V_g}\right)^2}{2\left(1-V_g\left(3-2\sqrt{V_g}\right)\right)} = \frac{1}{2}\frac{1+3\sqrt{V_g}}{1+2\sqrt{V_g}} \quad . \tag{4.116}$$

In a similar way, however with more effort, the spectra W_g and W_{gd} can be calculated. The results are:

$$W_g = 2kT\frac{(\omega C_g)^2}{g_m}\tilde{R}(V_g) \tag{4.117}$$

with

$$\tilde{R}(V_g) = \frac{\gamma^2(1 - V_g) - \frac{4}{3}\gamma(1 + \gamma)\left(1 - V_g^{3/2}\right) + \frac{1}{2}(1 + 4\gamma + \gamma^2)(1 - V_g^2)}{F_1(1)\left(1 - \sqrt{V_g}\right)\left(\gamma - \sqrt{V_g}\right)^2}$$

$$- \frac{\frac{4}{5}(1 + \gamma)(1 - V_g^{5/2}) - \frac{1}{3}(1 - V_g^3)}{F_1(1)\left(1 - \sqrt{V_g}\right)\left(\gamma - \sqrt{V_g}\right)^2}$$

$$= \frac{1}{10} \cdot \frac{1 + 7\sqrt{V_g}}{1 + 2\sqrt{V_g}} \qquad (4.118)$$

and

$$W_{gd} = -2kTj\omega C_g \tilde{Q}(V_g) \qquad (4.119)$$

with

$$\tilde{Q}(V_g) = \frac{\gamma(1 - V_g) - \frac{2}{3}(1 + 2\gamma)(1 - V_g^{3/2})}{F_1(1)\left(1 - \sqrt{V_g}\right)\left(\gamma - \sqrt{V_g}\right)}$$

$$+ \frac{\frac{1}{2}(2 + \gamma)(1 - V_g^2) - \frac{2}{5}(1 - V_g^{5/2})}{F_1(1)\left(1 - \sqrt{V_g}\right)\left(\gamma - \sqrt{V_g}\right)}$$

$$= \frac{1}{10} \cdot \frac{1 + 3\sqrt{V_g}(2 + \sqrt{V_g})}{(1 + \sqrt{V_g})(1 + 2\sqrt{V_g})}. \qquad (4.120)$$

Finally, the correlation between i_g and i_d is best described by the normalized cross-spectrum k_{gd}:

$$k_{gd} = \frac{W_{gd}}{\sqrt{W_g W_d}} = -j\frac{\tilde{Q}(V_g)}{\sqrt{\tilde{P}(V_g)\tilde{R}(V_g)}} = -j\tilde{C}(V_g) \qquad (4.121)$$

Problem

4.5   Calculate the function $\tilde{C}(V_g)$. Determine the upper and lower bounds of the normalized cross-spectrum in the range $0 \leq V_g \leq 1$.

Figure 4.20 shows the quantities $\tilde{P}, \tilde{Q}, \tilde{R}$ and $\tilde{C}$, which have been introduced above as a function of the normalized voltage V_g. All parameters only weakly depend on the gate-voltage and can even be considered to be constant for a first-order approximation. For low noise amplifiers, operating points at low drain currents are preferred. The corresponding V_g values typically range from 0.5 to 1. Therefore, a good approximation for the noise parameters is

$$\tilde{P}(V_g) \approx \frac{2}{3} , \qquad (4.122)$$

$$\tilde{R}(V_g) \approx \frac{1}{4} , \qquad (4.123)$$

$$\tilde{Q}(V_g) \approx \frac{1}{6} , \qquad (4.124)$$

$$\tilde{C}(V_g) \approx 0.4 . \qquad (4.125)$$

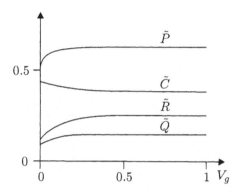

Fig. 4.20   FET noise parameters $\tilde{P}, \tilde{Q}, \tilde{R}$ and $\tilde{C}$ as a function of the normalized gate voltage V_g.

The above calculation of the noise spectra is based on the Shockley model. As was already mentioned, this model cannot be applied to GaAs FETs. For these devices the electric field in the channel can be so high that the electron mobility can no longer be considered to be constant. For a more realistic transistor model, it is assumed that Ohm's law holds in a part of the channel only and that the electrons move with a constant saturated drift velocity in the remaining channel region.

The strong electrical fields in the channel also require a more complicated noise model. In the relations for the thermal noise, e.g. Eq. (4.109), the physical temperature of the device can not be used anymore, because the electrons are no longer in thermal equilibrium with the crystal lattice. The actual electron temperature is typically higher than the lattice temperature (hot electrons). In the region of the saturated drift velocity, the current fluctuations cannot be considered as thermal noise but must be treated more generally as so-called diffusion noise.

The general equivalent circuit of Fig. 4.17 is valid for any linear circuit, i.e. also for GaAs FETs. Equations (4.115), (4.117) and (4.119) for the noise spectra can be used for GaAs FETs as well, except that the functions $\tilde{P}$, $\tilde{Q}$ and $\tilde{R}$ are different from the relations given by the constant mobility model. Among others, an important difference from the results obtained so far is that the noise currents i_g and i_d are almost completely correlated. Thus, the values of the quantity $\tilde{C}$ in Eq. (4.121) are close to 1, compared with 0.4 for the constant mobility model.

4.6.3 Noise figure of the complete FET

In addition to the inner FET, several parasitic resistances contribute to the noise of the complete FET. These are in particular the gate resistance R_g and the source resistance R_s. A more detailed noise equivalent circuit of the complete FET is shown in Fig. 4.21. The thermal noise of the parasitic

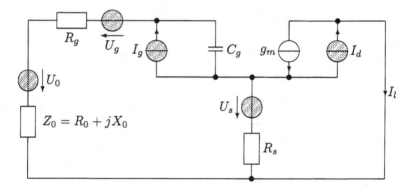

Fig. 4.21 Noise equivalent circuit of the complete FET.

resistances is described by noise voltage sources with the complex phasors U_g and U_s. The complex impedance $Z_0 = R_0 + jX_0$ is the source impedance of the signal source. The noise of the signal source is designated by the voltage phasor U_0.

The different noise phasors cause a short circuit noise current at the output with the complex amplitude I_l, which can easily be determined from the equivalent circuit of Fig. 4.21:

$$I_l = \frac{-U_0 - U_g + U_s - I_g(Z_0 + R_g + R_s)}{R_s + \dfrac{1}{g_m}(1 + j\omega C_g(Z_0 + R_g + R_s))}$$

$$+ \frac{\dfrac{I_d}{g_m}(1 + j\omega C_g(Z_0 + R_g + R_s))}{R_s + \dfrac{1}{g_m}(1 + j\omega C_g(Z_0 + R_g + R_s))} . \tag{4.126}$$

The squared magnitude of the phasor I_l yields the corresponding power spectrum at the output. Only the correlation between the currents I_g and I_d must be taken into account. Then, the noise figure F of the FET is given by

$$F = 1 + \frac{R_g + R_s}{R_0} + \frac{\tilde{P}}{g_m R_0} - \frac{2\omega C_g X_0}{g_m R_0}(\tilde{P} - \tilde{Q})$$
$$+ \frac{(\omega C_g)^2((R_0 + R_g + R_s)^2 + X_0^2)}{g_m R_0}(\tilde{P} + \tilde{R} - 2\tilde{Q}) \ . \quad (4.127)$$

Apart from the transistor properties, the noise figure also depends on the source impedance $Z_0 = R_0 + jX_0$. For an optimum source impedance $Z_{opt} = R_{opt} + jX_{opt}$ the minimum noise figure F_{min} is obtained. The partial derivatives of Eq. 4.127 yield the real and imaginary parts of the optimum source impedance:

$$R_{opt} = \sqrt{(R_g + R_s)^2 + \frac{\tilde{P}\tilde{R}(1 - \tilde{C}^2) + g_m(R_g + R_s)(\tilde{P} + \tilde{R} - 2\tilde{Q})}{((\omega C_g)(\tilde{P} + \tilde{R} - 2\tilde{Q}))^2}} \ ,$$
$$(4.128)$$

$$X_{opt} = \frac{1}{\omega C_g}\frac{\tilde{P} - \tilde{Q}}{\tilde{P} + \tilde{R} - 2\tilde{Q}} \ . \quad (4.129)$$

The resulting minimum noise figure is given by

$$F_{min} = 1 + 2\frac{(\omega C_g)^2}{g_m}(R_{opt} + R_g + R_s)(\tilde{P} + \tilde{R} - 2\tilde{Q}) \quad (4.130)$$

or

$$F_{min} = 1 + 2(\tilde{P} + \tilde{R} - 2\tilde{Q})\frac{(\omega C_g)^2}{g_m}(R_g + R_s)$$
$$\cdot \left(1 + \sqrt{1 + \frac{\tilde{P}\tilde{R}(1 - \tilde{C}^2) + g_m(R_g + R_s)(\tilde{P} + \tilde{R} - 2\tilde{Q})}{(\omega C_g(R_g + R_s)(\tilde{P} + \tilde{R} - 2\tilde{Q}))^2}}\right) \ .$$
$$(4.131)$$

Problem

4.6 Derive the expressions Eqs. (4.128) and (4.131) for the minimum noise figure and the optimum source impedance.

For frequencies that are small compared to the transit frequency of the transistor, we have $\omega C_g \ll g_m$ and thus the quotient under the square root in Eq. (4.131) is much larger than one. This leads to a simplified relation for the minimum noise figure:

$$F_{min} = 1 + 2\frac{\omega C_g}{g_m}\sqrt{\tilde{P}\tilde{R}(1 - \tilde{C}^2) + g_m(R_g + R_s)(\tilde{P} + \tilde{R} - 2\tilde{Q})}$$

$$+ 2\frac{(\omega C_g)^2}{g_m}(R_g + R_s)(\tilde{P} + \tilde{R} - 2\tilde{Q}) .$$ (4.132)

Equation (4.132) shows that the minimum noise figure far below the transit frequency increases linearly with frequency. Approaching the transit frequency, F_{min} increases faster due to the quadratic term in Eq. (4.132). The parasitic resistances R_g and R_s contribute substantially to this frequency behavior. If only the inner FET is considered, i.e. for $R_g = R_s = 0$, we get the simple expression

$$F_{min} = 1 + 2\sqrt{\tilde{P}\tilde{R}(1 - \tilde{C}^2)} \frac{\omega C_g}{g_m} .$$ (4.133)

As has already been mentioned, the two noise currents of the inner FET are strongly correlated for GaAs-FETs. With $\tilde{C} \approx 1$, the expression $\tilde{P}\tilde{R}(1 - \tilde{C}^2)$ under the square root of Eq. (4.132) can be neglected with respect to the second term. If, furthermore the quadratic term in frequency is omitted, then

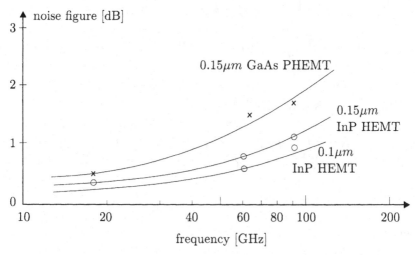

Fig. 4.22   Noise figure F of GaAs and InP field effect transistors.

the noise figure is given by

$$F_{min} = 1 + K\omega C_g\sqrt{\frac{R_g + R_s}{g_m}}$$ (4.134)

with

$$K = 2\sqrt{\tilde{P} + \tilde{R} - 2\tilde{Q}} \ . \tag{4.135}$$

The general dependence of the noise figure on the bias point can be derived from Eq. (4.134). If U_g becomes more negative, then C_g as well as g_m decrease. Also the ratio $\omega C_g/\sqrt{g_m}$ and thus the noise figure decreases at first. But if the transconductance finally approaches zero close to the pinch-off point, the noise figure increases again. Therefore, an optimum for the noise figure is observed at a certain drain current. For a GaAs FET the optimum gate voltage typically is close to the pinch-off voltage.

Problem

4.7 Determine the normalized gate voltage which yields the minimum noise figure, if g_m and C_g are described by the relations derived from the Shockley model.

Figure 4.22 shows typical noise figure values versus frequency curves of state of the art FETs, fabricated on GaAs or indium phosphide (InP) material. The best devices today achieve minimum noise figures close to 1 dB at 100 GHz.

5

Parametric Circuits

For a very important class of non-linear circuits the drive signal of the non-linear device (or devices) is a time-periodic pump voltage $u_p(t)$ of a relatively large amplitude and with the fundamental frequency f_p. Furthermore, there are a number of signals, $\Delta u(t)$, superimposed on the non-linear device, which are much smaller in amplitude than the pump signal and which normally have frequencies different from f_p. The so-called parametric approach is based on the assumption that the instantaneous properties of the non-linear device are determined exclusively by the periodic pump signal. It is assumed that the small signals do not influence the device behavior. As a first example, we will treat the Schottky diode with its unambiguous non-linear current-voltage characteristic. Then, we will transfer the parametric approach to other non-linear devices such as the field effect transistor or the varactor diode with its non-linear charge-voltage characteristic. We will see that parametric circuits can be treated quasi-linearly which simplifies the derivation of a noise model.

5.1 PARAMETRIC THEORY

The non-linear device relates the current I and the voltage U by an unambiguous current-voltage characteristic $I = I(U)$. The non-linear device is driven by a periodic signal $u_p(t)$ of a relatively large amplitude. A small signal $\Delta u(t)$ is superimposed on the large signal. If the amplitudes of $\Delta u(t)$ are very small compared with the amplitudes of $u_p(t)$, then a good approximation is given by the parametric approach:

$$I\left[u_p(t) + \Delta u(t)\right] \;=\; I\left(u_p(t)\right) + \left.\frac{dI}{dU}\right|_{u_p(t)} \cdot \Delta u(t)$$

$$=\; I\left(u_p(t)\right) + \Delta i(t) \;. \tag{5.1}$$

The small signal voltage $\Delta u(t)$ causes a small signal current $\Delta i(t)$ and both are related via the time dependent admittance $g(u_p(t))$:

$$\Delta i(t) \;=\; g(u_p(t)) \cdot \Delta u(t) \;,$$

$$g(t) \;=\; \frac{dI\left(u_p(t)\right)}{dU} \;. \tag{5.2}$$

The instantaneous admittance only depends on the large signal $u_p(t)$ and is the **parameter**, which is changed by the large signal. It should be noted that the small signal currents and voltages Δi and Δu are related linearly, because they do not influence the function $g(t)$. Therefore, the superposition principle holds for the small signal quantities. However, because the admittance $g(t)$ is time dependent, new frequency components will appear. The relationship between the different frequency components may be seen more clearly by using a phasor description. We will assume in the following derivation that the pump or local oscillator signal is periodic with the angular frequency $\omega_p = 2\pi f_p$. Then, the instantaneous admittance $g(t)$ is also periodic with ω_p and can be developed into a Fourier series:

$$g(u_p(t)) \;=\; \sum_{n=-\infty}^{+\infty} G_n \cdot \exp\left(j\,n\,\omega_p\,t\right)$$

$$G_n \;=\; \frac{1}{2\,\pi} \int_{-\pi}^{+\pi} g(u_p(t))\,\exp(-j\,n\,\omega_p\,t)\,d(\omega_p t) \;. \tag{5.3}$$

Because $g(t)$ is a real function, we have

$$G_{-n} = G_n^* \tag{5.4}$$

and G_0 is real. If we assume for the moment that $\Delta u(t)$ is sinusoidal with the angular frequency $\omega_s = 2\pi f_s$, then we conclude from Eq. (5.2) that $\Delta i(t)$ appears at all combination frequencies $|f_s \pm n \cdot f_p|$, $n = 0, 1, 2, 3$ etc. The small signal approximation has the consequence that harmonic frequencies of f_s can not appear. The small signal currents at the different combination frequencies lead to small signal voltages at these frequencies if the load impedances at these frequencies are not equal to zero.

At the moment, we will assume for the small signals that apart from the current at the signal frequency f_s only a current at the intermediate frequency

$f_i = f_s - f_p$ flows through the two-terminal non-linear device. At all other combination frequencies the load impedance is high enough to prevent any current flow. In a phasor description with the complex current phasors I_s, I_i at the signal and the intermediate frequency, respectively, we then may write for the small signal current

$$\Delta i(t) = \frac{1}{2} \left(I_s\, e^{j\omega_s t} + I_s^*\, e^{-j\omega_s t} + I_i\, e^{j\omega_i t} + I_i^*\, e^{-j\omega_i t} \right) \; . \tag{5.5}$$

Only the voltage phasors at the frequencies f_s and f_i together with the corresponding current phasors give rise to a power flow at the non-linear two-terminal element. We therefore introduce components only at the frequencies f_s and f_i also for $\Delta u(t)$:

$$\Delta u(t) = \frac{1}{2} \left(U_s\, e^{j\omega_s t} + U_s^*\, e^{-j\omega_s t} + U_i\, e^{j\omega_i t} + U_i^*\, e^{-j\omega_i t} \right) \; . \tag{5.6}$$

If we insert Eqs. (5.3), (5.5) and (5.6) into Eq. (5.2) and arrange according to frequency components, we obtain the following two equations in matrix form

$$\begin{bmatrix} I_s \\ I_i \end{bmatrix} = \begin{bmatrix} G_0 & G_1 \\ G_1^* & G_0 \end{bmatrix} \cdot \begin{bmatrix} U_s \\ U_i \end{bmatrix} = [G] \cdot \begin{bmatrix} U_s \\ U_i \end{bmatrix} \tag{5.7}$$

for $f_s > f_p$.

We can see that the complex current and voltage phasors are linearly related via an admittance matrix. However, in contrast to time-invariant linear two-ports, the phasors belonging to different indexes s and i also belong to different frequencies, f_s and f_i.

The amplitude of the pump signal does not enter into the equations explicitly. However, it determines the magnitude of the matrix elements G_0, G_n.

In contrast to the upper sideband conversion discussed so far (Fig. 5.1a), for the lower sideband conversion we have $f_i = f_p - f_s$ (Fig. 5.1b). For the

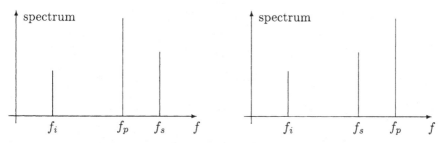

Fig. 5.1   (a) Upper sideband conversion;   (b) Lower sideband conversion.

case of the lower sideband conversion we obtain the following matrix equation for the current and voltage phasors:

$$\begin{bmatrix} I_s \\ I_i^* \end{bmatrix} = \begin{bmatrix} G_0 & G_1 \\ G_1^* & G_0 \end{bmatrix} \cdot \begin{bmatrix} U_s \\ U_i^* \end{bmatrix} = [G] \cdot \begin{bmatrix} U_s \\ U_i^* \end{bmatrix} \tag{5.8}$$

for $f_s < f_p$.

If the large signal time function $u_p(t)$, for a proper choice of the time origin, is an even function around $t = 0$, which e.g. is true for a cosine function, then the Fourier coefficients G_n in Eq. (5.3) are real. In this case, the admittance matrix $[G]$ is symmetrical and the corresponding two-port is reciprocal. Up- and down-conversion lead to the same conversion gain or conversion loss. If the large signal time function $u_p(t)$ is an arbitrary function of time, then the Fourier coefficients G_n in Eq. (5.3) are normally complex. Due to the special form of the admittance matrix $[G]$ the corresponding two-port is quasi-reciprocal and up- and down-conversion lead to the same conversion gain or conversion loss.

The size of the admittance matrix increases with the number of combination frequencies allowed. For instance, a 3×3 matrix is obtained if the image frequency $f_{im} = f_p \pm f_i$ is additionally taken into account.

5.2 DOWN CONVERTERS WITH SCHOTTKY DIODES

Schottky diodes are particularly well suited for the heterodyne reception at high frequencies, because they have a high cut-off frequency. As Schottky diodes are passive components, the stability of the mixer circuits is practically always guaranteed. As we will see in this chapter, frequency converters with Schottky diodes also have low noise figures. In general, circuits of down-

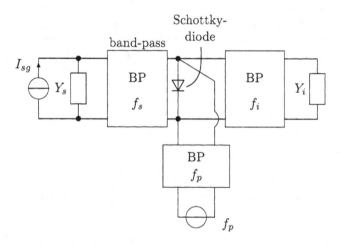

Fig. 5.2 Basic down converter circuit.

converters have a structure as shown in Fig. 5.2. For the band-pass filters BP shown, it is assumed that only the specified frequency is passed while all other frequencies are suppressed. Also, the input impedance of the band-pass filter is assumed to be very high at these other frequencies. In particular, it is assumed that all band-pass filters have a high input impedance at the image frequency f_{im}, at least for the circuit of Fig. 5.2. As a consequence no current will flow through the Schottky diode at the image frequency.

In a dual manner, for a Schottky diode connected in series, the band-pass filters should have a short circuit as input impedance at all frequencies outside the pass band. It is evident from these assumptions that only currents and voltages at the signal frequency f_s and the intermediate frequency f_i have to be taken into account. Therefore, the matrix relation Eq. (5.7) can be applied. Let Y_s be the source admittance, Y_i the load admittance and I_{sg} the signal generator current source, then we obtain the equations:

$$\begin{bmatrix} I_s \\ I_i \end{bmatrix} = \begin{bmatrix} G_0 & G_1 \\ G_1^* & G_0 \end{bmatrix} \cdot \begin{bmatrix} U_s \\ U_i \end{bmatrix} \quad \text{for} \quad f_s > f_p \tag{5.9}$$

$$I_{sg} = U_s \cdot Y_s + I_s \tag{5.10}$$

$$0 = U_i \cdot Y_i + I_i \ . \tag{5.11}$$

Figure 5.3 shows a two-port equivalent circuit in an admittance description with current and voltage phasors and a source and a load impedance. The

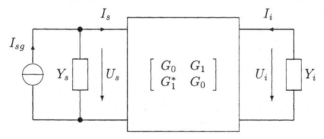

Fig. 5.3 Two-port equivalent circuit of a down converter.

elements of the admittance matrix G_0, G_1 are the Fourier coefficients of the periodic time function of $g(t)$ of the Schottky diode according to Eq. (5.3). We will assume that a sinusoidal voltage U_p from the pump signal or local oscillator drives the Schottky diode at the angular frequency $\omega_p = 2\pi f_p$. In addition, a d.c. bias voltage U_0 is applied to the Schottky diode. Then, the voltage $u_p(t)$ across the diode is given by

$$u_p(t) = U_0 + \hat{U}_p \cdot \cos \omega_p t \ . \tag{5.12}$$

Because $u_p(t)$ is an even function of time, all Fourier coefficients are real. For an exponential current-voltage characteristic according to

$$I = I_{ss} \cdot \left(\exp\left(\frac{U}{U_T} \right) - 1 \right) \tag{5.13}$$

and with

$$U_T = \tilde{n}kT/q \qquad (5.14)$$

the Fourier coefficients are obtained as

$$
\begin{aligned}
G_n &= \frac{I_{ss}}{U_T} \exp(U_0/U_T) \\
&\quad \cdot \frac{1}{2\pi} \int_{-\pi}^{+\pi} \exp\left(\frac{\hat{U}_p}{U_T} \cdot \cos(\omega_p t)\right) \cos(n\,\omega_p t)\, d(\omega_p t) \;. \qquad (5.15)
\end{aligned}
$$

The integral in Eq. (5.15) represents a modified Bessel function of the n-th order, $\tilde{I}_n(\hat{U}_p/U_T)$. Therefore, the Fourier coefficients may also be written as

$$G_n = \frac{I_{ss}}{U_T} \cdot \exp(U_0/U_T) \cdot \tilde{I}_n(\hat{U}_p/U_T) \;. \qquad (5.16)$$

According to this model the Fourier coefficients only depend on the peak value of the pump signal $\hat{U}_p$. Figure 5.4 shows a typical time characteristic of the admittance $g(t)$. For the following discussions, we will assume that G_0 is real

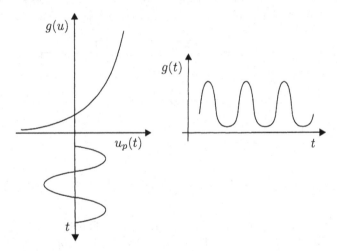

Fig. 5.4   Time characteristic of the admittance $g(t)$.

and positive and that G_1 and possibly G_2 etc. are allowed to be complex but have a positive real part. All G_0, G_1, G_2 etc. are known, because the pump level is assumed to be known. Normally, for a fundamental frequency converter, the relation $G_0 > G_1 > G_2$ holds. With known G_0 and G_1 and with known source and load admittances Y_s and Y_i we can determine the power gain or simply gain G_p, the available gain G_{av} and the maximum available gain G_m of the equivalent mixer circuit of Fig. 5.3. The gain of the down-converter is the ratio of the output power $|U_i|^2 \cdot \mathrm{Re}(Y_i)/2$ to the available source power

$|I_{sg}|^2/(8 \cdot \text{Re}(Y_s))$. The ratio U_i/I_{sg} can conveniently be determined from the extended matrix $[\tilde{G}]$, in which the Eqs. (5.9) to (5.11) are combined:

$$\begin{bmatrix} I_{sg} \\ 0 \end{bmatrix} = \begin{bmatrix} \tilde{G} \end{bmatrix} \cdot \begin{bmatrix} U_s \\ U_i \end{bmatrix} = \begin{bmatrix} G_0 + Y_s & G_1 \\ G_1^* & G_0 + Y_i \end{bmatrix} \cdot \begin{bmatrix} U_s \\ U_i \end{bmatrix} \qquad (5.17)$$

or

$$\begin{bmatrix} U_s \\ U_i \end{bmatrix} = \begin{bmatrix} \tilde{G} \end{bmatrix}^{-1} \cdot \begin{bmatrix} I_{sg} \\ 0 \end{bmatrix} . \qquad (5.18)$$

Equation (5.18) yields

$$\frac{U_i}{I_{sg}} = -\frac{G_1}{(G_0 + Y_s)(G_0 + Y_i) - |G_1|^2} , \qquad (5.19)$$

which leads to an expression for the gain G_p :

$$G_p = \left| \frac{U_i}{I_{sg}} \right|^2 \cdot 4 \cdot \text{Re}(Y_s) \cdot \text{Re}(Y_i)$$

$$= \frac{4 \cdot \text{Re}(Y_s) \cdot \text{Re}(Y_i) \cdot |G_1|^2}{|(G_0 + Y_s)(G_0 + Y_i) - |G_1|^2|^2} . \qquad (5.20)$$

For mixer or frequency converter circuits the available gain G_{av} is often required, which is the ratio of the available output power to the available source power. The available output power is obtained, if the load admittance Y_i is chosen equal to the complex conjugate of the input admittance Y_{ii} of the mixer circuit as seen from the load side at the intermediate frequency f_i :

$$Y_i = Y_{ii}^* . \qquad (5.21)$$

Conveniently, the input admittance Y_{ii} is computed from admittance matrix $[G_i]$ extended by the source admittance Y_s only:

$$\begin{bmatrix} 0 \\ I_i \end{bmatrix} = \begin{bmatrix} G_i \end{bmatrix} \cdot \begin{bmatrix} U_s \\ U_i \end{bmatrix} = \begin{bmatrix} G_0 + Y_s & G_1 \\ G_1^* & G_0 \end{bmatrix} \cdot \begin{bmatrix} U_s \\ U_i \end{bmatrix} \qquad (5.22)$$

or

$$\begin{bmatrix} U_s \\ U_i \end{bmatrix} = \begin{bmatrix} G_i \end{bmatrix}^{-1} \cdot \begin{bmatrix} 0 \\ I_i \end{bmatrix} . \qquad (5.23)$$

From Eq. (5.23) the input admittance Y_{ii} at the intermediate frequency or load side is obtained as

$$Y_{ii} = \frac{I_i}{U_i} = \frac{(G_0 + Y_s) G_0 - |G_1|^2}{(G_0 + Y_s)} = G_0 - \frac{|G_1|^2}{(G_0 + Y_s)} . \qquad (5.24)$$

With Y_{ii} from Eq. (5.24) and $Y_i = Y_{ii}^*$ the available gain is calculated in the same way as in Eq. (5.20), with Y_i replaced by Y_{ii}^* :

$$G_{av} = \frac{4 \cdot \text{Re}(Y_s) \cdot \text{Re}(Y_{ii}^*) \cdot |G_1|^2}{|(G_0 + Y_s)(G_0 + Y_{ii}^*) - |G_1|^2|^2} . \qquad (5.25)$$

The available gain does not depend on the load admittance Y_i. The reciprocal value of the available gain G_{av} will be denoted as conversion loss L_{av}:

$$L_{av} = \frac{1}{G_{av}} \ . \tag{5.26}$$

The maximum available gainavailable gain, maximum available gain $G_m = 1/L_m$ is obtained when an impedance match is provided at the mixer input as well as at the mixer output side. Since the gain is symmetrical with respect to the quantities Y_s and Y_i, we must have $Y_s = Y_i$ for the maximum available gain. For the input admittance at the source side, Y_{is}, we get, by similarity to Eq. (5.24),

$$Y_{is} = G_0 - \frac{|G_1|^2}{(G_0 + Y_i)} \ . \tag{5.27}$$

With impedance match at the source side, i.e. $Y_s = Y_{is}^*$ we obtain

$$Y_s = Y_i = Y_{is}^* = G_0 - \frac{|G_1|^2}{(G_0 + Y_i^*)} \tag{5.28}$$

or

$$|Y_s|^2 = |Y_i|^2 = G_0^2 - |G_1|^2 \tag{5.29}$$

and

$$\text{Im}\{Y_s\} = \text{Im}\{Y_i\} = 0 \tag{5.30}$$

and therefore,

$$Y_s = Y_i = \sqrt{G_0^2 - |G_1|^2} \ , \tag{5.31}$$

and Y_s and Y_i are real quantities. If these results for Y_s and Y_i are inserted into the relation Eq. (5.20) for the gain, then, after some manipulations the following expression is obtained for the maximum available gain G_m:

$$G_m = \left(\frac{|G_1|}{G_0}\right)^2 \left[\frac{1}{1 + \sqrt{1 - |G_1|^2/G_0^2}}\right]^2 \ . \tag{5.32}$$

The maximum available gain G_m only depends on the ratio of $|G_1|/G_0 < 1$. Therefore, the maximum available gain is smaller than 1.

In the extreme case of a mixer driven by very narrow pulses, so-called Dirac impulses, $|G_1| \approx G_0$ and thus $G_m \approx 1$. However, under these conditions the input admittance Y_{is} approaches zero and thus an impedance match is no longer possible. Practical mixers with Schottky diodes exhibit conversion losses in the range of 5 dB to 10 dB. The series resistance or bulk resistance of the Schottky diode, which has been neglected so far, is the reason for the somewhat higher conversion losses. Figure 5.5 shows an equivalent mixer circuit that has been supplemented by a series resistance R_b.

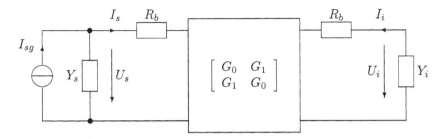

Fig. 5.5   Equivalent circuit of a down converter with a series resistance R_b.

Problem

5.1   Calculate the power gain, the available gain and the maximum available gain for the mixer circuit of Fig. 5.5 by taking into account the series resistance R_b .

For the down-converter, which has been discussed so far, we implicitly assumed that at the image frequency $f_{im} = f_p - f_i$ the load admittance is a short circuit, because we assumed the voltage U_{im} at f_{im} to be zero. Frequently, mixers do not have a short circuit at the image frequency, in particular, if the intermediate frequency f_i is low. In this case, the image frequency f_{im} and the signal frequency f_s are so close to each other that in practice it is difficult to supply a match at the signal frequency and simultaneously a short circuit at the image frequency. These practical difficulties become even more severe, if the pump frequency has to be tuned over a wide frequency range. For such a so-called broadband mixer, the load impedance at the signal frequency and the image frequency are nearly equal, at the price of a higher conversion loss. Figure 5.6 shows the equivalent circuit of a down-converter, where the load admittance at the image frequency, Y_{im}, has the same value as the load admittance Y_s at the signal frequency, i.e. $Y_{im} = Y_s$. For $f_s > f_p$ the currents and voltages are related via a 3-port admittance matrix $[G]$ as given by the following equation:

$$\begin{bmatrix} I_s \\ I_i \\ I_{im}^* \end{bmatrix} = \begin{bmatrix} G_0 & G_1 & G_2 \\ G_1^* & G_0 & G_1 \\ G_2^* & G_1^* & G_0 \end{bmatrix} \cdot \begin{bmatrix} U_s \\ U_i \\ U_{im}^* \end{bmatrix} \quad \text{for} \quad f_s > f_p \ . \quad (5.33)$$

Again the gains for up- and down-conversion are equal. For a feed signal from the intermediate frequency side, due to symmetry the signal power is evenly divided between the signal and the image port. However, because the power at the image load admittance Y_{im} is useless power with respect to the intended mode of operation, the conversion loss for the up- and down-conversion is at least 3 dB. Therefore, in practice, the conversion loss of a broadband mixer

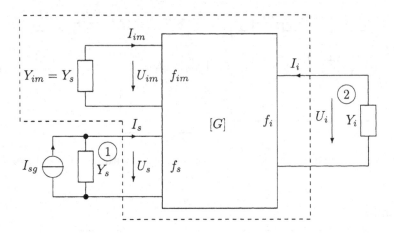

Fig. 5.6   3-port equivalent circuit of a down-converter with a load admittance Y_{im} at the image frequency.

is higher than the conversion loss of a mixer which is terminated by a lossless load admittance at the image frequency, e.g. a short circuit.

Problem

5.2   Calculate the power gain and the available gain for the case that the image frequency is terminated by the same complex load admittance as the signal frequency, $Y_{im} = Y_s$.

5.3   MIXER CIRCUITS

5.3.1   Single diode mixer

The single diode or one-diode mixer is particularly simple in its design and can achieve a broad bandwidth. A typical realization is shown in Fig. 5.7. The signal and the pump signal are combined via a coupler. The details of the coupler are not important. In Fig. 5.7 a transmission line coupler is drawn. The circuit has losses for the RF signal as well as for the pump signal. By the choice of the coupling factor, the losses can be shifted between the RF signal and the pump signal. If a 10 dB coupling is selected, then the pump signal is attenuated by 10 dB while the signal is attenuated by 0.46 dB. The high-pass filter HP must pass the high-frequency signal and the pump signal, and suppress the intermediate frequency, which is assumed to be much lower than the RF and the pump frequencies. The low-pass filter LP must pass the

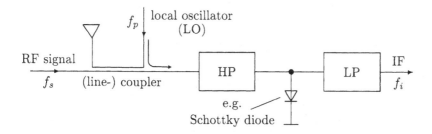

Fig. 5.7   Block diagram of a single diode mixer.

intermediate frequency and stop the RF signal and the pump frequency. On the diode side the low-pass filter must have a high impedance for the high-frequency components, while the high pass filter at the diode side must show a high impedance at the intermediate frequency.

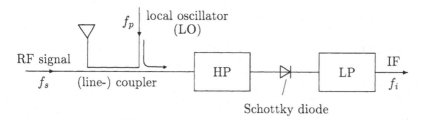

Fig. 5.8   Block diagram of a series connected one-diode mixer.

Instead of operating the diode in shunt configuration, one may connect the diode in series, as shown in Fig. 5.8. For a series connected diode the low-pass filter on the diode side must act as a short circuit against ground for the high-frequency signals and the high-pass filter as a short circuit against ground for the intermediate signal. Due to the rectification of the pump signal, a direct current will flow through the diode, if the current path is closed, driving the diode into a low impedance state. Therefore, normally, an additional bias voltage needs not to be applied to the diode.

5.3.2   Two-diode mixer or balanced mixer

The balanced mixer needs a 3 dB coupler of the 90°- or 180°-type and two diodes. Figure 5.9 shows a circuit with shunt diodes.

Similarly, the circuit can be built with series connected diodes. The balanced mixer circuit does not show additional losses, because for lossless couplers and filters the full power gets to the diodes. As an example we will consider a 3 dB-90° coupler. At the input of the coupler the signal and the

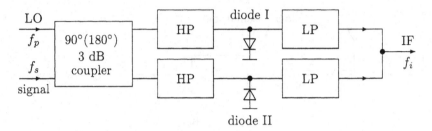

Fig. 5.9   Principal block diagram of a balanced mixer with two diodes.

pump signal are assumed to have the following time behavior:

$$\begin{aligned} u_s(t) &= \hat{U}_s \cdot \cos\left(\omega_s t + \phi_s\right) \\ u_p(t) &= \hat{U}_p \cdot \cos\left(\omega_p t\right) . \end{aligned} \qquad (5.34)$$

Then, we will observe the following signals at the diode I or diode II, respectively:

diode I : $\dfrac{1}{\sqrt{2}} \hat{U}_s \cdot \cos\left(\omega_s t + 90° + \phi_s\right) + \dfrac{1}{\sqrt{2}} \hat{U}_p \cdot \cos\left(\omega_p t\right)$

diode II : $\dfrac{1}{\sqrt{2}} \hat{U}_s \cdot \cos\left(\omega_s t + \phi_s\right) + \dfrac{1}{\sqrt{2}} \hat{U}_p \cdot \cos\left(\omega_p t + 90°\right)$. (5.35)

With the conversion factor κ we obtain the intermediate frequency signal $u_i(t)$ for the diode I:

$$\text{diode I :}\quad u_i^I = \kappa\, G_1\, \hat{U}_s \cos\left(\omega_i t + 90° + \phi_s\right) . \qquad (5.36)$$

Because the polarity is reversed for the diode II, the admittance function $g(t)$ has been shifted by half a period or 180° at the pump frequency f_p. Thus, we obtain for the diode II the expression:

$$\begin{aligned} \text{diode II :}\quad u_i^{II} &= \kappa\, G_1\, \hat{U}_s \cos\left(\omega_i t - 90° + 180° + \phi_s\right) \\ &= \kappa\, G_1\, \hat{U}_s \cos\left(\omega_i t + 90° + \phi_s\right) . \end{aligned} \qquad (5.37)$$

We assumed that the diodes are well paired, i.e. that they are equal. Then, both intermediate frequency signals are equal with respect to the phase and amplitude and they can be added, as it is shown in the circuit of Fig. 5.9. In total, the mixer circuit does not have additional signal and pump losses.

Problem

5.3   Show that a mixer with two diodes and a 180°-coupler does not exhibit additional signal or pump signal losses. What is the difference between a mixer with a 90°-coupler and a 180°-coupler?

The statement that a mixer is balanced has the following meaning. A pump oscillator signal normally shows irregular amplitude fluctuations, which are also denoted as amplitude noise. Such amplitude fluctuations are of stochastic nature and typically cover a broad frequency spectrum, which may also include spectral components in the intermediate frequency range. The origin of the amplitude noise will be discussed in detail in Chapters 6 and 7. The Schottky diodes of the mixer rectify the pump oscillator signal. Therefore, a direct voltage appears across the diodes, which also shows small irregular fluctuations, similar to the amplitude fluctuations of the pump oscillator. Spectral components of these fluctuations may fall into the frequency range of the intermediate frequency signal. Normally, the amplitude noise spectrum decreases with increasing offset frequency and, therefore, its contribution to the total noise may be negligible at high offset frequencies.

A mixer with two nearly identical diodes with opposite polarity and a superposition of the output signals as shown in Fig. 5.9 allows one to cancel the amplitude fluctuations independent of time. An identical small irregular fluctuation signal appears at the second diode, but due to the reversal of the second diode, this fluctuation signal has the opposite polarity versus time. When added these noise contributions of the pump oscillator signal will cancel. We will denote this balancing effect also as a noise balance. The radio frequency signal, in contrast, does not contribute substantially to the amplitude fluctuations, because it is much smaller than the pump signal and, therefore, all spectral contributions are transferred linearly without being rectified. On the other hand, the two intermediate frequency signals will add, as has been explained before.

For practical mixers the balancing effect is in the order of 20 to 40 dB. This is normally sufficient to eliminate the influence of the pump signal amplitude noise to the noise figure of the mixer. Then, the measured noise figure of the mixer as determined by its intrinsic noise is in agreement with the theory, which will be shown in the next sections.

A single diode mixer is not noise balanced, of course. A so-called double balanced mixer or ring modulator with four diodes arranged in a ring shows a similar noise balancing effect as the two-diode mixer.

5.3.3 Four-diode double balanced mixer

An example of a so-called double balanced mixer is the ring mixer as shown in Fig. 5.10. The name stems from the fact that the four diodes are arranged in a ring configuration, if one follows their topology in the direction of e.g. forward conduction. The two transformers cause a ground-symmetrical excitation of the four diodes for the radio frequency signals and the pump signal. The center tapping of the transformers acts as a direct galvanic connection to the diodes for the low intermediate frequency. We may introduce directed arrows to indicate the polarity of the radio frequency (RF) signal and the pump signal or local oscillator (LO) signal. We will make the assumption that a LO directed arrow in the forward conduction direction of a diode, i.e. in the direction of the tip of the diode symbol, represents an admittance time characteristic with a phase of 0° relative to the LO-signal, a directed arrow against the diode tip for a phase of 180°. The phase of the intermediate frequency signal (IF) results from the difference of the RF-phase and the phase of the admittance time characteristic. With these representations the resulting directed arrows for the IF-signals give the result shown in Fig. 5.11. We notice that all four arrows for the IF-signal are parallel and thus sum up. Due to the bridge arrangement of the diodes the RF port and the LO port are isolated, provided that the diodes are well paired. We may thus speak of a signal balancing of these ports. Similarly, for symmetry reasons, the IF port is isolated from the RF and LO ports, which may also be called a signal balancing. The double balanced mixer is noise balanced, because the rectified LO-signals cancel, due to their opposite polarities.

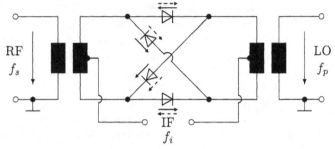

Fig. 5.10   Double balanced mixer or ring mixer.

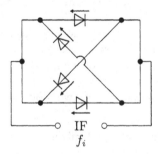

Fig. 5.11   Directed arrows for the IF-signals of the double balanced mixer of Fig. 5.10.

At higher frequencies, the transformers of Fig. 5.10 might be replaced by 180° − 3 dB couplers as shown in Fig. 5.12. In this figure, the 180° − 3 dB couplers are realized by 90° branch-line couplers with additional 90° phase shifters (PS). The low-pass filters should have a high impedance for the high-frequency signals at the diodes side. The RF and LO ports are isolated.

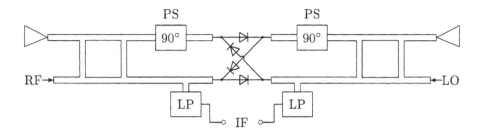

Fig. 5.12   Double balanced mixer with two 90° branch-line couplers.

5.4   NOISE EQUIVALENT CIRCUIT OF PUMPED SCHOTTKY DIODES

The sensitivity of a down-converter for the reception of weak high-frequency signals depends on the conversion loss, ultimately however, it is the noise figure which determines the deterioration of the signal-to-noise ratio at the intermediate frequency output of the mixer. In order to calculate the noise figure of the mixer, we need a noise equivalent circuit of the mixer. We will choose an equivalent circuit with a parallel noise current source $I_{ns} = I_{n1}$ for the signal input port at the frequency f_s and a second noise current source $I_{ni} = I_{n2}$ for the intermediate frequency output port at the frequency f_i (Fig. 5.13). The down converter is assumed to have a short circuit as a load admittance at the image frequency and all other relevant combination frequencies. The two-port with the admittance matrix $[G]$ itself is noise-free, G_1 is assumed to be real for the initial discussion. Later on, it may also be complex. The difference to equivalent circuits of normal linear circuits is

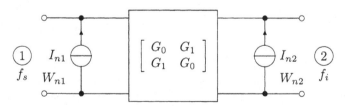

Fig. 5.13   Noise equivalent circuit of a down converter with an ideal Schottky diode.

that the ports 1 and 2 refer to different frequencies. In order to simplify the following discussion, we will neglect the noise contribution of the small reverse saturation current I_{ss}. Furthermore, we will neglect the thermal noise of the series resistance of the Schottky diode, but will add this contribution later on.

If we neglect the contribution of the reverse saturation current, then the Fourier coefficients of the current, I_μ, and of the admittance, G_μ, are proportional to each other, because the corresponding time characteristics are proportional to each other. We get

$$i(t) \;=\; \sum_{\mu=-\infty}^{+\infty} I_\mu \cdot \exp\left(j\,\mu\,\omega_p\,t\right) \approx U_T \cdot g(t)$$

and

$$I_\mu \;=\; U_T \cdot G_\mu \;. \tag{5.38}$$

The power spectra of the noise current sources are easily calculated. With a short circuit at port 2, the input admittance at port 1 is just G_0, and the current source shows shot-noise according to the mean value of the current, I_0, or the mean admittance, G_0. With the ideality factor $\tilde{n}$ for the Schottky diode the two-sided spectra W_{n1} and W_{n2} are given by

$$W_{n1} = |I_{n1}|^2 = q \cdot I_0 = 2\,k \cdot \left(\frac{1}{2}\,\tilde{n}\,T\right) \cdot G_0 = W_{n2} = |I_{n2}|^2 \;. \tag{5.39}$$

Here, as before, k is the Boltzmann constant and T is the absolute temperature.

The considerations concerning the cross-spectrum are more difficult. If the pump amplitude is equal to zero, then G_1 is also equal to zero and the ports 1 and 2 are isolated. According to the mathematical definition of correlation, signals of different, non-overlapping frequency bands are always uncorrelated. However, in the context of noise in mixers, the evaluation of the correlation requires to shift one of the two frequency bands in frequency such that both frequency bands perfectly overlap. For the unmodulated shot noise of the Schottky diode, i.e. $G_1 = 0$, there is no correlation even after a frequency translation for the purpose of achieving a frequency coincidence. This will be shown in problem (5.4).

Problem

5.4   Prove for white unmodulated noise that the noise signals at the output of two band-pass filters are uncorrelated, as long as the pass bands do not overlap. For the calculation of the correlation, first a frequency translation for the purpose of frequency coincidence should be performed.

We will normally observe a correlation if the shot noise of the Schottky diode is modulated by the pump signal, because the modulation generates noise sidebands at corresponding frequencies. After the frequency translation towards frequency coincidence, there are noise components from the same origin, which may lead to a correlation.

The unmodulated white shot noise is assumed to have the time characteristic $s(t)$. The autocorrelation function of $s(t)$, i.e. ρ, is assumed to be a Dirac function $\delta(\tau)$:

$$\rho(\tau) = \text{E}\{s(t) \cdot s(t+\tau)\} = \rho_0 \cdot \delta(\tau) \ . \tag{5.40}$$

The modulated shot noise is assumed to have the time characteristic $s_m(t)$. For the following considerations, we will make the fundamental assumption that the instantaneous power $s_m^2(t)$ of the modulated noise is proportional to the instantaneous current $i(t)$ through the diode or admittance of the diode $g(t)$:

$$
\begin{aligned}
s_m^2(t) &= s^2(t) \frac{i(t)}{I_0} \\
&= s^2(t) \cdot \left[1 + \frac{2\,I_1}{I_0}\cos(\omega_p t) + \frac{2\,I_2}{I_0}\cos(2\omega_p t) + \cdots\right] \\
&= s^2(t) \cdot \left[1 + \frac{2\,G_1}{G_0}\cos(\omega_p t) + \frac{2\,G_2}{G_0}\cos(2\omega_p t) + \cdots\right]
\end{aligned} \tag{5.41}
$$

or

$$s_m(t) = s(t) \cdot \sqrt{1 + \frac{2\,G_1}{G_0}\cos(\omega_p t) + \frac{2\,G_2}{G_0}\cos(2\omega_p t) + \cdots} \tag{5.42}$$

As shown in Fig. 5.14, we will discuss the correlation between the rectangularly band-pass filtered noise signals $X_i(t)$ at the frequency f_i and $X_s(t)$ at the frequency f_s, with $f_s - f_i = f_p$. We assume that the bandwidths Δf of the band-pass filters are equal and small. In the following, only the term G_1 will

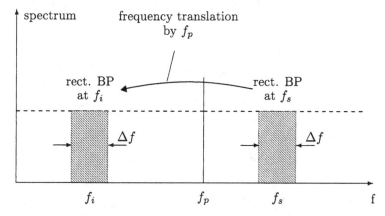

Fig. 5.14 Illustration of the correlation between two frequency bands around f_i and f_s.

be taken into account, because further terms, e.g G_2, do not contribute to the correlation, as may be shown. The functions $h_i(t)$ and $h_s(t)$ denote the impulse responses of the ideal band-pass filters with the center frequencies f_i and f_s. Then, we obtain the band-pass filtered signals $X_i(t)$ and $X_s(t)$ from $s_m(t)$ by a convolution of the impulse responses $h_i(t)$ and $h_s(t)$:

$$
\begin{aligned}
X_i(t) &= \int_{-\infty}^{+\infty} h_i(t') \cdot s_m(t - t')\, dt' \\
&= \int_{-\infty}^{+\infty} h_i(t') \cdot s(t - t') \cdot \sqrt{1 + \frac{2\,G_1}{G_0} \cos[\omega_p(t - t')]}\, dt' \\
X_s(t) &= \int_{-\infty}^{+\infty} h_s(t'') \cdot s_m(t - t'')\, dt'' \\
&= \int_{-\infty}^{+\infty} h_s(t'') \cdot s(t - t'') \cdot \sqrt{1 + \frac{2\,G_1}{G_0} \cos[\omega_p(t - t'')]}\, dt'' \ .
\end{aligned}
$$

$$(5.43)$$

The signal $X_s(t)$ includes spectral components around f_s. In order to calculate the correlation, a frequency translation or frequency shift by the amount or distance f_p from the frequency band at f_s to the frequency band at f_i must be performed. As illustrated in Fig. 5.15, such a frequency translation can be realized with the help of an ideal analog multiplier and a band-pass filter. We obtain the signal $\tilde{X}_s(t)$ after the ideal frequency shift of the signal $X_s(t)$ by the frequency distance f_p. The analog multiplier creates the sum and the

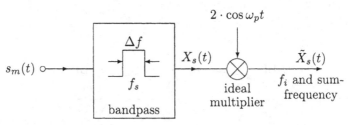

Fig. 5.15   Ideal frequency translator.

difference frequency. The difference frequency falls into the frequency band at f_i, as intended, while the sum frequency $f_p + f_s$ does not contribute to the correlation. The signals $\tilde{X}_s(t)$ and X_i both have spectral components in the vicinity of f_i and can be considered as the input signals of a correlator as shown in Fig. 3.15. Then, the correlation $\mathrm{E}\{X_i(t) \cdot \tilde{X}_s(t)\}$ is obtained from the following expression, in which the integration and time averaging have already been exchanged:

$$\langle X_i(t) \cdot \tilde{X}_s(t) \rangle \;=\; \mathrm{E}\{X_i(t) \cdot \tilde{X}_s(t)\} \;=\; \int\!\!\!\int\limits_{-\infty}^{+\infty} h_i(t') \cdot h_s(t'')$$

$$\cdot\, \mathrm{E}\Bigg\{ 2 \cdot \cos\omega_p t \cdot s(t - t') \cdot \sqrt{1 + \frac{2G_1}{G_0} \cos[\omega_p(t - t')]}$$

$$\cdot\, s(t - t'') \cdot \sqrt{1 + \frac{2G_1}{G_0} \cos[\omega_p(t - t'')]} \Bigg\} \, dt' dt'' \; .$$

$$(5.44)$$

We note that the expected value of the expression in the curly brackets, denoted by $E\{\tilde{x}\}$, becomes identical to zero for $t' \neq t''$, for similar reasons as $\rho(\tau)$ in Eq. (5.40) for $\tau \neq 0$. We therefore analyze the expression $E\{\tilde{x}\}$ for $t' = t''$:

$$E\{\tilde{x}\} \;=\; \mathrm{E}\Bigg\{ 2 \cdot \cos\omega_p t \cdot s^2(t - t') \cdot \left(1 + \frac{2G_1}{G_0} \cos[\omega_p(t - t')]\right)\Bigg\}\Bigg|_{t'=t''}$$

$$=\; \mathrm{E}\{2 \cdot \cos\omega_p t \cdot s^2(t - t')\} +$$

$$\mathrm{E}\Bigg\{ \frac{4G_1}{G_0} \cdot \cos\omega_p t \cdot \cos[\omega_p(t - t')] \cdot s^2(t - t') \Bigg\}\Bigg|_{t'=t''}$$

$$=\; \mathrm{E}\{2 \cdot \cos\omega_p t \cdot s^2(t - t')\} + \mathrm{E}\Bigg\{ \frac{2G_1}{G_0} \cdot \cos\omega_p t' \cdot s^2(t - t') \Bigg\}$$

$$+\, \mathrm{E}\Bigg\{ \frac{2G_1}{G_0} \cdot \cos[2\omega_p t - \omega_p t'] \cdot s^2(t - t') \Bigg\}\Bigg|_{t'=t''}$$

$$E\{\tilde{x}\} \;=\; 0 \qquad \text{for} \quad t' \neq t'' \; . \tag{5.45}$$

The first and the third expression on the right hand side of Eq. (5.45) are zero, because $s^2(t - t')$ is multiplied by an alternating and limited weighting function. Therefore, the final result is

$$E\{\tilde{x}\} = \mathrm{E}\Bigg\{ \frac{2G_1}{G_0} \cdot \cos\omega_p t' \cdot s^2(t - t') \Bigg\} = \frac{2G_1}{G_0} \cdot \cos\omega_p t' \cdot \mathrm{E}\{s^2(t - t')\}\Bigg|_{t'=t''}$$

$$E\{\tilde{x}\}=0 \qquad \text{for} \quad t' \neq t'' \tag{5.46}$$

or with Eq. (5.40):

$$E\{\tilde{x}\} \;=\; \frac{2G_1}{G_0} \cdot \cos\omega_p t' \cdot \rho_0 \cdot \delta(t'' - t') \; . \tag{5.47}$$

We therefore can write for Eq. (5.44):

$$\mathrm{E}\{X_i(t) \cdot \tilde{X}_s(t)\} \;=\; \int\!\!\!\int\limits_{-\infty}^{+\infty} h_i(t') \cdot h_s(t'')$$

$$\cdot \frac{2\,G_1}{G_0} \cdot \cos(\omega_p t') \cdot \rho_0 \cdot \delta(t'' - t')\, dt'\, dt''$$

$$= \quad \rho_0 \cdot \frac{2\,G_1}{G_0} \cdot \int\limits_{-\infty}^{+\infty} h_i(t') \cdot h_s(t') \cos(\omega_p t')\, dt' \quad . \quad (5.48)$$

In the following, we will calculate the normalized correlation coefficient, i.e. the correlation coefficient normalized to the autocorrelation coefficient $E\{X_i(t) \cdot X_i(t)\}$ in the band with the center frequency f_i. This ratio is equal to the ratio of the cross-spectrum to the auto spectrum, $(I_{n1}^* \cdot I_{n2})/|I_{n1}|^2$, because, according to the assumption made, the crossspectrum of the current sources is real due to the real $[G]$-matrix. Thus the imaginary part of $I_{n1}^* I_{n2}$ is zero (cf. problem (5.5)). Therefore, we can write:

$$\frac{E\{X_i(t) \cdot \tilde{X}_s(t)\}}{E\{X_i^2(t)\}} \quad = \quad \frac{\mathrm{Re}\{I_{n1}^* I_{n2}\}}{|I_{n1}|^2} = \frac{I_{n1}^* \cdot I_{n2}}{|I_{n1}|^2}$$

$$= \quad \frac{\rho_0 \cdot \dfrac{2\,G_1}{G_0} \cdot \displaystyle\int_{-\infty}^{+\infty} h_i(t') \cdot h_s(t') \cos(\omega_p t')\, dt'}{\rho_0 \displaystyle\int_{-\infty}^{+\infty} h_i^2(t')\, dt'} \quad .$$

$$(5.49)$$

For further evaluation of Eq. (5.49) we will use the explicit expressions for the impulse responses $h_i(t)$ and $h_s(t)$ for a rectangular band-pass filter, as already known from a similar calculation in problem 1.7:

$$h_{i,s} = 2\,\Delta f \cdot \cos(2\pi f_{i,s} \cdot t) \cdot \mathrm{si}(\pi\,\Delta f \cdot t) \quad . \qquad (5.50)$$

For a small value of Δf the integrals in Eq. (5.49) are almost zero, if the products of cosine functions in front of the term $\mathrm{si}^2(\pi\,\Delta f \cdot t')$ are alternating weighting functions. A contribution of the integrals only occurs if the products of cosine functions results in a constant term (cf. problem 5.5). With this knowledge Eq. (5.49) yields:

$$\frac{E\{X_i(t) \cdot \tilde{X}_s(t)\}}{E\{X_i^2(t)\}} \quad = \quad \frac{I_{n1}^* \cdot I_{n2}}{|I_{n1}|^2}$$

$$= \quad \frac{2\,G_1}{G_0} \cdot \frac{\dfrac{1}{4} \displaystyle\int_{-\infty}^{+\infty} \mathrm{si}^2(\pi\,\Delta f \cdot t')\, dt'}{\dfrac{1}{2} \displaystyle\int_{-\infty}^{+\infty} \mathrm{si}^2(\pi\,\Delta f \cdot t')\, dt'} = \frac{G_1}{G_0} \quad . \quad (5.51)$$

From Eqs. (5.39) and (5.51) we finally get the requested cross-spectrum $I_{n1}^* \cdot I_{n2} \equiv I_{ni}^* \cdot I_{ns}$:

$$I_{n1}^* \cdot I_{n2} = 2\,k \left(\frac{1}{2}\, \tilde{n}\, T \right) \cdot G_1 \quad . \qquad (5.52)$$

We notice that the cross-spectrum of two frequency bands that are related by the pump frequency f_p, are proportional to the Fourier coefficient G_1. This is for instance true for the frequency ranges f_i and f_s or f_i and f_{im}. Similarly, the cross-spectrum of two frequency ranges that are related by twice the pump frequency $2f_p$, as for instance f_s and f_{im}, are proportional to the Fourier coefficient G_2 :

$$I_{ns}^* \cdot I_{n,im} = 2\,k\left(\frac{1}{2}\tilde{n}\,T\right) \cdot G_2 \ . \tag{5.53}$$

Thus, the so-called correlation matrix for the equivalent noise current sources of an ideally pumped Schottky diode has the same structure as that of a passive noisy two-port with only thermal noise at a homogeneous temperature T. This equivalent passive two-port is also described by its admittance matrix $[Y]$ and with current sources (see also Eq. (2.47) and Eq. (2.48)). The Fourier coefficient G_0 corresponds to the admittance matrix element Y_{11} and the element G_1 corresponds to the admittance matrix element $Y_{21} = Y_{12}$. However, instead of the temperature T for the passive circuit with thermal noise, we have to use the temperature $\tilde{n}\,T/2$, i.e. the effective temperature already known from the Schottky diode with a bias current but without series resistance. The correlation matrix which corresponds to Eq. (2.32) is:

$$\begin{bmatrix} I_{ns}^* \cdot I_{ns} & I_{ns}^* \cdot I_{ni} & I_{ns}^* \cdot I_{n,im}^* \\ I_{ni}^* \cdot I_{ns} & I_{ni}^* \cdot I_{ni} & I_{ni}^* \cdot I_{n,im}^* \\ I_{n,im} \cdot I_{ns} & I_{n,im} \cdot I_{ni} & I_{n,im} \cdot I_{n,im}^* \end{bmatrix} = 2\,k\,(\tilde{n}\,T/2)\cdot \begin{bmatrix} G_0 & G_1 & G_2 \\ G_1^* & G_0 & G_1 \\ G_2^* & G_1^* & G_0 \end{bmatrix} \ . \tag{5.54}$$

We can state that the mixer noise correlation matrix is proportional to the correlation matrix of a passive time-invariant thermally noisy multi-port of a homogeneous temperature and the same admittance matrix. The proportionality constant is $\tilde{n}\,T/2$. This statement remains valid if some of the elements of the G-matrix are complex, due to a non-even pump-drive of the mixer, as will be shown in problem 5.5. The Eq. (5.54) has been given in complex form.

Problem

5.5   Determine the correlation spectra of an ideally pumped Schottky diode for the case of complex elements G_1, G_2 of the G-matrix, due to a non-even pump-drive.

The noise model of the mixer of Fig. 5.13 can be extended in such a way that it includes the thermal noise of the time invariant series or bulk resistance R_b (Fig. 5.16). Since the series resistance is time invariant, the noise sources W_{u1} and W_{u2} are uncorrelated. For the noise sources I_{n1} and I_{n2}, the relations Eq. (5.39) and Eq. (5.40) remain valid. The noise sources I_{n1} and I_{n2} are not correlated with the noise sources U_{b1} and U_{b2}.

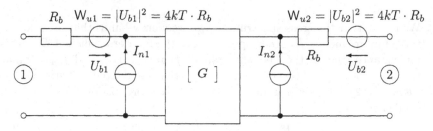

Fig. 5.16 Noise equivalent circuit of a pumped Schottky diode with a series resistance R_b.

5.5 NOISE FIGURE OF DOWN-CONVERTERS WITH SCHOTTKY DIODES

For the moment, we will neglect the series or bulk resistance R_b. In this case, the calculation of the noise figure is quite straightforward. Because the equivalent circuit with current sources and an admittance matrix has the same structure and the same form of the correlation matrix as a thermally noisy two-port at a homogeneous temperature T, we can adopt the results of time-invariant passive noisy two-ports with thermal noise only. The difference for the Schottky diode mixer is that we have to use the effective temperature $T_{ef} = \tilde{n}T/2$ and not the physical temperature T of the semiconductor. Here, $\tilde{n}$ is again the ideality factor of the diode.

With Eq. (2.90) from chapter 2 and $T_1 = T_{ef} = \tilde{n}T/2$, with the available gain $G_{av} = 1/L_{av}$, and the ambient temperature T_0, the noise figure F_m of the mixer is obtained as

$$F_m = 1 + \frac{\tilde{n}}{2} \cdot \frac{T}{T_0} \cdot \frac{1 - G_{av}}{G_{av}} = 1 + \frac{\tilde{n}}{2} \frac{T}{T_0} (L_{av} - 1) \ . \qquad (5.55)$$

One may determine the available gain from Eq. (5.25). The relation for the noise figure Eq. (5.55) applies, if the image frequency and all further combination frequencies of secondary importance are terminated by a short circuit or an open circuit or more generally by a lossless admittance. The relation for the noise figure is valid for a down-converter as well as for an up-converter, for a lower sideband up-converter as well as for an upper sideband up-converter.

Equation (5.55) may also be proven directly, i.e. with the equivalent circuit of Fig. 5.13 and Eqs. (5.39) and (5.52). One has to express the noise figure as a function of the available gain (problem 5.6).

Problem

5.6 Calculate the noise figure of a down-converter with a short circuit at the image frequency with the help of the equivalent circuit of Fig. 5.13 and derive Eq. (5.55).

For the maximum available gain the noise figure attains its minimum value F_{min}. Theoretically, for a narrow-band mixer and in the limit of an ideal impulse pump drive, the noise figure may approach the value of 1 or 0 dB.

The noise figure increases if the series resistance of the diode is taken into account, as will be shown in problem 5.7.

Problem

5.7 Calculate the noise figure of a down-converter with a series resistance of the Schottky diode with the help of the equivalent circuit of Fig. 5.16.

The mixer with a Schottky diode and a series resistance can also be described by a two-temperature model, i.e. with a temperature $\tilde{n}T/2$ of the ideal Schottky junction and the temperature T of the series or bulk resistance R_b. In general, we may write for the noise figure of a two-temperature two-port:

$$F = 1 + \frac{\beta_s\,T + \beta_j\,\frac{\tilde{n}}{2}\,T}{T_0 \cdot G_{av}}\,. \tag{5.56}$$

Here, β_s and β_j are the relative dissipated powers in the series resistance and junction, respectively, when feeding from the load side and assuming reciprocity.

Problem

5.8 Calculate the noise figure of a down-converter with a series resistance as in problem 5.7 via the relation Eq. (5.56) and with the help of the dissipation theorem.

The broadband mixer according to Fig. 5.6 can also be treated as a two-temperature problem. It is assumed that the same load admittance is effective at the image frequency f_{im} and at the signal frequency f_s. For the moment, the series resistance will be neglected. The part in the dashed box of the circuit in Fig. 5.6 contains the noisy two-port of the mixer with the ports 1 and 2. This two-port consists of two temperature regions, i.e. the load admittance $Y_{im} = Y_s$ at the image frequency with the ambient temperature T_0 and the Schottky diode junction with the temperature $\tilde{n}T/2$. Therefore, the noise figure expression is similar to Eq. (5.56):

$$F = 1 + \frac{\beta_{im}\,T_0 + \beta_j\,\frac{\tilde{n}}{2}\,T}{T_0 \cdot G_{av}}\,. \tag{5.57}$$

Here, β_{im} is the relative power dissipated in the admittance Y_{im}, when feeding from the load side, i.e. from the intermediate frequency side. If $L_{av} = 1/G_{av}$

is the conversion loss from port 1 to port 2, then, due to reciprocity, the conversion loss from port 2 to port 1 is the same. For reasons of symmetry, we have

$$\beta_{im} = \frac{1}{L_{av}}. \tag{5.58}$$

The part of the total power which is not absorbed in the load admittance Y_{im} at the image frequency or the generator admittance Y_s remains in the diode junction. According to the dissipation theorem, the feeding occurs from the intermediate frequency port. We therefore conclude that

$$\beta_j = 1 - \frac{1}{L_{av}} - \frac{1}{L_{av}} = 1 - \frac{2}{L_{av}}. \tag{5.59}$$

With this result for β_j we obtain for the noise figure of the broadband down-converter from Eqs. (5.57) and (5.58)

$$F = 1 + \frac{\frac{1}{L_{av}}T_0 + \left(1 - \frac{2}{L_{av}}\right) \cdot \frac{\tilde{n}}{2}T}{T_0 \cdot 1/L_{av}} = 2 + (L_{av} - 2) \cdot \frac{\tilde{n}}{2}\frac{T}{T_0}. \tag{5.60}$$

Because $L_{av} \geq 2$, the lower limit for the minimum noise figure of the broadband mixer is 3 dB.

5.6   MIXERS WITH FIELD EFFECT TRANSISTORS

The circuit structure of a mixer with a field effect transistor (FET) as the non-linear element is particularly simple, because the FET as a three-terminal component already provides an inherent isolation between the RF and LO port or the gate and drain, respectively. The pump signal (LO) is fed to the gate. The LO-signal periodically alters the value of the drain-source channel admittance and thus determines the admittance time function $g(t)$. Normally, the field-effect transistor is operated passively, i.e. without a drain-source bias voltage and within the ohmic part of the current-voltage characteristic. This type of operation has the advantage that the mixer has good large signal properties, is unconditionally stable and, to a first approximation, does not exhibit $1/f$-noise. The bias voltage together with the pump signal amplitude at the gate typically are chosen such that the modulation of the channel admittance varies between the open and the closed channel conditions. The parametric mixer theory of this chapter can be applied to the FET mixer without change. In Section 4.6 it was described how the admittance time characteristic $g(t)$ can be determined for a given gate signal (LO-signal) according to the approximate Shockley model. Figure 5.18 shows the cross-section through the idealized FET for the case that the drain-source bias voltage is zero, i.e. a 'cold' operation of the FET mixer. In contrast to Fig. 4.14, now the interface of the space charge region and the conducting channel is a horizontal straight

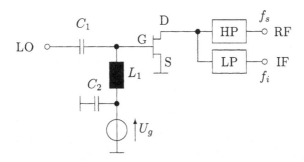

Fig. 5.17   Basic equivalent circuit of a single FET mixer.

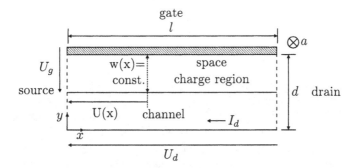

Fig. 5.18   Cross-section through the inner FET for a zero drain-source bias voltage.

line, which moves up and down with the applied gate voltage. In the equivalent circuit of Fig. 5.17 the capacitance C_1 blocks the gate bias voltage, the inductance L_1 and the capacitance C_2 separate the gate bias voltage U_g from the LO-circuit. The low-pass filter should have a high input impedance for the RF-signal, the high pass filter should have a high input impedance for the intermediate frequency signal.

A single FET mixer is inherently noise balanced, because the LO-signal at the gate is not rectified. This, however, is no longer true for very high frequencies, because a part of the LO-signal may couple into the drain-source conducting channel via the gate-drain capacitance and may be rectified there by means of non-linear current voltage relations. Therefore, at high frequencies, the noise balance effect will decrease by 20 dB/decade with increasing frequency.

A mixer with two FETs has additional degrees of freedom and enables, e.g. by the use of symmetries, a better isolation of the LO port and the RF port, as shown for the mixer-circuit in Fig. 5.19, which employs, as an example, a 3 dB-0° and a 3 dB-180° coupler. Field effect transistors are unipolar devices

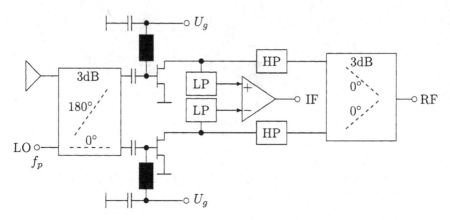

Fig. 5.19   Example of an equivalent circuit of a two-FET mixer.

and, therefore, it is not possible, as for Schottky diodes, to change the polarity by reversing the device. Within the circuit of Fig. 5.19 it is thus necessary to combine the intermediate frequency signals by a differential amplifier. There is a twofold noise balance. First, we get a reduction of the pump signal amplitude noise by taking the difference of both FET intermediate frequency signals. Second, the FET is inherently noise balanced, as has been discussed before. Both noise reduction effects add up.

The conversion loss of a FET mixer is comparable to that of a Schottky diode mixer. Typical values are 5 to 10 dB for broadband mixers.

5.7   NOISE FIGURE OF DOWN CONVERTERS WITH FIELD EFFECT TRANSISTORS

If a FET mixer is operated in the ohmic or cold mode, i.e. without a drain-source bias voltage or quasi-passively, then we can assume that the channel admittance generates thermal noise only, according to its physical temperature T. Because the value of the admittance $g(t)$ of the channel is changed periodically by the pump-signal at the gate, we can adopt the parametric Schottky diode mixer theory for the FET mixer, provided that the signals at the channel admittance are small enough. In particular, the correlation matrix is proportional to the admittance matrix. In the FET case, however, the proportionality constant is given by the physical channel temperature T. Thus one obtains for the noise figure F_{fet} the same expressions as for the Schottky diode mixer. Only the temperature $\tilde{n}T/2$ for the Schottky diode must be replaced by the temperature T of the FET channel. With the source temperature T_0 and the mixer conversion loss $L_{av} = 1/G_{av}$, which is by definition the inverse of the available gain, we obtain for the noise figure of the

FET mixer:

$$F_{fet} = 1 + \frac{T}{T_0}(L_{av} - 1) \qquad \begin{array}{c} \text{narrow-band mixer,} \\ \text{filter at the image frequency} \end{array} \qquad (5.61)$$

$$F_{fet} = 2 + \frac{T}{T_0}(L_{av} - 2) \quad \text{broadband mixer .} \qquad (5.62)$$

If the channel temperature T is equal to the generator source temperature T_0, then the relation for the noise figure further simplifies to the expression

$$F_{fet} = L_{av} = \frac{1}{G_{av}} . \qquad (5.63)$$

This latter relation holds for a mixer with a filter at the image frequency (narrow-band mixer) as well as for a broadband mixer. The relation even remains valid if additional resistive losses occur in the FET and the surrounding mixer circuit, provided that these losses also relate to the temperature T_0. Equation (5.63) thus applies to FET-mixers quite generally, in agreement with measurements.

The quasi-passive or cold operation of a FET-mixer, i.e. the operation of the mixer without a drain-source bias current, has the important advantage, as has been mentioned before, that $1/f$-noise is practically not induced. $1/f$-noise is rather pronounced in GaAs devices if a continuous current flows in the channel. The $1/f$-noise power increases approximately quadratically with an increasing continuous current.

5.8 HARMONIC MIXERS

For the harmonic mixer or sampling mixer a pump signal is employed which consists of a periodic sequence of small pulses, i.e. nearly δ-pulses, with the pulse repetition frequency f_p. Then, the admittance time function $g(t)$, which relates the small signals according to Eq. (5.2), also has a periodic and pulse-like behavior with the pump signal frequency f_p. The Fourier series coefficients of the admittance time function $g(t)$ are approximately constant up to an upper cut-off frequency f_h and approximately zero above the cut-off frequency. Thus we have

$$g(t) = \sum_{n=-N}^{N} G_n \cdot \exp(jn\omega_p t) \qquad (5.64)$$

$$\begin{array}{lll} \text{with} & |G_n| \approx \text{constant} = G_0 & \text{for} \quad n \leq N \\ \text{and} & |G_n| \approx 0 & \text{for} \quad n > N . \end{array}$$

In Fig. 5.20, a sketch of the line spectrum of such a periodic admittance time characteristic $g(t)$ is shown.

The upper frequency limit f_h only depends on the shape of the pulse, but not on the repetition frequency f_p. However, the value of $|G_n|$ decreases proportional to $1/N$, while the conversion loss as expressed by power ratios even decreases proportional to $1/N^2$. On the other hand, for a constant pulse shape, the maximum harmonic number N increases proportional to $1/f_p$. The conversion efficiencies at all harmonic spectral lines up to the maximum harmonic number N are approximately equal. One obtains an intermediate frequency signal whenever the equation

$$|f_s - n \cdot f_p| = f_i, \qquad n = 1, 2, 3 \ldots N \tag{5.65}$$

is fulfilled.

The conversion loss of a harmonic mixer is higher than of a fundamental mixer and increases with an increasing maximum harmonic number N. Thus, the conversion loss also increases for a given shape of the pulse but a lowered repetition frequency of the pump signal.

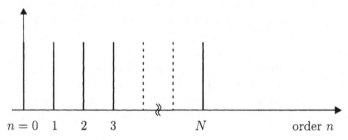

Fig. 5.20   Line spectrum of the admittance time characteristics for a harmonic mixer.

A possible circuit of a harmonic mixer is shown in Fig. 5.21 in the form of a two-wire equivalent circuit. Over a certain length l_p the ground wire is split into two parallel lines that may serve as a two-wire transmission line. This two-wire transmission line is marked with thick lines in Fig. 5.21.

The two-wire transmission line or double line forms a section of a symmetrical transmission line which is short circuited at both ends. The leading edge of the pump pulse drives the diodes into a conducting state. The pump pulse travels along the two-wire transmission line and is reflected at the short circuits and travels back with the opposite pulse polarity. By the reversed pump pulse the diodes are switched off. The travel time and thus the length l_p must be adapted to the pulse width. The signal line is isolated from the LO-pulse line over a large bandwidth. Therefore, the circuit in Fig. 5.21 is signal- and noise-balanced. The capacitors C are filter elements and separate the low intermediate frequency signals from the high frequency signals and pulses.

We should add the remark that such a harmonic mixer circuit may also serve as a sampler or sampling unit in a sampling oscilloscope. The two-wire

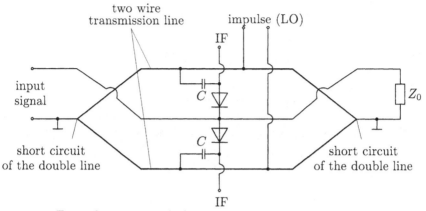

Fig. 5.21 Equivalent circuit of a harmonic mixer with two Schottky diodes.

transmission line is often realized as a slot line in the metallized ground plane of a microstrip circuit.

The realization of a harmonic mixer is even simpler with a field effect transistor (FET) as the non-linear device, because the FET is a three-terminal device. Figure 5.22 shows a harmonic mixer circuit with one FET. The pump

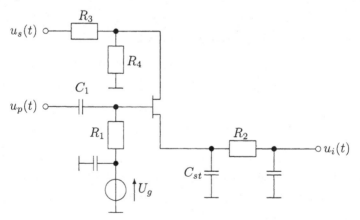

Fig. 5.22 Equivalent circuit of a harmonic mixer with one FET.

signal is applied to the gate, the RF-signal and the intermediate signal are applied to the drain or taken from the gate, respectively. The pump signal is inherently isolated from the RF-signal and the intermediate signal. The RF-signal and the intermediate signal are isolated by RC-filters. Typically, the gate bias voltage is adjusted to pinch off in the channel when no pump pulse is applied, while the pump pulses have a polarity to open the channel for a short time. This mode of operation leads to a conveniently high impedance level at the intermediate frequency.

The circuit of Fig. 5.23 utilizes two identical FETs, of which one FET is employed as a dummy component for the purpose of compensating any residual direct voltages. This occurs at the expense of an additional 3 dB insertion loss.

For the quantitative description of a harmonic mixer, we will start from an idealized equivalent circuit similar to the one in Fig. 5.6, but with a much higher number of ports to be taken into account. For the calculation of the

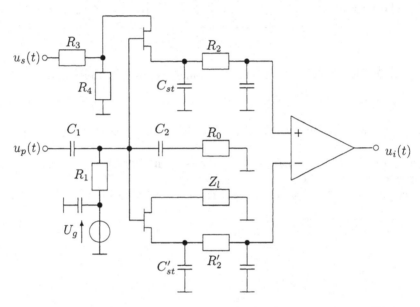

Fig. 5.23   Equivalent circuit of a harmonic mixer with two FETs.

insertion loss of a harmonic mixer we may proceed in a similar way as for the broadband mixer and start from an equation similar to Eq. (5.33), but with a higher order of the extended admittance matrix:

$$
\begin{bmatrix} I_{sg} \\ 0 \\ 0 \\ \cdots \\ 0 \end{bmatrix} = \begin{bmatrix} G_0 + Y_s & G_1 & G_2 & \cdots & G_N \\ G_1^* & G_0 + Y_i & G_1 & \cdots & \cdots \\ G_2^* & G_1^* & G_0 + \cdots & \cdots & \cdots \\ \cdots & \cdots & \cdots & \cdots & \cdots \\ G_N^* & \cdots & \cdots & \cdots & G_0 + \cdots \end{bmatrix} \cdot \begin{bmatrix} U_s \\ U_i \\ \cdots \\ \cdots \\ \cdots \end{bmatrix} \quad (5.66)
$$

Equation (5.66) is not well suited for the analytical determination of the conversion loss or gain of a harmonic mixer because the order of the matrix may be very high, up to several hundred or even a few thousand. However, the expression Eq. (5.66) is quite convenient for a numerical evaluation.

If all small signal components are terminated with the same real load admittance Y_0, then we may find a much simpler solution for the harmonic mixer in the time domain. The assumption of a common load at all frequencies might be a good approximation for a broadband harmonic mixer. The equivalent circuit for the mixer shown in Fig. 5.24 consists of only one loop.

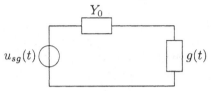

Fig. 5.24 Equivalent circuit for a harmonic mixer in the time domain.

The admittance Y_0 is the source admittance as well as the load admittance for all small signal frequency components. The admittance time function $g(t)$ is periodic with the angular pump frequency or pulse repetition frequency ω_p. We thus obtain the following relationship between the signal generator voltage u_{sg} and the small signal currents $\Delta i(t)$:

$$\Delta i(t) = \frac{g(t)\,Y_0}{g(t) + Y_0} \cdot u_{sg} = \tilde{g}(t) \cdot u_{sg} \tag{5.67}$$

with

$$\tilde{g}(t) = \frac{g(t)\,Y_0}{g(t) + Y_0} \ . \tag{5.68}$$

Because $\tilde{g}(t)$ is also a periodic function of time, it may be expanded into a Fourier series. For simplicity we assume that $\tilde{g}(t)$ is an even function of time and can be written as a Fourier series with cosine terms only.

$$\begin{aligned}
\tilde{g}(t) &= \tilde{G}_0 + \tilde{G}_1 \cdot \cos(\omega_p t) + \ldots + \tilde{G}_n \cdot \cos(n\,\omega_p t) \\
&= \sum_{n=0}^{N} \tilde{G}_n \cdot \cos(n\,\omega_p t) \ .
\end{aligned} \tag{5.69}$$

Without loss of generality we can write for the source voltage $u_{sg}(t)$

$$u_{sg}(t) = V_{sg} \cdot \cos(\omega_s t) \tag{5.70}$$

and then obtain the small signal current $\Delta i(t)$ by inserting Eq. (5.70) into Eq. (5.69)

$$\begin{aligned}
\Delta i(t) &= V_{sg} \cdot \cos(\omega_s t) \cdot \sum_{n=0}^{N} \tilde{G}_n \cdot \cos(n\,\omega_p t) \\
&= \frac{1}{2} V_{sg} \cdot \sum_{n=0}^{N} \tilde{G}_n \left(\cos[(\omega_s - n\,\omega_p)\,t] + \cos[(\omega_s + n\,\omega_p)\,t] \right) \ .
\end{aligned}$$

$$\tag{5.71}$$

If we are only interested in the spectral current component at the intermediate frequency $\omega_i = \omega_s - n\omega_p$, then one term of the sum of Eq. (5.71) is sufficient to determine the intermediate frequency current $i_i(t)$:

$$i_i(t) = \frac{1}{2} V_{sg} \, \tilde{G}_n \, \cos(\omega_i t) \; . \tag{5.72}$$

With the knowledge of the intermediate frequency current $i_i(t)$, the Fourier coefficient $\tilde{G}_n$ and the source voltage E_{sg}, it is straightforward to calculate the conversion loss or gain. One may wonder why the calculation of the gain of the harmonic mixer is so much simpler than the calculation in the frequency domain via Eq. (5.66). The answer is that the values of the Fourier coefficients G_n and $\tilde{G}_n$ are different. These Fourier coefficients are related via Eq. (5.68). Taking the fundamental mixer as an example, it can be demonstrated in which way the Fourier coefficients differ from each other.

The fundamental mixer described in the frequency domain needs one Fourier coefficient G_1 only, which for simplicity is assumed to be real:

$$g(t) = G_0 + 2\,G_1 \cdot \cos(\omega_p t) \; . \tag{5.73}$$

The Fourier series for the mixer described in the time domain is given by

$$\tilde{g}(t) = \tilde{G}_0 + 2\,\tilde{G}_1 \cdot \cos(\omega_p t) + \cdots \; . \tag{5.74}$$

According to the relation Eq. (5.68) we may formulate two equations by the method of harmonic balance, linking the Fourier coefficients $\tilde{G}_1$ and G_1:

$$\begin{aligned}
\tilde{G}_0 \,(G_0 + Y_0) &= G_0\,Y_0 - G_1\,\tilde{G}_1 \\
\tilde{G}_1 \,(G_0 + Y_0) &= 2\,Y_0\,G_1 - 2\,G_1\,\tilde{G}_0 \; .
\end{aligned} \tag{5.75}$$

These two equations can be solved for $\tilde{G}_1$:

$$\tilde{G}_1 = \frac{2\,Y_0^2\,G_1}{(G_0 + Y_0)^2 - 2\,G_1^2} \; . \tag{5.76}$$

We obtain for the voltage U_i at the intermediate frequency:

$$U_i = -\frac{I_i}{Y_0} = V_{sg}\,Y_0 \cdot \frac{G_1}{2\,G_1^2 - (G_0 + Y_0)^2} \; , \tag{5.77}$$

which is the same expression as for a calculation in the frequency domain, given by Eq. (5.18).

5.9   NOISE FIGURE OF HARMONIC MIXERS

We will assume that the harmonic mixer has the same real load admittance Y_0 at all frequencies involved. Furthermore, in this section we will neglect any bulk resistance R_b and employ an effective noise temperature $T_{ef} = \tilde{n}T/2$ for the Schottky diode. For the calculation of the noise figure we make use of the dissipation theorem and the reciprocity of the available gain. For the application of the dissipation theorem we have to feed the harmonic mixer from the intermediate frequency side, although we consider a down converter. Feeding the harmonic mixer with a sinusoidal signal at the intermediate frequency will produce a spectrum as shown in Fig. 5.25, with pairs of spectral signal lines below and above all harmonics of the pump or pulse repetition frequency ω_p. We assume that the Fourier series coefficients of the admittance time function $g(t)$ are approximately constant up to an upper cut-off frequency f_h and are approximately zero above the cut-off frequency. With constant real Fourier

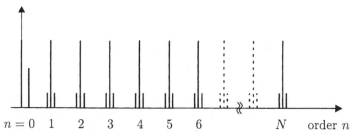

Fig. 5.25 Spectrum of a harmonic mixer fed with a sinusoidal signal from the intermediate frequency port.

coefficients up to a maximum harmonic number of N and a total of $2N - 1$ spectral signal lines we can also assume that the available conversion gain from the intermediate frequency side to all spectral signal lines is approximately the same and equal to G_{av}. Then, we obtain for the noise figure F_t of a Schottky diode harmonic mixer with a junction temperature T in accordance with Eq. (5.57):

$$F_t = 1 + \frac{(2N-1) \cdot G_{av} \cdot T_0 + (1 - 2NG_{av}) \frac{\tilde{n}}{2} T}{G_{av} \cdot T_0} . \qquad (5.78)$$

The corresponding expression for the noise figure F_t of the harmonic mixer using a field effect transistor (FET) with the channel temperature T is

$$F_t = 1 + \frac{(2N-1) \cdot G_{av} \cdot T_0 + (1 - 2NG_{av})T}{G_{av} \cdot T_0} . \qquad (5.79)$$

Finally, for the case that the channel temperature of the field effect transistor is at the ambient temperature T_0, the expression for the noise figure of the harmonic mixer simplifies to

$$F_t = \frac{1}{G_{av}} , \qquad (5.80)$$

as expected.

5.10   NOISE FIGURE MEASUREMENTS OF DOWN CONVERTERS

Since a down-converter has loss and not gain, a low noise figure of the first intermediate frequency amplifier is very important for a low overall noise figure. An alternative solution is a low noise amplifier already in front of the mixer.

For the noise figure measurement of the down-converter there is a peculiarity which has to be noted. Because the noise sources which are employed are typically broadband sources, a broadband down-converter will receive noise power from the noise generator not only at the signal frequency but also at the image frequency. Thus, the noise power of the noise generator seems to have doubled. Therefore, the measured noise figure value, measured e.g. with the Y-factor method, has to be increased by 3 dB. Except for the addition of 3 dB, the measurement of the noise figure of the mixer and a post-amplifier does not differ from the noise figure measurement of an ordinary linear two-port, apart from the fact that the input and output frequencies are not the same. The effective bandwidth is normally determined by a band-pass filter with a center frequency f_i and a bandwidth Δf at the intermediate frequency. Then, for a broadband mixer two frequency bands above and below the pump frequency at $f_p + f_i$ and $f_p - f_i$ with a bandwidth of Δf enter into the measurement. If the mixer properties are somewhat frequency dependent, then the intermediate frequency should not be too high in order to ensure that the noise figure does not change significantly between the upper and lower sideband. If the amplifier-converter stage already includes a sufficiently narrow radio frequency (RF) band-pass filter, then the measured noise figure may be the correct one and 3 dB must not be added to the measured noise figure.

5.11   NOISE FIGURE OF A PARAMETRIC AMPLIFIER

Reverse biased pn-diodes have a capacitance which is not constant but depends on the reverse bias voltage U_{rv}. These so called varactorvaractor diodes or simply varactorsvaractor, varactor diode can be applied for a number of functions in the high frequency area, e.g. for frequency up-converters and frequency multipliers of relatively high powers. Also parametric amplifiers have been realized with low noise figures. Although parametric amplifiers are no longer of practical importance, their principle of operation is of general interest. Often a voltage controlled oscillator with a varactor diode as the frequency controlling device shows instabilities due to parametric amplification. A good comprehension of the mechanism of the parametric instability in conjunction with the varactor diode may help to avoid this effect.

5.11.1 Characteristics and parameters of depletion layer varactors

The voltage swing of the varactor is assumed to occur only in the depletion region. Then, the pn-junction is mainly a variable capacitance that depends on the bias voltage U_0. We will limit the discussion to the simple case of an abrupt pn-junction with a constant doping of the p- and n-regions. For the case of a piecewise constant donor and acceptor doping, N_D and N_A, the space charge $\rho(x)$, the electric field $E(x)$ and the potential $\Phi(x)$ are shown in Fig. 5.26. Here, U_D denotes the diffusion voltage, x is the spacial coordinate in the one-dimensional model, q is the elementary charge, and ϵ is the dielectric constant in the semiconductor. Twofold integration of the Poisson equation,

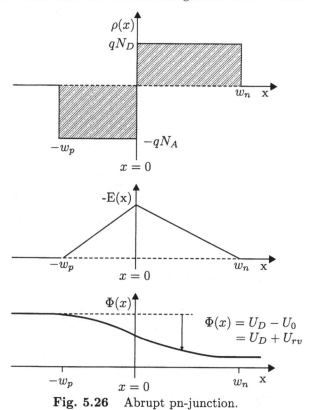

Fig. 5.26 Abrupt pn-junction.

steady field and potential behavior at $x = 0$ and vanishing electrical field in the neighboring bulk regions yields with $U_{rv} = -U_0$ a relationship between the reverse bias voltage U_{rv}, the doping levels N_D and N_A and the space charge widths w_n and w_p:

$$U_{rv} + U_D = \frac{q}{2\,\epsilon}\left(N_D \cdot w_n^2 + N_A \cdot w_p^2\right) = w_n^2 \cdot \frac{q}{2\,\epsilon}\left(N_D + \frac{N_D^2}{N_A}\right)$$

$$w_n = \sqrt{\frac{2\,\epsilon}{q}(U_{rv} + U_D)\frac{N_A}{N_D\,(N_A + N_D)}}$$

$$w_p = \sqrt{\frac{2\,\epsilon}{q}(U_{rv} + U_D)\frac{N_D}{N_A\,(N_A + N_D)}} \ . \tag{5.81}$$

From this expression one obtains for the total space charge width $w_t = w_n + w_p$:

$$w_t = w_n + w_p = \sqrt{\frac{2\,\epsilon}{q}(U_{rv} + U_D)\left(\frac{1}{N_A} + \frac{1}{N_D}\right)} \ . \tag{5.82}$$

In order to determine the differential junction capacitance $C(U_{rv})$ one has to know the charge of one polarity of the total charge, for instance the charge Q_n of the n-region:

$$Q_n = q \cdot A \cdot w_n \cdot N_D \ . \tag{5.83}$$

In the latter equation, A is the area of the semiconductor device. The differential capacitance $C(U_{rv})$ follows from a differentiation of the charge Q_n with respect to the voltage U_{rv}:

$$C(U_{rv}) = \frac{\mathrm{d}\,Q_n}{\mathrm{d}\,U_{rv}} = q \cdot A \cdot N_D \cdot \frac{\mathrm{d}\,w_n}{\mathrm{d}\,U_{rv}}$$

$$= A \cdot \sqrt{\frac{q \cdot \epsilon \cdot N_A \cdot N_D}{2\,(N_A + N_D)(U_{rv} + U_D)}} = \frac{A \cdot \epsilon}{w_t} \ . \tag{5.84}$$

Let $Q_D = Q_n(U_{rv} = 0)$ be the charge of one polarity without an applied reverse bias voltage:

$$Q_D = q \cdot A \cdot N_D \cdot \sqrt{\frac{2\,\epsilon}{q}\,U_D\,\frac{N_A}{N_D\,(N_A + N_D)}} \ . \tag{5.85}$$

Let Q_{rv} be the additional charge $Q_n - Q_D$ at a reverse bias voltage U_{rv} with $Q_{rv} \geq 0$ and $U_{rv} \geq 0$. Then we get for the additional charge Q_{rv}:

$$Q_{rv} = Q_n - Q_D = Q_D \left[\sqrt{1 + \frac{U_{rv}}{U_D}} - 1\right] \ . \tag{5.86}$$

Rearranged, this equation can be written as

$$\frac{U_{rv} + U_D}{U_D} = \left(\frac{Q_{rv} + Q_D}{Q_D}\right)^2 \ . \tag{5.87}$$

Except for additive constants this last equation establishes a quadratic relationship between the charge Q_{rv} and the voltage U_{rv}. For the differential capacitance $C(U_{rv})$ we obtain from Eq. (5.86):

$$C(U_{rv}) = \frac{\mathrm{d}\,Q_{rv}}{\mathrm{d}\,U_{rv}} = \frac{Q_D}{2\,U_D} \cdot \frac{1}{\sqrt{1 + U_{rv}/U_D}} \ . \tag{5.88}$$

Instead of the differential capacitance $C(U_{rv})$ it may sometimes be more convenient to perform the mathematical calculations with the inverse of the capacitance, the so called differential elastance $S(U_{rv})$ or $S(Q_{rv})$, because then one obtains a linear relationship between the elastance S and the charge Q_{rv} :

$$S(U_{rv}) = \frac{1}{C(U_{rv})} = 2 \cdot \left(\frac{U_D}{Q_D}\right) \cdot \left(\frac{Q_{rv} + Q_D}{Q_D}\right) = S(Q_{rv}) \ . \qquad (5.89)$$

For the reverse bias operation the capacitance of the space charge region follows any voltage changes nearly instantaneously, because the growth and the decay of the space charge region is a majority carrier effect. The maximum voltage swing is limited by the breakdown voltage U_B. The charge corresponding to U_B may be Q_B. Figure 5.27 shows the capacitance and elastance as a function of the reverse bias voltage U_{rv} or charge Q_{rv}, respectively. Losses are

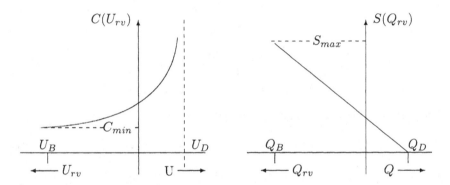

Fig. 5.27 Capacitance versus voltage and elastance versus charge for an abrupt pn-junction.

introduced by the constant, i.e. voltage independent bulk or series resistance R_b. We define a cut-off frequency f_c as the frequency, at which the maximum capacitive reactance, as given by the minimum capacitance C_{min}, is equal to R_b :

$$f_c = \frac{1}{2\pi} \frac{1}{C_{min} \cdot R_b} = \frac{S_{max}}{2\pi R_b} \ . \qquad (5.90)$$

5.11.2 Parametric operation of a varactor

Often a varactor is operated parametrically, i.e. a strong pump signal at the frequency f_p determines the instantaneous operating point for a variety of small signals. If the pump signal as described by the charge as a function of time, $Q_p(t)$, is a periodic function of time, then also the elastance $S(t)$ is a periodic function of time and can be developed into a Fourier series. For an abrupt pn-junction, $S(t)$ and $Q_p(t)$ are linearly related according to

Eq. (5.89). If $Q_p(t)$ is an even function of time, which we will always assume in this section, then $S(t)$ can be expressed as a Fourier series with cosine terms only:

$$S(t) = S_0 + 2\,S_1 \cos \omega_p\,t + 2\,S_2 \cos 2\,\omega_p\,t + \cdots \; . \tag{5.91}$$

We will assume small signal phasors for the voltage U and the charge Q at three different frequencies, namely the upper sideband at the frequency f_u with the phasors U_u and Q_u, the lower sideband at the frequency f_d with the phasors U_d and Q_d and the intermediate frequency at the frequency f_i with the phasors U_i and Q_i. The following relationships hold between the different frequencies:

$$\begin{aligned} f_u &= f_p + f_i \\ f_d &= f_p - f_i \; . \end{aligned} \tag{5.92}$$

As for the mixer, the small signal phasors are related to the real Fourier coefficients S_0, S_1, S_2 via a linear set of equations.

$$\begin{bmatrix} U_u \\ U_i \\ U_d^* \end{bmatrix} = \begin{bmatrix} S_0 & S_1 & S_2 \\ S_1 & S_0 & S_1 \\ S_2 & S_1 & S_0 \end{bmatrix} \begin{bmatrix} Q_u \\ Q_i \\ Q_d^* \end{bmatrix} \; . \tag{5.93}$$

The current and charge phasors are related by

$$I_u = j\,\omega_u \cdot Q_u \; ; \; I_i = j\,\omega_i \cdot Q_i \; ; \; I_d^* = -j\,\omega_d \cdot Q_d^* \; . \tag{5.94}$$

Then, the relation of voltage and current phasors is given by

$$\begin{bmatrix} U_u \\ U_i \\ U_d^* \end{bmatrix} = \begin{bmatrix} S_0/j\,\omega_u & S_1/j\,\omega_i & -S_2/j\,\omega_d \\ S_1/j\,\omega_u & S_0/j\,\omega_i & -S_1/j\,\omega_d \\ S_2/j\,\omega_u & S_1/j\,\omega_i & -S_0/j\,\omega_d \end{bmatrix} \begin{bmatrix} I_u \\ I_i \\ I_d^* \end{bmatrix} \; . \tag{5.95}$$

The relation between currents and voltages at the varactor is no longer reciprocal. Among others, one result from this fact is that down- and up-conversion do not have the same conversion loss or gain.

Problem

5.9   What does Eq. (5.93) look like if the pump drive is a general periodic function of time?

5.11.3 Parametric amplifier

The parametric amplifier requires the same frequency scheme as a lower sideband up-converter (Fig. 5.28).

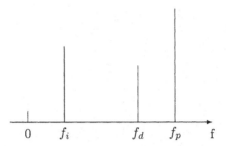

Fig. 5.28   Frequency scheme of the parametric amplifier.

The amplification occurs at the frequency f_i. One needs an auxiliary resonant circuit at the frequency f_d, which needs not be accessible from outside. At the frequency f_u one must provide a high impedance termination. Then the current at this frequency is negligibly small, i.e. $I_u = 0$. The current and voltage phasors U_i, U_d and I_i, I_d are related via an impedance matrix $[Z]$ in accordance with the matrix of Eq. (5.95):

$$\begin{bmatrix} U_i \\ U_d^* \end{bmatrix} = \begin{bmatrix} S_0/j\,\omega_i & -S_1/j\,\omega_d \\ S_1/j\,\omega_i & -S_0/j\,\omega_d \end{bmatrix} \begin{bmatrix} I_i \\ I_d^* \end{bmatrix} = [Z] \cdot \begin{bmatrix} I_i \\ I_d^* \end{bmatrix} .$$

(5.96)

The equivalent circuit of a parametric amplifier is shown in Fig. 5.29. The

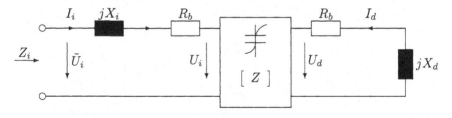

Fig. 5.29   Equivalent circuit of a parametric amplifier.

large signal drive of the elastance is assumed to be cosinusoidal at the pump frequency f_p, i.e. $S_2 = 0$. The lower sideband with the frequency f_d serves as an auxiliary circuit and is terminated by the inductive reactance jX_d. Furthermore, the input circuit is extended by the inductive reactance jX_i. Both the inductive reactances jX_d and jX_i are chosen in such a way that the capacitive reactances $S_0/j\omega_d$ and $S_0/j\omega_i$ are tuned or compensated in the form of a series resonance by jX_d and jX_i, respectively. This proves to be expedient in order to optimize the gain of the parametric amplifier.

The $\tilde{Z}$-matrix extended by jX_d and jX_i and twice R_b reads:

$$\begin{bmatrix} \tilde{U}_i \\ 0 \end{bmatrix} = \begin{bmatrix} R_b + S_0/j\,\omega_i + jX_i & -S_1/j\,\omega_d \\ S_1/j\,\omega_i & R_b - S_0/j\,\omega_d - jX_d \end{bmatrix} \begin{bmatrix} I_i \\ I_d^* \end{bmatrix}$$

$$= \begin{bmatrix} R_b & -S_1/j\,\omega_d \\ S_1/j\,\omega_i & R_b \end{bmatrix} \begin{bmatrix} I_i \\ I_d^* \end{bmatrix} . \qquad (5.97)$$

The input impedance Z_i at the frequency f_i is given by

$$Z_i = \frac{\tilde{U}_i}{I_i} = R_b \left[1 - \frac{S_1^2}{\omega_i\,\omega_d\,R_b^2} \right] . \qquad (5.98)$$

The input impedance Z_i is real and for a sufficiently large S_1 it becomes negative. The possible amplification for a negative resistance Z_i can be expressed by the reflection coefficient ρ:

$$\rho = \frac{Z_i - Z_0}{Z_i + Z_0} . \qquad (5.99)$$

In the above equation Z_0 is the characteristic impedance of the reference transmission line. For $\mathrm{Re}(Z_i) < 0$ the reflection coefficient becomes greater than one, i.e. $|\rho| > 1$, which means that the reflected wave is larger than the incident wave. One may employ a lossless transformer in order to increase the negative impedance. This, however may reduce the bandwidth. The amplification grows beyond all limits, i.e. the amplifier oscillates, if $Z_i = Z_0$. Practically, a gain of 15 dB to 20 dB is possible. Normally, the pump signal amplitude must be controlled in order to keep S_1 and thus the gain constant. The separation of the incident and reflected waves may be accomplished by a circulator, as outlined in Fig. 5.30.

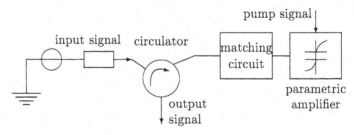

Fig. 5.30 Parametric amplifier with a circulator.

One can estimate the minimum cut-off frequency of the varactor, which is necessary to balance the negative and positive resistance. For this purpose, the expression in the parentheses of Eq. (5.98) must be negative, corresponding to

$$S_1^2 > \omega_i\omega_d \cdot R_b^2 . \qquad (5.100)$$

For a cosinusoidal charge pump at the frequency f_p, the elastance time function has a cosine form also and S_2 and S_3 are zero. The Fourier coefficient S_1 becomes maximum for the largest possible excitation, i.e. when $S(t)$ approaches the values zero and S_{max}. Then, this maximum S_1 is

$$S_{1\,max} = \frac{1}{4} S_{max} = \frac{1}{4\,C_{min}} \ . \tag{5.101}$$

The inequality Eq. (5.100) can be expressed by the cut-off frequency f_c:

$$f_c > 4\,\sqrt{f_d \cdot f_i} \ . \tag{5.102}$$

In reality, the cut-off frequency must even be higher than described by equation (5.102) in order to compensate for circuit losses and also guarantee a positive gain.

5.11.4   Noise figure of the parametric amplifier

The equivalent circuit of a parametric amplifier is extended by a thermally noisy generator with the real source impedance R_g as shown in Fig. 5.31. In this equivalent circuit, the real load impedance R_l is identical to the source impedance R_g, i.e. $R_l = R_g$. In a practical circuit, the corresponding signals must be separated, e.g. by a circulator as shown in Fig. 5.30. The only noise

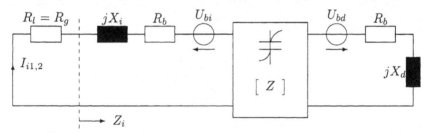

Fig. 5.31   For the explanation of the noise figure of a parametric amplifier.

considered in the parametric amplifier is the thermal noise of the series or bulk resistor R_b, namely at the auxiliary or image frequency f_d and the intermediate or generator frequency f_i. The noise contributions of the thermally noisy bulk resistance as expressed by the noise voltages U_{bi} and U_{bd} at these two frequencies are uncorrelated, because the bulk resistance is assumed to be constant versus time, i.e. not modulated by the pump signal. The non-linear capacitance is a lossless pure reactance and thus it is noise-free. With these assumptions the calculation of the noise figure does not pose any particular problem. We will again make the assumption that the two inductive reactances compensate for the capacitive reactances of the varactor so that the input impedance of the varactor is real. With Z_i from Eq. (5.103) one obtains for the squared magnitude of the current I_{i1}, caused by the noise voltage U_{bi}:

$$|I_{i1}|^2 = \frac{|U_{bi}|^2}{(R_l + Z_i)^2} .$$ (5.103)

The noise current I_{i2}, caused by the noise voltage U_{bd}, is conveniently evaluated via the following extended matrix:

$$\begin{bmatrix} 0 \\ U_{bd}^* \end{bmatrix} = \begin{bmatrix} R_l + R_b & -S_1/j\omega_d \\ S_1/j\omega_i & R_b \end{bmatrix} \cdot \begin{bmatrix} I_{i2} \\ I_d^* \end{bmatrix} .$$ (5.104)

From this extended matrix we obtain:

$$I_{i2} = \frac{S_1/j\omega_d}{(R_l + R_b) R_b - \dfrac{S_1^2}{\omega_i \omega_d}} \cdot U_{bd}^* ,$$

$$|I_{i2}|^2 = \frac{S_1^2/\omega_d^2}{\left[(R_l + R_b) R_b - \dfrac{S_1^2}{\omega_i \omega_d} \right]^2} \cdot |U_{bd}|^2$$

$$= \frac{S_1^2/\omega_d^2}{R_b^2 \cdot (R_l + Z_i)^2} \cdot |U_{bd}|^2 .$$ (5.105)

Thus, we find for the spectrum ΔW_2 of the noisy two-port at the load resistance R_l with the contributions of the noisy bulk resistance R_b at the temperature T:

$$\Delta W_2 = R_l \cdot |I_{i1}|^2 + R_l \cdot |I_{i2}|^2$$

$$= 4 k T \cdot R_b \cdot R_l \frac{1 + S_1^2 \cdot \omega_d^{-2} \cdot R_b^{-2}}{(R_l + Z_i)^2} .$$ (5.106)

The spectrum W_{20} is the amplified source spectrum supplied to the load resistance R_l via the circulator. For this calculation the parametric amplifier is assumed to be noise-free. Regarding R_g as the real reference impedance Z_0, then we have

$$W_{20} = k \cdot T_0 \cdot \left(\frac{Z_i - R_g}{Z_i + R_g} \right)^2 .$$ (5.107)

With $R_l = R_g$ the Eqs. (5.106) and (5.107) yield the noise figure F_t:

$$F_t = 1 + \frac{\Delta W_2}{W_{20}} = 1 + 4 \frac{T}{T_0} \cdot R_b R_g \left[1 + \frac{S_1^2}{\omega_d^2 \cdot R_b^2} \right] \cdot \frac{1}{(Z_i - R_g)^2} .$$ (5.108)

For a low noise figure it is apparently favorable to choose the auxiliary frequency f_d as high as possible.

Problem

5.10 Show that the gain and the noise figure become optimum, if the inductive reactances jX_i and jX_d just compensate the capacitive reactances $S_0/j\omega_i$ and $S_0/j\omega_d$.

For a correct operation, the parametric amplifier only shows the relatively low thermal noise of the bulk resistance R_b with the temperature T. By cooling the amplifier, the temperature T and thus the noise figure can be reduced. Uncooled parametric amplifiers achieve system temperatures $T_s = (F_t - 1) T_0$ of about 150 to 200 K at several GHz. Cooled amplifiers, cooled down to the temperature of liquid helium at 4.2 K, may have system temperatures as low as 5 to 10 K. The varactor diodes must be made of GaAs, because for silicon at very low temperatures there are not enough electrons available in the conduction band. For low noise figures also the circuit around the varactor diode should be cooled, in order to reduce circuit losses and thermal noise. Parametric amplifiers have been installed e.g. in satellite and astronomical receivers and some are still in operation, but are no longer considered in modern systems due to their only moderately low noise figures, their complexity and narrow bandwidth.

An oscillator tuned in frequency by a varactor diode, may show additional spurious oscillations, because unintentionally one may have realized an unstable parametric amplifier. This, however, must strictly be avoided because then additional spurious spectral lines can appear and the noise spectrum may deteriorate. For a large tuning range of a voltage controlled oscillator (VCO), the varactor diode must tightly be coupled to the oscillator, which may lead to a large voltage excursion at the diode. This is a situation which is favorable for the excitation of parametric self oscillations. Among others, stability can be improved by also terminating the sum frequency $f_u = f_p + f_i$ by a suitable impedance, which helps to reduce the gain of the parametric circuit.

5.12 UP-CONVERTERS WITH VARACTORS

The frequency scheme of the upper sideband up-converter is the same as in Fig. 5.1a. Feeding takes place at the intermediate frequency f_i and for reasons of stability the output power is taken at the upper sideband only. Small signals may be present at the intermediate frequency f_i and the upper sideband frequency f_u, while at the pump frequency $f_p = f_u - f_i$ a large voltage excursion is necessary. An equivalent circuit of the upper sideband up-converter is shown in Fig. 5.32.

Also for this circuit it proves to be advantageous to choose the inductive reactances jX_i and jX_u in such a way that the capacitive reactances of the

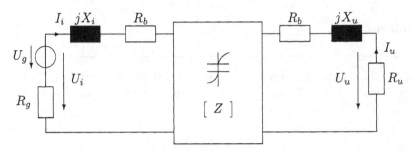

Fig. 5.32   Equivalent circuit of an upper sideband up-converter.

varactor diode $S_0/j\omega_i$ and $S_0/j\omega_u$ are compensated. From the extended matrix with the real load and source or generator impedances R_u and R_g

$$\left[\begin{array}{c} U_g \\ 0 \end{array}\right] = \left[\begin{array}{cc} R_g + R_b & S_1/j\omega_u \\ S_1/j\omega_i & R_u + R_b \end{array}\right] \cdot \left[\begin{array}{c} I_i \\ I_u \end{array}\right] \ , \qquad (5.109)$$

we can determine the gain G_p . We get

$$G_p = \frac{\left(\dfrac{S_1}{\omega_i}\right)^2 \cdot 4\,R_g \cdot R_u}{\left[(R_g + R_b)(R_u + R_b) + S_1^2(\omega_i\,\omega_u)\right]^2} \ . \qquad (5.110)$$

Up-converters with varactors show good efficiencies and may even have gain and handle large powers. They have been used when post-amplification was difficult. Also the noise figure is low. However, the circuit is narrow-band and an up-converter with Schottky diodes or field effect transistors (FETs) is usually preferred.

Problem

5.11   What is the maximum gain G_m and the noise figure F_t for the above upper sideband up-converter with a varactor diode?

6

Noise in Non-linear Circuits

6.1 INTRODUCTION

The two-ports that have been treated so far were linear with respect to the input and output signals. The parametric systems like mixers and parametric amplifiers of Chapter 5 are based on devices operated in a non-linear mode, but the relationship between the input and output phasors, although assigned to different frequencies, is still quasi linear. For all linear two-ports the noise figure or noise factor is the most commonly used quantity for the characterization of the noise behavior. The noise figure can be determined from the signal and noise powers at the input and the output of the two-port under investigation. The situation is illustrated in Fig. 6.1. A signal with the signal power

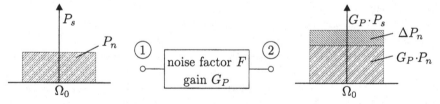

Fig. 6.1 Illustration of signal and noise powers of linear two-ports.

P_s is fed to the input of a two-port. Furthermore, the signal source supplies the noise power P_n. The operating bandwidth is assumed to be small enough that the properties of the two-port do not change within this frequency band. For a noise-free two-port with power gain G_p we get a signal power $G_p \cdot P_s$

and a noise power $G_p \cdot P_n$ at the output. Therefore, the ratio of the signal power to the noise power has not changed while passing through the linear two-port. For a noisy two-port the noise power at the output is raised by the noise contribution ΔP_n, which stems from the internal noise sources of the two-port. The noise figure is defined by

$$F = \frac{G_p P_n + \Delta P_n}{G_p P_n} = 1 + \frac{\Delta P_n}{G_p P_n} \ . \tag{6.1}$$

Here, the noise temperature of the signal source has the fixed value T_0, e.g. 290 K.

According to Eq. (6.1) the noise figure does not depend on the signal power. This statement, however, is only valid on the premise that the quantities ΔP_n and G_p are independent of the signal power. This condition is met by linear two-ports but not by non-linear ones. Together with some further effects this leads to the situation that the noise figure and similar quantities from the linear regime are not suitable for the description of non-linear networks. This will be discussed in this chapter.

6.2   PROBLEMS WITH THE NOISE CHARACTERIZATION OF NON-LINEAR TWO-PORTS

A very important practical example of a non-linear two-port is an amplifier under large signal conditions. If the input power, starting from small values, is steadily increased then the output power cannot grow forever at the same rate but will finally reach a saturation value, which mainly depends on the characteristics of the active components used. This means that the power gain will eventually decrease, once the input power has passed a certain threshold value. Then also the noise figure will depend on the input power. Generally, however, not only the power gain will change under large signal conditions but also the noise power ΔP_n generated within the two-port. There are two different reasons for this situation. Firstly, under large signal conditions, parameters may vary which directly influence the physical reason for the noise, e.g. temperatures may change and thus the thermal noise or dc currents and thus the shot noise. Secondly, a non-linear operation can lead to frequency conversions, similar to that in mixers. Generally, for a large signal drive at an angular frequency Ω_0 the noise can be shifted in frequency by $\pm N\Omega_0$ with $N = 1, 2, 3, \ldots$. Therefore, it is theoretically possible to observe noise at the output of a non-linear two-port at an angular frequency Ω_0, even without any physical noise sources existing at this frequency.

This discussion shows that the interrelations between signals and noise are much more complicated for non-linear two-ports than for linear circuits. Nevertheless, it might seem possible to use the concept of noise figures for the characterization of non-linear networks as well, if in each case the input and output signal power levels are specified, too. However, the noise signal at the output of a non-linear two-port shows two additional particularities, which are not directly covered by the noise power ΔP_n and, therefore, do not enter into the noise figure. Firstly, due to the frequency translations, spectral components at different frequencies are not necessarily uncorrelated, as is the case in linear circuits. This in particular is true for spectral components that are located symmetrically with respect to the signal frequency Ω_0. It is thus possible that a low frequency noise signal at a frequency ω with $\omega \ll \Omega_0$ gives rise to noise components at the frequencies $\Omega_0 - \omega$ and $\Omega_0 + \omega$ at the output of the two-port by mixing with the carrier signal. These noise components are fully correlated, since they have their origin in the same physical source. The detailed and quantitative knowledge of this correlation is very important for the assessment of how this noise interferes with other signals. The second particularity concerns the frequency dependence of the noise power spectral density. So far, we have dealt with noise processes with nearly white spectra, i.e. nearly frequency independent power spectral densities. This in most cases is true for the output noise of linear two-ports, if the operating bandwidth is so small that the frequency dependence within this range can be neglected. For non-linear two-ports, however, one often observes noise spectra as in Fig. 6.2, where the power spectral density close to the high-frequency or RF signal changes rapidly. For many systems the disturbing effect of the noise also depends upon the frequency difference to the RF signal, so that again a specification of the total noise power alone is not sufficient for a proper characterization of the noisy system.

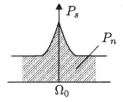

Fig. 6.2 Typical spectrum of a RF signal plus noise at the output of a non-linear two-port.

In conclusion, we can state that the noise figure is a meaningful quantity for the description of the noise behavior of linear two-ports only. For non-linear two-ports additional properties of the noise have to be considered that require a different kind of characterization. Before we introduce new parameters for this purpose we first shall discuss the reasons for the frequency dependent noise spectra as depicted schematically in Fig. 6.2.

6.3   1/F-NOISE

The physical noise processes that have been treated so far, can be described by almost white i.e. nearly frequency independent spectra. However, in many materials carrying a direct current, in particular in all semiconductor components, an additional noise mechanism occurs with a spectrum approximately inversely proportional to the frequency. All these noise mechanisms are collectively called $1/f$-**noise** or **flicker-noise**. A combination of $1/f$-noise and white noise may lead to a spectrum as in Fig. 6.3. At low frequencies the

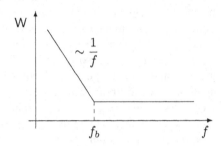

Fig. 6.3   Noise spectrum with a $1/f$-component (logarithmic scale).

$1/f$-component dominates, while at high frequencies only the white noise is apparent. The boundary between both regions, where both contributions are equal, is designated as the noise corner frequency f_b. The corner frequencies of semiconductor components range from below 1 kHz up to approximately 100 MHz. A particularly strong $1/f$-noise is observed in compound semiconductors, e.g. gallium arsenide. This semiconductor is very important for microwave applications because, e.g. amplifiers with GaAs field effect transistors can operate up to mm-wave frequencies.

For all linear applications of electronic components, e.g. small signal amplifiers, the $1/f$-noise is of no concern, if the frequency range of interest lies above the corner frequency. The situation changes drastically when non-linear effects have to be taken into account, as for power amplifiers, mixers or oscillators. By the non-linear interaction between the low frequency noise and the large high-frequency signal, the low frequency noise is upconverted resulting in lower and upper noise sidebands around the RF signal. Another view is that the RF signal is modulated by the $1/f$-noise. In this way, we can obtain spectra like the one in Fig. 6.2.

The physical origin of the $1/f$-noise still is not completely understood. Only in special cases, it has been possible to develop satisfactory models for the $1/f$-noise. The $1/f$-noise does not seem to have one unique origin. It appears that quite a number of different fluctuation processes may lead to a $1/f$ type spectrum. In particular, this frequency dependence is difficult to model. An analysis of physical fluctuation phenomena often leads to spectra of the form

$$W \sim \frac{1}{1 + (f/f_{lp})^2},\tag{6.2}$$

where f_{lp} is a characteristic corner frequency. It is possible to show, however, that in a limited frequency range the superposition of many spectra of the type of Eq. (6.2) with different corner frequencies f_{lp} may approximately lead to a $1/f$-spectrum. It is not known if there is a lower limit of the $1/f$-noise spectrum. In experimental investigations, the $1/f$-law of the noise spectrum could be confirmed even at frequencies as low as 10^{-6} Hz.

Apart from models for certain electronic components there are also theories to explain the $1/f$-noise as a universal phenomenon. An experimental observation for the power spectrum $W_i(f)$ of the low frequency fluctuations of a current I through a homogeneous metal or semiconductor sample is the approximate relation

$$W_i(f) = \frac{\alpha_H I^2}{N f} \; . \tag{6.3}$$

Here, N is the total number of free electrons in the sample, f is the frequency and α_H is a proportionality factor (Hooge's constant), which approximately equals $2 \cdot 10^{-3}$ for a number of materials. However, it does not seem to be possible to extend this relation, which is approximately valid for homogeneous samples, to more complicated structures like semiconductor components.

6.4   AMPLITUDE AND PHASE NOISE

Because the usual description of linear two-ports is not well suited for non-linear circuits, new quantities will be introduced now that are better suited to noise in non-linear circuits.

6.4.1   Noise modulation

The spectrum of a sinusoidal carrier signal with an underlying small noise spectrum, which may also include small coherent signals, typically has the form as shown in Fig. 6.2. It will be assumed throughout this chapter that the noise spectrum is band limited to the range $\Omega_0 - \Delta\Omega_0$ to $\Omega_0 + \Delta\Omega_0$. It is further assumed that the bandwidth $2\Delta\Omega_0$ of the small-signal noise-band is smaller than the carrier frequency Ω_c. If an ideal sinusoidal carrier passes through a linear or non-linear two-port, either high-frequency noise may add or by some sort of modulation, low-frequency noise may be up-converted to the carrier, e.g. by amplitude or phase modulation or both. The same may happen with small coherent signals which may also add to the noise signals. A signal $x(t)$ is supposed to consist of a sinusoidal carrier signal $X_0 \cos(\Omega_0 t + \phi_0)$ and a superimposed small noise-signal $n(t)$. The noise-signal $n(t)$ may also include spurious coherent signals but, nevertheless, we shall refer to it as a noise signal. We thus can write:

$$\begin{aligned} x(t) &= X_0 \cos(\Omega_0 t + \phi_0) + n(t) \\ &= [X_0 + \Delta x(t)] \cdot \cos[\Omega_0 t + \phi_0 + \Delta\phi(t)] \; . \end{aligned} \tag{6.4}$$

Here, the influence of the noise signal $n(t)$ on the carrier is described by random amplitude and phase fluctuations of the sinusoidal carrier. This may also be regarded as amplitude noise and phase noise of the carrier signal $x(t)$. The mean values of the amplitude and phase fluctuations are supposed to be zero:

$$\overline{\Delta x(t)} = \overline{\Delta \phi(t)} = 0 \ . \tag{6.5}$$

In complex form Eq. (6.4) reads:

$$x(t) = \text{Re}\left\{ X_0 \left[1 + \frac{\Delta x(t)}{X_0} \right] e^{j[\Omega_0 t + \phi_0 + \Delta \phi(t)]} \right\} . \tag{6.6}$$

The amplitude and phase fluctuations normally are very small with $\Delta x(t)/X_0 \ll 1$ and $\Delta \phi(t) \ll 1$. Therefore, $\Delta x(t) \cdot \Delta \phi(t)/X_0 \approx 0$, because this product is small of higher order, and also:

$$e^{j\Delta\phi(t)} \approx 1 + j\Delta\phi(t) \ . \tag{6.7}$$

With these approximations the equation (6.6) simplifies to

$$x(t) = \text{Re}\left\{ X \left[1 + \frac{\Delta x(t)}{X_0} + j\Delta\phi(t) \right] e^{j\Omega_0 t} \right\} , \tag{6.8}$$

where

$$X = X_0 e^{j\phi_0} \tag{6.9}$$

is the complex phasor of the carrier. In order to find a relation between the high-frequency noise signal $n(t)$ and the amplitude and phase fluctuations $\Delta x(t)$ and $\Delta \phi(t)$ we shall first analyze the case of small sinusoidal fluctuation signals. Later on this can easily be extended to arbitrary signal waveforms, because all fluctuation quantities are related by linear equations.

6.4.2   Sinusoidal amplitude and phase modulation

Sinusoidal amplitude and phase fluctuations with a frequency $\omega \ll \Omega$ can be described by complex phasors ΔX and $\Delta \Phi$:

$$\Delta x(t) = \text{Re}\left\{ \Delta X e^{j\omega t} \right\} = \frac{1}{2}\left(\Delta X e^{j\omega t} + \Delta X^* e^{-j\omega t} \right) , \tag{6.10}$$

$$\Delta \phi(t) = \text{Re}\left\{ \Delta \Phi e^{j\omega t} \right\} = \frac{1}{2}\left(\Delta \Phi e^{j\omega t} + \Delta \Phi^* e^{-j\omega t} \right) . \tag{6.11}$$

Inserting the equations (6.10) and (6.11) into Eq. (6.8) we obtain:

$$x(t) = \text{Re}\left\{ X e^{j\Omega_0 t} + \frac{1}{2}X\left(\frac{\Delta X^*}{X_0} + j\Delta\Phi^* \right) e^{j(\Omega_0 - \omega)t} \right.$$
$$\left. + \frac{1}{2}X\left(\frac{\Delta X}{X_0} + j\Delta\Phi \right) e^{j(\Omega_0 + \omega)t} \right\} . \tag{6.12}$$

We note that the amplitude and phase modulation leads to two sideband signals at a distance ω from the carrier. For the complex phasors X_l ($l \hat{=}$ lower sideband) and X_u ($u \hat{=}$ upper sideband) we get the following equations from Eq. (6.12):

$$X_l = \frac{X}{2}\left(\frac{\Delta X^*}{X_0} + j\Delta\Phi^*\right) \ , \quad X_u = \frac{X}{2}\left(\frac{\Delta X}{X_0} + j\Delta\Phi\right) \ , \qquad (6.13)$$

or, in matrix notation,

$$\left[\begin{array}{c} X_l^* \\ X_u \end{array}\right] = \frac{X_0}{2}\left[\begin{array}{cc} e^{-j\phi_0} & -j\,e^{-j\phi_0} \\ e^{j\phi_0} & j\,e^{j\phi_0} \end{array}\right]\left[\begin{array}{c} \dfrac{\Delta X}{X_0} \\ \Delta\Phi \end{array}\right] \ . \qquad (6.14)$$

By inversion of the matrix Eq. (6.14) the amplitude and phase fluctuations may be expressed as a function of the lower and upper sideband phasors:

$$\left[\begin{array}{c} \dfrac{\Delta X}{X_0} \\ \Delta\Phi \end{array}\right] = \frac{1}{X_0}\left[\begin{array}{cc} e^{j\phi_0} & e^{-j\phi_0} \\ e^{j\phi_0} & j\,e^{-j\phi_0} \end{array}\right]\left[\begin{array}{c} X_l^* \\ X_u \end{array}\right] \ . \qquad (6.15)$$

Generally, the relations (6.14) and (6.15) can not directly be transferred to the corresponding spectra by forming the squared magnitude of the phasors. This is due to the fact that the phasors belong to different frequencies. The problem can be avoided by assigning equivalent baseband phasors at the frequency ω to the sideband phasors at the frequencies $\Omega_0 \pm \omega$. With an electronic circuit consisting of ideal filters and an ideal mixer or analog multiplier the sideband signals can be converted to baseband signals and vice versa. Figure 6.4 shows

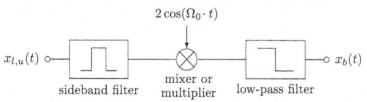

$$2\cos(\Omega_0 \cdot t)$$

$x_{l,u}(t)$ — sideband filter — mixer or multiplier — low-pass filter — $x_b(t)$

Fig. 6.4   Ideal single sideband converter.

a possible realization of a single sideband converter.single sideband converter The circuit can operate in both directions. When operated as a single sideband receiver, a high-frequency signal $x_{l,u}(t)$ is fed from the left-hand side, which may or may not include both sidebands:

$$x_{l,u}(t) = \text{Re}\left\{X_l e^{j(\Omega_0 - \omega)t} + X_u e^{j(\Omega_0 + \omega)t}\right\} \ . \qquad (6.16)$$

The sideband or bandpass filter suppresses either the lower or the upper sideband. By mixing with the local oscillator signal $2\cos\Omega_0 t$ in an ideal mixer or analog multiplier and subsequent filtering with a low-pass filter, the resulting output signal in the baseband is either

$$x_{lb}(t) = \mathrm{Re}\{X_l^* e^{j\omega t}\} \tag{6.17}$$

or

$$x_{ub}(t) = \mathrm{Re}\{X_u e^{j\omega t}\} \ , \tag{6.18}$$

depending on the sideband that is passed by the filter. In complex phasor notation we thus can write

$$X_{lb} = X_l^* \tag{6.19}$$

and

$$X_{ub} = X_u \ . \tag{6.20}$$

If the low frequency baseband signal $x_b(t)$ is the input signal to the circuit of Fig. 6.4, then the circuit operates as a single sideband modulator. The output signal is a sinusoidal signal at the frequency $\Omega_0 - \omega$ or $\Omega_0 + \omega$. The corresponding image sideband is suppressed by the band-pass filter. The relation between the complex phasors of the corresponding signals still is given by Eq. (6.19) and Eq. (6.20).

Problem

6.1   Prove the equations (6.17) through (6.20).

If in Eqs. (6.14) and (6.15) the sideband phasors X_l, X_u are replaced by the equivalent baseband phasors X_{lb}, X_{ub}, then we only deal with phasors of the same frequency ω, which are related by linear expressions. We therefore can apply the known methods for linear circuits and the corresponding transformation rules for the noise spectra.

6.4.3   Spectra of the amplitude and phase noise

Let W_{lb} and W_{ub} be the spectra of the equivalent baseband signal and let W_ϕ and W_α be the spectra of the phase noise phasor $\Delta\Phi$ and of the normalized amplitude noise phasor $\Delta X/X_0$, respectively. Then, with the corresponding cross spectra W_{lub} and $W_{\alpha\phi}$, Eqs. (6.14) and (6.15) yield:

$$X_{lb}^* \cdot X_{lb} \quad \rightarrow W_{lb} = \frac{X_0^2}{4}\left[W_\alpha + W_\phi + 2\,\mathrm{Im}\{W_{\alpha\phi}\}\right] \ , \tag{6.21}$$

$$X_{ub}^* \cdot X_{ub} \quad \rightarrow W_{ub} = \frac{X_0^2}{4}\left[W_\alpha + W_\phi - 2\,\mathrm{Im}\{W_{\alpha\phi}\}\right] \ , \tag{6.22}$$

$$\frac{\Delta X^* \cdot \Delta X}{X_0^2} \quad \rightarrow W_\alpha = \frac{1}{X_0^2}\left[W_{lb} + W_{ub} + 2\,\mathrm{Re}\left\{e^{-j2\phi_0} W_{lub}\right\}\right] \ , \tag{6.23}$$

$$\Delta\Phi^* \cdot \Delta\Phi \quad \rightarrow W_\phi = \frac{1}{X_0^2}\left[W_{lb} + W_{ub} - 2\,\mathrm{Re}\left\{e^{-j2\phi_0} W_{lub}\right\}\right] \ . \tag{6.24}$$

Notice: Although in the above equations (6.21) through (6.24) double-sided spectra have been used, the relations are valid for positive frequencies only. For negative frequencies some signs would have to be changed. This is due to the fact that for negative frequencies the matrix elements in the equations (6.14) and (6.15) must be replaced by their complex conjugates. In order to simplify the equations, only the version for positive frequencies will be given in the following.

Problem

6.2   Prove the equations (6.21) through (6.24).

Finally, the relation between the baseband spectra W_{lb}, W_{ub} and the spectrum W_n of the high-frequency noise signal $n(t)$ is of interest. Similar to the description of the sinusoidal sideband signals it is also appropriate to split the noise signal $n(t)$ into a lower and an upper sideband signal:

$$n(t) = n_l(t) + n_u(t) \ . \tag{6.25}$$

The corresponding spectra W_{lb}, W_{ub} and W_n are related by

$$W_l(\Omega) \quad = \quad \begin{cases} W_n(\Omega) & \text{for } \Omega < \Omega_0 \\ 0 & \text{else} \end{cases} \ , \tag{6.26}$$

$$W_u(\Omega) \quad = \quad \begin{cases} W_n(\Omega) & \text{for } \Omega > \Omega_0 \\ 0 & \text{else} \end{cases} \ . \tag{6.27}$$

The relation between the high-frequency spectra W_l, W_u and the corresponding baseband spectra W_{lb}, W_{ub} can be calculated by means of the ideal single sideband converter of Fig. 6.4. Instead of a rigorous derivation only an illustrative explanation will be given here. The single sideband converter links the baseband components at the frequency ω to the sideband signals at the frequencies $\Omega_0 - \omega$ or $\Omega_0 + \omega$ in a quasi linear way. Therefore, we may expect the relationships:

$$W_l(\Omega_0 - \omega) = W_{lb}(\omega) \tag{6.28}$$

and

$$W_u(\Omega_0 + \omega) = W_{ub}(\omega) \ . \tag{6.29}$$

This is confirmed by a more rigorous treatment. The relations between the different spectra are shown schematically in Fig. 6.5.

By inserting the equations (6.26) to (6.29) into (6.21) to (6.24) we obtain the required relationships between the high-frequency noise and the amplitude and phase noise:

$$W_n(\Omega_0 - \omega) \quad = \quad \frac{X_0^2}{4} \left[W_\alpha(\omega) + W_\phi(\omega) + 2\operatorname{Im}\{W_{\alpha\phi}(\omega)\} \right] \ , \tag{6.30}$$

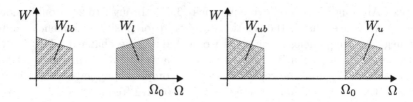

Fig. 6.5 High-frequency noise spectra W_l, W_u and the corresponding baseband spectra W_{lb}, W_{ub}.

$$W_n(\Omega_0 + \omega) = \frac{X_0^2}{4} \left[W_\alpha(\omega) + W_\phi(\omega) - 2 \,\mathrm{Im}\{W_{\alpha\phi}(\omega)\} \right] , \quad (6.31)$$

$$W_\alpha(\omega) = \frac{1}{X_0^2} \left[W_n(\Omega_0 - \omega) + W_n(\Omega_0 + \omega) \right.$$

$$\left. + 2 \,\mathrm{Re}\left\{ e^{-j2\phi_0} W_{lub}(\omega) \right\} \right] , \quad (6.32)$$

$$W_\phi(\omega) = \frac{1}{X_0^2} \left[W_n(\Omega_0 - \omega) + W_n(\Omega_0 + \omega) \right.$$

$$\left. - 2 \,\mathrm{Re}\left\{ e^{-j2\phi_0} W_{lub}(\omega) \right\} \right] . \quad (6.33)$$

Only for the cross-spectrum W_{lub} there is no equivalent expression, because signals with different and non overlapping frequency ranges, in this case the lower and upper high-frequency noise sidebands, are always uncorrelated according to the mathematical definition. A comparison of the equations (6.30) to (6.33) with the equations (6.14) and (6.15) shows, that again we can describe the carrier amplitude and phase noise or the carrier sideband noise by complex phasors. The squared magnitudes of these phasors are equivalent to the corresponding noise spectra. We may observe, however, that in contrast to linear systems we often calculate spectra by combing phasors of different frequency bands. Furthermore, as an example, the phasor product $X_l X_u$ and not $X_l^* X_u$ stands for the cross-spectrum W_{lub}. With the power spectra $W_\alpha(\omega)$ and $W_\phi(\omega)$ of the amplitude and phase noise as well as their cross-spectrum $W_{\alpha\phi}$ the noise properties of a non-linear two-port are described completely and much more in detail than by the concept of the noise figure. The frequency ω is named the baseband or offset frequency. Normally, all spectra will change, if one parameter of the input signal is changed, e.g. the frequency, the amplitude, or the form of the signal.

6.5   NORMALIZED SINGLE SIDEBAND NOISE POWER DENSITY

As has been shown in the preceding sections, the noise behavior of a non-linear two-port with a noise-free and precisely specified periodic input signal or an input signal with known noise fluctuations is fully characterized by the noise spectra W_α, W_ϕ and $W_{\alpha\phi}$. However, these quantities can not be measured directly by a sensitive spectrum analyzer, as is possible for the highfrequency noise spectra. According to the equations (6.30) and (6.31) the noise sidebands depend upon the amplitude and phase noise in a rather complicated way. The relations simplify essentially, if one of the two kinds of fluctuations can be neglected. Quite often the phase noise dominates and the amplitude noise can be neglected. In this case, the equations (6.30) and (6.31) simplify to

$$W_n(\Omega_0 - \omega) = W_n(\Omega_0 + \omega) = \frac{X_0^2}{4}W_\phi(\omega) \qquad (6.34)$$

or for a one-sided noise spectrum to

$$\mathsf{W}_n(\Omega_0 \pm \omega) = \frac{X_0^2}{4}\mathsf{W}_\phi(\omega) \ . \qquad (6.35)$$

The term $X_0^2/2$ is a measure of the power P_c of the carrier, $\mathsf{W}_n(\Omega_0 \pm \omega)$ is a measure of the noise power P_{ssb} of one sideband in 1 Hz bandwidth and at an offset-frequency ω from the carrier. Figure 6.6 schematically shows the powers

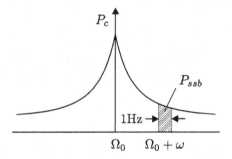

Fig. 6.6   Illustration of the normalized single sideband noise power.

P_c and P_{ssb} within the spectrum of a large sinusoidal signal plus noise. The ratio P_{ssb}/P_c is the normalized single sideband noise power. With Eq. (6.35) we obtain for a carrier with phase noise:

$$\left(\frac{P_{ssb}}{P_c}\right)_\phi = \frac{1}{2}\mathsf{W}_\phi(\omega) = W_\phi(\omega) \ . \qquad (6.36)$$

A similar relation may be given for the case where the noise sidebands are caused by amplitude fluctuations only:

$$\left(\frac{P_{ssb}}{P_c}\right)_\alpha = \frac{1}{2}\mathsf{W}_\alpha(\omega) = W_\alpha(\omega) \ . \qquad (6.37)$$

The equations (6.36) and (6.37) may also be regarded as definitions for the equivalent normalized single sideband noise power of the amplitude and phase noise, when both kinds of fluctuations are present simultaneously. The noise powers then are no longer identical to the physical high-frequency noise sidebands. However, the ratios $(P_{ssb}/P_c)_\alpha$ and $(P_{ssb}/P_c)_\phi$ can be used for the characterization of the carrier noise instead of the spectra W_α and W_ϕ, as is often done in practice. The power ratios are usually given logarithmically in dBc/Hz as a function of the offset frequency ω. The correlation between amplitude and phase noise usually is not specified in data sheets. The cross-spectrum $W_{\alpha\phi}$ is less important and is also difficult to measure.

6.6   AMPLITUDE AND PHASE NOISE OF AMPLIFIERS

In the following, we will discuss the amplitude noise and the phase noise of amplifiers for various drive levels. If the noise-free sinusoidal input signal with the angular frequency Ω_0 and the amplitude X_0 is small enough so that non-linear effects will not occur in the amplifier, then only a white noise signal with the constant spectral density W_0 is superimposed to the carrier. Since there is no interaction between the carrier and the noise signal, the noise spectrum remains unchanged and we obtain for the noise sidebands

$$W_n(\Omega_0 - \omega) = W_n(\Omega_0 + \omega) = W_0 = \text{const.} \qquad (6.38)$$

Furthermore, both sidebands are independent of each other, because no coupling occurs via the carrier. Therefore, the equivalent baseband signals are uncorrelated:

$$W_{lub}(\omega) = 0 \ . \qquad (6.39)$$

With the equations (6.32), (6.33) and (6.38), (6.39) we obtain for the spectra of the amplitude and phase fluctuations the simple result

$$W_\alpha(\omega) = W_\phi(\omega) = \frac{2}{X_0^2} W_0 \ . \qquad (6.40)$$

Thus we see that the superposition of white noise to a sinusoidal carrier leads to amplitude and phase noise with identical white spectra. Starting from Eq. (6.15) we are also able to calculate the cross-spectrum of the amplitude and phase noise. We obtain:

$$\left(\frac{\Delta X}{X_0}\right)^* \Delta\Phi = \frac{1}{X_0^2}\left[j|X_l|^2 - j|X_u|^2 - je^{-j2\phi_0}X_lX_u + je^{j2\phi_0}X_l^*X_u^*\right] \ . \qquad (6.41)$$

With the equations (6.38) and (6.39) the result for the cross-spectrum from Eq. (6.41) is

$$W_{\alpha\phi}(\omega) = \frac{1}{X_0^2}\left[jW_n(\Omega_0 - \omega) - jW_n(\Omega_0 + \omega)\right] = 0 \ . \qquad (6.42)$$

We thus conclude that, for the assumptions made, the phase noise and the amplitude noise are completely uncorrelated.

Within the limits of linear amplification the amplitude and phase noise may also be expressed by the noise figure F. If both the carrier and the noise signals are referred to the amplifier input port, then $X_0^2/2$ corresponds to the carrier input power P_{in} and FkT_0 is the one-sided spectral power density of the white noise, which is referred to the carrier at the input port. Then

$$W_\alpha(\omega) = W_\phi(\omega) = \frac{1}{2}\frac{FkT_0}{P_{in}} \qquad (6.43)$$

or with the equations (6.36), (6.37)

$$\left(\frac{P_{ssb}}{P_c}\right)_\alpha = \left(\frac{P_{ssb}}{P_c}\right)_\phi = \frac{1}{2}\frac{FkT_0}{P_{in}} \ . \qquad (6.44)$$

For example, for a noise figure of $F = 2 \hat{=} 3$ dB and $P_{in} = 1\mu\,\mathrm{W} \hat{=} -30$ dBm Eq. (6.44) yields -144 dBc/Hz, independent of the offset frequency ω. Apparently, the amplitude and phase noise can be reduced by increasing the input power. A limit is reached when the amplifier is driven into saturation and thus the linear range is left.

In the strongly non-linear region, the amplifier may approximately be treated as if the carrier is modulated in amplitude and phase by a single low frequency noise process. Therefore, one may write for the fluctuations in complex form:

$$\frac{\Delta X}{X_0} = m_\alpha M \ , \qquad (6.45)$$

$$\Delta\Phi = m_\phi M \ . \qquad (6.46)$$

Here, the phasor M describes the low frequency noise process, which is linked to the amplitude and phase noise of the carrier by the complex frequency-independent modulation factors m_α and m_ϕ. With $|M^2| \hat{=} W_m$ the spectra are given by

$$W_\alpha(\omega) = |m_\alpha|^2 W_m(\omega) \ , \qquad (6.47)$$

$$W_\phi(\omega) = |m_\phi|^2 W_m(\omega) \ . \qquad (6.48)$$

Since both kinds of fluctuations are caused by the same noise source, we expect that the amplitude and phase noise are perfectly correlated. This is confirmed by a direct calculation via the equations (6.45) and (6.46). From $(\Delta X/X_0)^* \Delta\Phi$ we conclude

$$W_{\alpha\phi}(\omega) = m_\alpha^* m_\phi W_m(\omega) \ , \qquad (6.49)$$

and the normalized cross-spectrum

$$\frac{W_{\alpha\phi}(\omega)}{\sqrt{W_\alpha(\omega)W_\phi(\omega)}} = \frac{m_\alpha^* m_\phi}{|m_\alpha||m_\phi|} \qquad (6.50)$$

has a magnitude of one.

With the equations (6.30) and (6.31) the spectra of the high-frequency noise sidebands are obtained as

$$W_n(\Omega_0 - \omega) = \frac{X_0^2}{4} \left[|m_\alpha|^2 + |m_\phi|^2 + 2\text{Im}\{m_\alpha^* m_\phi\} \right] W_m(\omega) , \quad (6.51)$$

$$W_n(\Omega_0 + \omega) = \frac{X_0^2}{4} \left[|m_\alpha|^2 + |m_\phi|^2 - 2\text{Im}\{m_\alpha^* m_\phi\} \right] W_m(\omega) . \quad (6.52)$$

The imaginary part of the product $m_\alpha^* m_\phi$ disappears if the amplitude and phase modulation have the same phase. Then, the high-frequency noise spectrum is symmetrical with respect to the carrier frequency Ω_0. If one of the two kinds of modulation dominates, the spectrum also is symmetrical. In general, however, the noise sidebands may differ in magnitude. With the help of Eq. (6.14) we are able to determine the cross-spectrum of the equivalent base band signals. From

$$X_l X_u = \frac{X_0^2}{4} \left[e^{j2\phi_0} \left| \frac{\Delta X}{X_0} \right|^2 - e^{j2\phi_0} |\Delta\Phi|^2 \right.$$
$$\left. + j\, e^{j2\phi_0} \left(\frac{\Delta X}{X_0} \right)^* \Delta\Phi + j\, e^{j2\phi_0} \frac{\Delta X}{X_0} \Delta\Phi^* \right] \quad (6.53)$$

we get for the cross-spectrum from the equations (6.47) to (6.49)

$$W_{lub}(\omega) = \frac{X_0^2}{4} e^{j2\phi_0} \left[|m_\alpha|^2 - |m_\phi|^2 + 2j\text{Re}\{m_\alpha^* m_\phi\} \right] W_m(\omega) . \quad (6.54)$$

From equations (6.51), (6.52) and (6.54) the normalized cross-spectrum can be calculated:

$$\frac{W_{lub}(\omega)}{\sqrt{W_n(\Omega_0 - \omega) W_n(\Omega_0 + \omega)}}$$
$$= e^{j2\phi_0} \frac{|m_\alpha|^2 - |m_\phi|^2 + 2j\text{Re}\{m_\alpha^* m_\phi\}}{\sqrt{(|m_\alpha|^2 + |m_\phi|^2)^2 - 4\text{Im}^2\{m_\alpha^* m_\phi\}}} . \quad (6.55)$$

A more detailed analysis of Eq. (6.55) shows that also the normalized cross-spectrum of the noise sidebands has a unit magnitude, as was to be expected from the analogy to the spectrum $W_{\alpha\phi}$.

Problem

6.3 Prove that the normalized cross-spectrum of Eq. (6.55) has a magnitude of one.

Finally, the normalized single sideband noise powers are obtained from equations (6.36), (6.37) and (6.47), (6.48):

$$\left(\frac{P_{ssb}}{P_c}\right)_\alpha = |m_\alpha|^2 W_m(\omega) , \tag{6.56}$$

$$\left(\frac{P_{ssb}}{P_c}\right)_\phi = |m_\phi|^2 W_m(\omega) . \tag{6.57}$$

Physically, the spectrum $W_m(\omega)$ often corresponds to the $1/f$-noise of semiconductor components within the amplifier. Then, also the amplitude and phase fluctuations show a $1/f$-spectrum. For a strongly non-linear mode of operation, the input power only has a minor influence on the noise of the output signal. A practical large signal amplifier is usually operated somewhere between the extreme cases regarded here. Therefore, the amplitude and phase noise in general will partly be correlated and the spectra will show a $1/f$-region as well as a frequency independent region at higher offset frequencies.

6.7 TRANSFORMATION OF AMPLITUDE AND PHASE NOISE IN LINEAR TWO-PORTS

Up to now we have assumed that the amplitude and phase noise fluctuations which appear at the output of a two-port are entirely caused by noise processes within the two-port, i.e. we assumed that the input carrier signal is noise-free. This situation will never occur in practice. At least a certain noise floor will be superimposed on the input signal, for instance the thermal noise from the signal source. Generally, amplitude and phase fluctuations are altered when the signal passes through the system, if this system is frequency dependent. Therefore, the amplitude and phase fluctuations of the output signal consist of a noise contribution from the two-port itself and a contribution by the input signal.

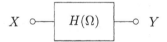

Fig. 6.7   Linear system with the transfer function $H(\Omega)$.

In this section, a linear system with the transfer function $H(\Omega)$ as depicted in Fig. 6.7 will be considered. For the derivation of a conversion matrix we start from Eq. (6.14), assuming that the system itself is noise-free. With the phase angle ϕ_0 and the magnitude X_0 of the complex input carrier signal $X_c = X_0 \exp(j\phi_0)$ the likewise complex amplitude and phase fluctuations $\Delta X/X_0$ and $\Delta\Phi$ can be transformed into the complex sideband phasors X_l

and X_u:

$$
\begin{bmatrix} X_l^* \\ X_u \end{bmatrix} = \frac{X_0}{2} \begin{bmatrix} e^{-j\phi_0} & -j\,e^{-j\phi_0} \\ e^{j\phi_0} & j\,e^{j\phi_0} \end{bmatrix} \begin{bmatrix} \dfrac{\Delta X}{X_0} \\ \Delta\Phi \end{bmatrix} . \tag{6.58}
$$

A similar relation holds for the complex carrier signal at the output $Y_c = Y_0 \exp(j\psi_0)$ with the complex amplitude and phase fluctuations $\Delta Y/Y_0$ and $\Delta\Psi$:

$$
\begin{bmatrix} Y_l^* \\ Y_u \end{bmatrix} = \frac{Y_0}{2} \begin{bmatrix} e^{-j\psi_0} & -j\,e^{-j\psi_0} \\ e^{j\psi_0} & j\,e^{j\psi_0} \end{bmatrix} \begin{bmatrix} \dfrac{\Delta Y}{Y_0} \\ \Delta\Psi \end{bmatrix} . \tag{6.59}
$$

Since the network with the transfer function H is linear, the corresponding sidebands are at the same frequency and are linked via the transfer functions H_l and H_u, while the carrier signals at the input and output of the network are linked by the transfer function H_c:

$$
Y_l = H(\Omega_0 - \omega)X_l = H_l X_l \ , \tag{6.60}
$$

$$
Y_u = H(\Omega_0 + \omega)X_u = H_u X_u \ , \tag{6.61}
$$

$$
Y_0 e^{j\psi_0} = H(\Omega_0)X_0 e^{j\phi_0} = H_c X_0 e^{j\phi_0} \ . \tag{6.62}
$$

From equations (6.58) to (6.61) we get

$$
\begin{bmatrix} \dfrac{\Delta Y}{Y_0} \\ \Delta\Psi \end{bmatrix} = \frac{1}{2}\frac{X_0}{Y_0} \begin{bmatrix} e^{j\psi_0} & e^{-j\psi_0} \\ je^{j\psi_0} & -je^{-j\psi_0} \end{bmatrix} \begin{bmatrix} H_l^* \left(e^{-j\phi_0}\dfrac{\Delta X}{X_0} - j\,e^{-j\phi_0}\,\Delta\Phi \right) \\ H_u \left(e^{j\phi_0}\dfrac{\Delta X}{X_0} + j\,e^{j\phi_0}\,\Delta\Phi \right) \end{bmatrix}
$$

and with Eq. (6.62)

$$
\begin{bmatrix} \dfrac{\Delta Y}{Y_0} \\ \Delta\Psi \end{bmatrix} = \frac{e^{j(\psi_0 - \phi_0)}}{2H_c} \Bigg[\left(H_l^*\, e^{j(\psi_0 - \phi_0)} + H_u\, e^{-j(\psi_0 - \phi_0)} \right)\frac{\Delta X}{X_0}
$$

$$
j\left(H_l^*\, e^{j(\psi_0 - \phi_0)} - H_u\, e^{-j(\psi_0 - \phi_0)} \right)\frac{\Delta X}{X_0}
$$

$$
+ j\left(-H_l^*\, e^{j(\psi_0 - \phi_0)} + H_u\, e^{-j(\psi_0 - \phi_0)} \right)\Delta\Phi
$$

$$
+ j\left(H_l^*\, e^{j(\psi_0 - \phi_0)} + H_u\, e^{-j(\psi_0 - \phi_0)} \right)\Delta\Phi \Bigg] . \tag{6.63}
$$

With

$$
e^{j2(\psi_0 - \phi_0)} = \left(\frac{H_c}{|H_c|} \right)^2 = \frac{H_c^2}{H_c H_c^*} = \frac{H_c}{H_c^*} \tag{6.64}
$$

one finally obtains:

$$
\begin{bmatrix} \dfrac{\Delta Y}{Y_0} \\ \Delta \Psi \end{bmatrix} = \frac{1}{2} \begin{bmatrix} \dfrac{H_u}{H_c} + \dfrac{H_l^*}{H_c^*} & j\left(\dfrac{H_u}{H_c} - \dfrac{H_l^*}{H_c^*} \right) \\ -j\left(\dfrac{H_u}{H_c} - \dfrac{H_l^*}{H_c^*} \right) & \dfrac{H_u}{H_c} + \dfrac{H_l^*}{H_c^*} \end{bmatrix} \begin{bmatrix} \dfrac{\Delta X}{X_0} \\ \Delta \Phi \end{bmatrix} . \qquad (6.65)
$$

In this way, all elements of the conversion matrix can be calculated from the transfer function. We may observe from Eq. (6.65) that the amplitude and phase fluctuations of the input signal are transferred to the output by the same factor. The two factors for the AM-PM- and PM-AM-conversion differ by the sign only. For a symmetrical transfer function with respect to the carrier frequency, i.e. for $H_u/H_c = (H_l/H_c)^*$, there is no mutual conversion with respect of the two kinds of fluctuations. The same is true for a frequency independent network.

6.8 TRANSFORMATION OF AMPLITUDE AND PHASE NOISE IN NON-LINEAR TWO-PORTS

6.8.1 Conversion matrix

In this section, the transformation of the amplitude and phase noise in a non-linear system will be discussed. Again, the noise fluctuations of the output signal consist of a noise contribution by the non-linear network in addition to the noise contribution from the input signal. However, by passing through the non-linear two-port, these fluctuations do not remain constant but are subject to a certain change, with a possible conversion of amplitude noise to phase noise and vice versa. If we denote the fluctuations of the input signal by $\Delta X/X_0$ and $\Delta \Phi$, and those of the output signal by $\Delta Y/Y_0$ and $\Delta \Psi$, then we can express the relation between these quantities by the following matrix equation:

$$
\begin{bmatrix} \dfrac{\Delta Y}{Y_0} \\ \Delta \Psi \end{bmatrix} = \begin{bmatrix} K_{\alpha\alpha} & K_{\alpha\phi} \\ K_{\phi\alpha} & K_{\phi\phi} \end{bmatrix} \begin{bmatrix} \dfrac{\Delta X}{X_0} \\ \Delta \Phi \end{bmatrix} + \begin{bmatrix} \dfrac{\Delta Y_n}{Y_0} \\ \Delta \Psi_n \end{bmatrix} . \qquad (6.66)
$$

The elements of the conversion matrix $[K]$ describe the way in which the fluctuations of the input signal are changed by passing through the non-linear network, while the quantities $\Delta Y_n/Y_0$ and $\Delta \Psi_n$ describe the contribution of the internal noise of the two-port. These contributions have been dealt with in a preceding section. In the following, we will discuss some of the properties of the conversion matrix. The discussion is not only valid for noisy signals but also includes the case that the fluctuations are caused by a deterministic modulation, intentional or not, as long as the fluctuations are small, i.e. as long as $|\Delta X/X_0| \ll 1$ and $\Delta \Phi \ll 1$.

While linear two-ports can be described by a complex transfer function, non-linear networks can be characterized by the so-called **describing function**. For the definition, we assume at the input a perfect sinusoidal signal of angular frequency Ω_0 with the complex amplitude X. Due to the non-linear properties of the two-port the output signal normally is no longer sinusoidal, however, it remains periodic with the angular frequency Ω_0 and it will typically include higher harmonics. The output signal can therefore be described as a Fourier series with a complex amplitude Y for the fundamental angular frequency Ω_0. We will again denote Y as a phasor quantity. The ratio of the complex phasors X and Y defines the describing function, denoted by D:

$$D = \frac{Y}{X} \ . \tag{6.67}$$

The describing function does not directly depend on the higher harmonics of the output signal. In contrast to the transfer function of linear two-ports, the describing function not only is a function of the frequency but also of the amplitude of the input signal. With $X_0 = |X|$ we can write for the describing function D with respect to the amplitude and phase:

$$D = D_0(X_0, \Omega_0) e^{j\varrho(X_0, \Omega_0)} \ . \tag{6.68}$$

D_0 is defined as a real-valued quantity.

Problem

6.4   A piecewise linear relation as shown in the figure below shall describe the relation between the input voltage $u_{in}(t)$ and the output voltage $u_{out}(t)$ of an amplifier. Calculate the describing function in dependence of the amplitude X_0 of the input signal.

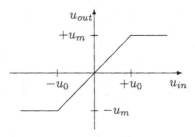

For the amplitude and phase noise, the interesting angular offset frequencies ω usually are small compared to the angular carrier frequency Ω_0. Because we consider active non-linear devices that have a relatively large bandwidth around the center frequency Ω_0 and due to the assumption that $\omega \ll \Omega_0$ we shall neglect the frequency dependence of the active non-linear two-port in the small frequency region of $\Omega_0 \pm \omega$. With $X = X_0 e^{j\phi_0}$ and $Y = Y_0 e^{j\psi_0}$ the amplitude and phase of the input signal and the amplitude and phase of the fundamental frequency component of the output signal are related by

$$Y_0 = D_0(X_0) \cdot X_0 , \tag{6.69}$$

$$\psi_0 = \phi_0 + \varrho(X_0) . \tag{6.70}$$

If the amplitude and phase of the input signal change by small amounts Δx and $\Delta \phi$, then the resulting changes Δy and $\Delta \psi$ of the output signal may be calculated by a linear Taylor approximation of the describing function:

$$Y_0 + \Delta y = \left[D_0(X_0) + \left. \frac{dD_0}{dX_0} \right|_{\hat{X}_0} \cdot \Delta x \right] (X_0 + \Delta x) , \tag{6.71}$$

$$\psi_0 + \Delta \psi = \phi_0 + \Delta \phi + \varrho(X_0) + \left. \frac{d\varrho}{dX_0} \right|_{\hat{X}_0} \cdot \Delta x . \tag{6.72}$$

With Eqs. (6.69), (6.70) and by neglecting the term $(\Delta x)^2$ in Eq. (6.71), which is of higher order small, we obtain

$$\Delta y = D_0(X_0) \left[1 + \left. \frac{X_0}{D_0(X_0)} \frac{dD_0}{dX_0} \right|_{\hat{X}_0} \right] \Delta x , \tag{6.73}$$

$$\Delta \psi = \Delta \phi + \left. \frac{d\varrho}{dX_0} \right|_{\hat{X}_0} \Delta x . \tag{6.74}$$

Assuming that the relations (6.73) and (6.74), which were derived for static amplitude and phase variations, are also valid for low-frequency time-variant fluctuations, the elements of the conversion matrix directly follow as

$$K_{\alpha\alpha} = 1 + \left. \frac{X_0}{D_0(X_0)} \frac{dD_0}{dX_0} \right|_{\hat{X}_0} , \tag{6.75}$$

$$K_{\alpha\phi} = 0 , \tag{6.76}$$

$$K_{\phi\alpha} = \left. X_0 \frac{d\varrho}{dX_0} \right|_{\hat{X}_0} , \tag{6.77}$$

$$K_{\phi\phi} = 1 . \tag{6.78}$$

With derivations with respect to amplitude and phase of the describing function, we can define an amplitude compression factor k_α and an AM-PM conversion factor k_ϕ:

$$k_\alpha = - \left. \frac{X_0}{D_0(X_0)} \frac{dD_0}{dX_0} \right|_{\hat{X}_0} , \tag{6.79}$$

$$k_\phi = \left. X_0 \frac{d\varrho}{dX_0} \right|_{\hat{X}_0} . \tag{6.80}$$

We can thus write the conversion matrix in the following form:

$$[K] = \begin{bmatrix} 1 - k_\alpha & 0 \\ k_\phi & 1 \end{bmatrix} .$$

$$(6.81)$$

In the linear case we have $k_\alpha = k_\phi = 0$. The amplitude and phase fluctuations of the input signal are not changed by the two-port, under the assumption, however, that the frequency dependence of the two-port can be neglected.

Problem

6.5   Determine the amplitude compression factor k_α for the amplifier example of the problem (6.4). What is the value of k_ϕ?

6.8.2   Large signal amplifiers

For an amplifier under large signal conditions, the amplitude compression is the main effect of the non-linear operation. The AM-PM conversion may be important since it can cause crosstalk, if the carrier is subject to a combined amplitude and phase modulation. For noise considerations the conversion factor k_ϕ is of minor importance. In practical systems the phase noise usually dominates. Then it is of no concern if a part of the much smaller amplitude noise is converted into phase noise. For this reason, the AM-PM conversion factor k_ϕ will be neglected in the following.

For a large number of amplifiers, the dependence of the power gain G_p as a function of the signal input power P_s may approximately be described by the following empirical equation:

$$G_p(P_s) = \frac{P_{sat}}{P_s} \left[1 - \exp\left(-\frac{G_0 P_s}{P_{sat}} \right) \right] .$$

$$(6.82)$$

Here, G_0 is the small signal power gain and P_{sat} the saturation value of the output power. For $P_s \ll P_{sat}/G_0$ we get $G_p = G_0$. With Eq. (6.82) we are able to determine the compression factor k_α. If we assume a real input impedance Z_i of the amplifier and a voltage amplitude X_i of the input signal, then $P_s = X_i^2/(2Z_i)$. The corresponding describing function of the amplifier follows from the equation (6.82):

$$D_0(X_i) = \sqrt{\frac{2 Z_i P_{sat}}{X_i^2} \left[1 - \exp\left(-\frac{G_0 X_i^2}{2 Z_i P_{sat}} \right) \right]} .$$

$$(6.83)$$

With Eqs. (6.79) and (6.83) the amplitude compression factor k_α is obtained as:

$$k_\alpha = \frac{\dfrac{2Z_i P_{sat}}{X_i^2}\left[1 - \exp\left(-\dfrac{G_0 X_i^2}{2Z_i P_{sat}}\right)\right] - G_0 \exp\left(-\dfrac{G_0 X_i^2}{2Z_i P_{sat}}\right)}{\dfrac{2Z_i P_{sat}}{X_i^2}\left[1 - \exp\left(-\dfrac{G_0 X_i^2}{2Z_i P_{sat}}\right)\right]} \tag{6.84}$$

or

$$k_\alpha = 1 - \frac{G_0}{G_p(P_s)} \exp\left(-\frac{G_0 P_s}{P_{sat}}\right) . \tag{6.85}$$

For small input powers, i.e. for $P_s \ll P_{sat}/G_0$, we see that $k_\alpha \approx 0$ and an amplitude compression does not occur. With increasing input power the exponential function in Eq. (6.85) converges to zero and k_α approaches the value one. For $k_\alpha = 1$ all amplitude fluctuations of the input signal will be perfectly suppressed, the amplitude of the output signal is constant.

In order to estimate typical values of k_α within the limits 0 and 1, it should be clarified which input power is a suitable or perhaps optimum choice for a large signal amplifier. For a power amplifier one often attempts to optimize the difference between the output power $G_p P_s$ and the input power P_s, i.e. the added power ΔP_s. With Eq. (6.82) we obtain for this power difference:

$$\Delta P_s = G_p P_s - P_s = P_{sat}\left[1 - \exp\left(-\frac{G_0 P_s}{P_{sat}}\right)\right] - P_s . \tag{6.86}$$

Differentiation with respect to P_s yields

$$\frac{d(\Delta P_s)}{dP_s} = G_0 \exp\left(-\frac{G_0 P_s}{P_{sat}}\right) - 1 . \tag{6.87}$$

From this result we obtain the optimum input power

$$P_{s,opt} = \frac{P_{sat}}{G_0} \ln G_0 . \tag{6.88}$$

The power gain at this point of operation is

$$G_p(P_{s,opt}) = \frac{G_0}{\ln G_0}\left(1 - \frac{1}{G_0}\right) = \frac{G_0 - 1}{\ln G_0} . \tag{6.89}$$

Finally, we get for the corresponding amplitude compression factor

$$k_\alpha = 1 - \frac{1}{G_p(P_{s,opt})} = 1 - \frac{\ln G_0}{G_0 - 1} . \tag{6.90}$$

In Fig. 6.8 we can see this relation in a graphical form. For a typical small signal gain usually between 6 dB and 20 dB, we observe a strong amplitude compression if the amplifier is operated as a power amplifier with optimum drive level. Because this compression partly also suppresses amplitude fluctuations that are generated in the amplifier itself, the amplitude noise is usually smaller than the phase noise. The phase noise is nearly uneffected by non-linear effects, due to $K_{\phi\phi} = 1$.

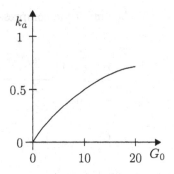

Fig. 6.8   Amplitude compression factor for an optimal operation point.

6.8.3 Frequency multipliers and dividers

Up to now, the input and output signals of the non-linear two-ports had the same frequency. Because of $K_{\phi\phi} = 1$ the phase fluctuations of the input signal are transferred to the output without change, except for internal noise and possibly additional noise contributions due to AM-PM conversion. This is no longer true if the output frequency changes due to frequency multiplication or frequency division.

If the input signal of a frequency multiplier is described by

$$x(t) = X_0 \cos[\Omega t + \Delta\phi(t)] \ , \tag{6.91}$$

then the output signal is given by

$$y(t) = Y_0 \cos[N\Omega t + N\Delta\phi(t)] \ , \tag{6.92}$$

with N as the multiplication factor. The ratio of the amplitudes Y_0 and X_0 depends upon the conversion loss or the gain, respectively, of the frequency multiplier. As a consequence of the frequency multiplication all phase excursions are amplified by the factor N, too. If the phase fluctuations of the output signal are denoted by $\Delta\psi(t)$, then Eq. (6.92) yields

$$\Delta\psi(t) = N \cdot \Delta\phi(t) \tag{6.93}$$

or

$$K_{\phi\phi} = N \ . \tag{6.94}$$

This linear relationship can be transformed directly into a relation of the corresponding spectra, i.e. the input spectrum $W_\phi(\omega)$ and the output spectrum $W_\psi(\omega)$:

$$W_\psi(\omega) = N^2 \, W_\phi(\omega) \ . \tag{6.95}$$

Due to Eq. (6.36) the factor N^2 is also valid for the normalized single sideband noise powers. Then, e.g. for a multiplication by a factor of 10, the ratio of the single sideband noise power to the carrier power deteriorates by 20 dB. This effect normally is crucial for the phase noise of the output signal. The contribution from the internal noise of the multiplier can often be neglected. Good practical frequency multipliers, e.g. multipliers with step recovery diodes, usually only show an internal noise slightly above thermal noise and also low $1/f$ noise cut-off frequencies.

For the generation of microwave signals with good long term frequency stability, the output signal of a quartz crystal oscillator is often multiplied by a chain of multipliers up to the required frequency. In this way, the total multiplication factor can be rather high and, therefore, a strong phase noise of the output signal can result. If, e.g. , the crystal oscillator has a frequency of 10 MHz and an output power of 10 dBm and if one assumes that -as the best possible situation- only thermal noise of -174 dBm/Hz is added to the carrier, then the phase noise of the crystal oscillator is $(P_{ssb}/P_c)_\phi = -187$ dBc/Hz. This value results from Eq. (6.44) with $F = 1$ and $P_{in} = 10$ dBm. By a multiplication to a frequency of e.g. 10 GHz the normalized single sideband noise power deteriorates by 60 dB to $(P_{ssb}/P_c)_\psi = -127$ dBc/Hz.

For frequency dividers with the division ratio N we observe just the inverse relationships. If, again, we denote the phase fluctuations of the input signal by $\Delta\phi(t)$ and those of the output signal by $\Delta\psi(t)$, then we have

$$\Delta\psi(t) = \frac{1}{N} \cdot \Delta\phi(t) \tag{6.96}$$

or

$$K_{\phi\phi} = \frac{1}{N} \ . \tag{6.97}$$

Again, this linear relationship can be transformed into a relation of the corresponding spectra:

$$W_\psi(\omega) = \frac{1}{N^2} W_\phi(\omega) \ . \tag{6.98}$$

Because the phase fluctuations of the input signal are reduced in magnitude at the output, the contribution of the internal noise of the divider to the output phase noise will typically be of relevance. Therefore, for frequency dividers the magnitude of the internal noise is of greater importance than for frequency multipliers and is more likely to contribute to the overall noise performance of a system.

Problem

6.6   The input signal of a frequency divider with a division ratio of N consists of two sinusoidal components with the small frequency difference $\Delta f = f_2 - f_1$ and the large amplitude ratio $A_1/A_2 \gg 1$ according to the figure below. What does the spectrum of the output signal look like? The same question should be answered for a frequency multiplier with a multiplication factor of N.

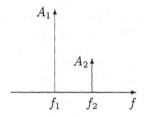

6.8.4   Frequency converters or mixers

For a shift in frequency of a carrier signal with small phase fluctuations by means of a frequency converter or mixer the situation is particularly simple, if we can assume that the mixer is operated quasi linearly or parametrically. Under this and the further assumption that the local oscillator signal is ideal and free of noise and neglecting the noise contribution of the mixer itself, it follows that the output phase noise $\Delta\psi(t)$ is equal to the input phase noise $\Delta\phi(t)$, independent of the frequencies involved:

$$\Delta\psi(t) = \Delta\phi(t) \tag{6.99}$$

and

$$K_{\phi\phi} = 1 \ . \tag{6.100}$$

From this result we conclude that the corresponding spectra are equal:

$$W_\psi(\omega) = W_\phi(\omega) \ . \tag{6.101}$$

Under the assumption of quasi-linearity of the mixer also the amplitude fluctuations will linearly be transposed in frequency. If, however, we do not neglect the phase fluctuations of the local oscillator $\Delta\Phi_c$ with the corresponding spectrum W_c, then the equation (6.101) will read as

$$W_\psi(\omega) = W_\phi(\omega) + W_c(\omega) \ , \tag{6.102}$$

provided that the local oscillator signal is free of amplitude fluctuations and that its phase noise is not correlated with the phase noise of the other signals. The last equation (6.102) is even valid under large signal conditions for all signals involved at the mixer, as long as the signals are free of amplitude fluctuations. Furthermore, we must require, as before, that the signals due to phase fluctuations are small relative to the carrier signals.

6.9   MEASUREMENT OF THE PHASE NOISE

The amplitude noise of a non-linear two-port is not measured very often, because it is not as important as the phase noise and because its measurement

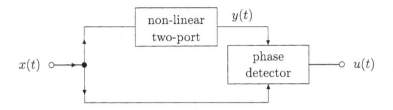

Fig. 6.9   Measurement of the phase noise of non-linear two-ports.

is less straightforward. The phase noise, on the other hand can be determined relatively easy. The measurement principle is demonstrated in Fig. 6.9.

The input signal $x(t)$ and the output signal $y(t)$ of a non-linear two-port are both connected to the input ports of a phase detector. With $x(t) = X_0 \cos[\Omega_0 t + \Delta\phi(t)]$ and $y(t) = Y_0 \cos[\Omega_0 t + \Delta\psi(t)]$ the voltage $u(t)$ at the output of the phase detector is proportional to the difference of the phase fluctuations of $x(t)$ and $y(t)$. The proportionality constant is named K_{PD}:

$$u(t) = K_{PD}[\Delta\psi(t) - \Delta\phi(t)] \ . \tag{6.103}$$

The phase fluctuations $\Delta\psi(t)$ of the output signal include the fluctuations $\Delta\phi(t)$ of the input signal and the noise contribution $\Delta\psi_n(t)$ of the noisy two-port under test:

$$\Delta\psi(t) = \Delta\phi(t) + \Delta\psi_n(t) \ . \tag{6.104}$$

We can thus write for the spectrum of $u(t)$:

$$W_u(\omega) = K_{PD}^2 \cdot W_{\psi n}(\omega) \ . \tag{6.105}$$

Therefore, the spectrum W_u is a replica of the spectrum $W_{\psi n}$ of the internal noise fluctuations of the two-port under test. After a sufficient amplification it can be displayed, e.g. with a spectrum analyzer.

For a quantitative measurement we need to know the product of the phase detector constant K_{PD} and the amplification factor between the phase detector and the display system. This product can be determined e.g. by a calibration measurement. For this purpose, two sinusoidal signals of the same frequency can be applied to the inputs of the phase detector. One of these signals is phase modulated in a well defined manner. The frequency and amplitude of the calibration signals must be identical to those of the measurement signals, because otherwise the detector constant K_{PD} may change.

For not too high frequencies, integrated digital circuits are available as phase detectors. For even higher frequencies one can employ balanced mixers for this purpose. The sensitivity then depends according to a sinus function on the phase difference of the two input signals. In order to adjust the system for maximum sensitivity, the circuit in Fig. 6.9 must be extended by a variable phase shifter at one of the input ports of the phase detector. In practice, however, and for convenience one should first try to perform the phase noise measurement directly with a sensitive high-frequency spectrum analyzer instead of using a measurement circuit as shown in Fig. 6.9. This requires that the amplitude fluctuations are small enough which often is the case. The spectrum analyzer offers the additional advantage of a spectrum power calibration.

Problem

6.7 A balanced mixer can be treated as an ideal analog multiplier. Determine the phase detector constant K_{PD}.

The circuit in Fig. 6.9 can only be applied if no frequency translation occurs within the non-linear two-port under test. In order to perform also measurements with frequency multipliers and frequency dividers we must extend the circuit according to Fig. 6.10. In this measurement setup instead of one circuit

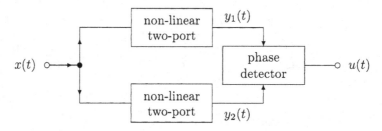

Fig. 6.10 Measurement circuit for the phase noise with two identical non-linear two-ports.

under test, two samples of the two-port under test are required with properties as similar as possible. The input signal $x(t)$ is applied to both two-ports. The phase fluctuations $\Delta\psi_1(t)$ and $\Delta\psi_2(t)$ of the output signals both contain a contribution of the input noise signal and the internal noises $\Delta\psi_{n1}(t)$ and $\Delta\psi_{n2}(t)$:

$$\Delta\psi_1(t) = N \cdot \Delta\phi(t) + \Delta\psi_{n1}(t) \ , \qquad (6.106)$$

$$\Delta\psi_2(t) = N \cdot \Delta\phi(t) + \Delta\psi_{n2}(t) \ . \qquad (6.107)$$

For frequency multipliers and frequency dividers, the frequency multiplication factor or division factor N is different from one, i.e. $N \neq 1$. For the spectrum W_u the equations (6.103), (6.106) and (6.107) yield

$$W_u(\omega) = K_{PD}^2\big[W_{\psi n1}(\omega) + W_{\psi n2}(\omega)\big] \ , \qquad (6.108)$$

because the noise contributions of both two-ports under test are uncorrelated and because the phase fluctuations $\Delta\phi(t)$ of the input signals cancel. As a result the measurement delivers the sum of both phase noises of the two-ports under test. For equal or nearly equal noise properties of the two-ports, we therefore obtain the result of the individual two-port simply by reducing the total measurement result of the spectrum by 3 dB. If the two-ports are unequal, then the spectrum of any of the two two-ports is smaller than the measured total spectrum. Furthermore, if available, the second measurement object may be replaced by a frequency multiplier or frequency divider with much lower noise. In this case, the total measured output noise directly yields the phase noise of the device under test. Finally, as will be shown in the next problem, one can use three possibly different devices under test and perform three paired measurements in order to determine all individual phase noise spectra.

Problem

6.8   Three samples of a frequency divider (frequency multiplier) with equal or unequal noise properties are available. Determine the individual noise properties of the dividers (multipliers) from the result of sequential measurements of all possible pairs of the samples.

Again, in practice, and for convenience, one will usually first try to perform the phase noise measurement directly with a sensitive high-frequency spectrum analyzer and not use a measurement circuit as in Fig. 6.10. Again the requirement for this procedure is that the amplitude fluctuations are small enough compared with the phase fluctuations, which very often is the case. Furthermore, it is required that the phase noises of the internal sources of the spectrum analyzer are smaller than the phase noise to be measured. The spectrum analyzer offers the additional advantage of an easy spectrum power calibration. Fig. 6.11 shows examples of typical measured results of the internal phase noise for an amplifier and a frequency divider in a schematic representation.

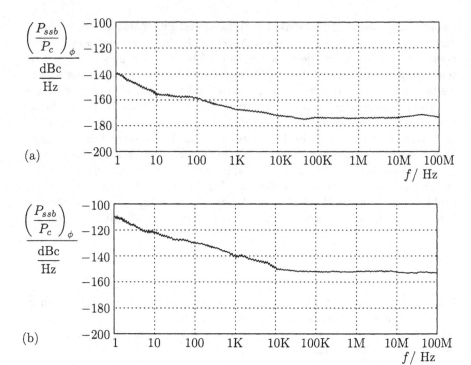

Fig. 6.11   Phase noise of (a) an amplifier and (b) of a frequency divider.

7

Noise in Oscillators

The output signal of an ideal oscillator has an exactly periodic, for high-frequency oscillators usually a sinusoidal, function of time. The frequency and the amplitude of the output signal are constant with time, except for possible slow drift effects, e.g. by temperature changes or aging effects. For a real oscillator, the noise processes within the circuit lead to small disturbances of the frequency or phase and the amplitude of the output signal. As for the nonlinear two-ports of Chapter 6, we will describe the disturbing effects of the noise on the output signal by amplitude and phase fluctuations of the carrier signal. Therefore, we can fully adopt the concepts of this former chapter, e.g. the concept of single sideband noise power.

In this chapter, we will discuss models that allow one to describe the influence of the different circuit parameters on the noise behavior of the oscillator. Furthermore, we will introduce possible concepts for the practical realization of low-noise signal sources. Finally, we will deal with the appropriate measurement techniques.

7.1 TWO-PORT AND ONE-PORT OSCILLATORS

Each oscillator consists of an active amplifying part and a passive network with a frequency selective feedback circuit that determines the oscillation frequency and a signal divider part to branch off the output signal. Principally, there are two different classes of oscillators. A two-port oscillator consists of an active two-port, that provides amplification, and a passive, resonant and thus frequency selective feedback network, which together with the output port may be described as a three-port (Fig. 7.1). One-port oscillators are based on an active one-port with a negative real part of its terminal impedance and a passive frequency selective circuit, which is usually a two-port. One-port oscillators are the typical choice in the mm-wave region, while at lower frequencies both types of oscillators are employed. Often it only depends on the point of view, whether the oscillator topology is considered as a one-port or a two-port oscillator. Most common active elements are bipolar transistors and field effect transistors. Gunn and Impatt diodes are sometimes found in existing mm-wave oscillators but are now more and more replaced by bipolar and field effect transistors. Vacuum tubes are still very important for oscillators with high powers and high frequencies but will not be discussed in this text.

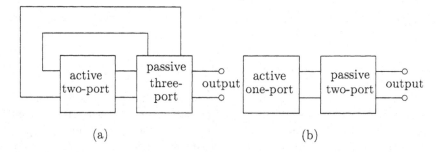

(a)   (b)

Fig. 7.1   Block diagram of (a) a two-port and (b) a one-port oscillator.

7.2   OSCILLATION CONDITION

Both kinds of oscillators can be described by the flow graph shown in Fig. 7.2. The active element is operated under large signal conditions and, therefore, must be treated as a non-linear network. The relationship between the complex input phasor X with $|X| = X_0$, a signal with a sinusoidal waveform, and the complex phasor of the output Y, a signal component at the fundamental frequency which, therefore, also has a sinusoidal wave form, is given by the describing function D

$$Y = D(X_0)X \tag{7.1}$$

We will assume throughout this chapter that the describing function of the active element of an oscillator, operating at a fixed frequency, only depends on the amplitude of the input signal and not on the frequency.

For oscillators this assumption is usually justified, because the frequency dependence of the active part normally is negligible compared with the strongly frequency dependent passive part of the circuit, which typically has the prop-

feedback network

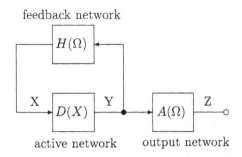

active network   output network

Fig. 7.2   General block diagram of oscillators.

erties of a resonator. For the two-port oscillator the phasors X and Y describe the input and output signals of the active two-port, e.g. the input and output signals of a bipolar or field effect transistor. With respect to the dimension these signals may be voltages, currents or waves. If quantities of the same dimension are used for the input and output signal, then the describing function is dimensionless. For a one-port oscillator the describing function for the active element usually has a dimension. If the input signal is chosen as a current and the output signal as a voltage at the terminals of the active one-port, then the describing function has the dimension of an impedance. A dimensionless describing function results for the one-port, if the signals at the interface between the active and passive parts of the circuit are introduced as waves propagating towards and back from the active impedance. Then, the describing function corresponds to the reflection coefficient of the active one-port.

The properties of the passive network of both the two-port oscillator and the one-port oscillator are described by frequency dependent transfer functions in the signal flow diagram. The transfer function $H(\Omega)$ designates the fraction of the output signal Y of the active element which is coupled back to its input as a signal X, in order to maintain a stationary oscillation. The transfer function $A(\Omega)$ specifies the relationship between the output signal Y and the signal Z at the external load impedance of the oscillator. Regarding the dimension of the transfer function, the same considerations hold as for the describing function. From Fig. 7.2 we get

$$X = H(\Omega)\,Y \tag{7.2}$$

and

$$Z = A(\Omega)\,Y \ . \tag{7.3}$$

Combining the equations (7.1) and (7.2) yields

$$D(X_0)\,H(\Omega) = 1 \ . \tag{7.4}$$

This last equation is known as the **oscillation condition**. This condition states that the oscillator can not oscillate with arbitrary amplitude and frequency but only with those combinations of X_{0i}, Ω_i, which meet the condition Eq. (7.4). For a given oscillator circuit with known functions $D(X_0)$ and $H(\Omega)$, all possible modes of oscillation can be found by solving Eq. (7.4). On the other hand, the oscillation condition can be used as a basis for the systematic design of the oscillator circuit. For this purpose, one may start with a proper selection of the signal amplitude X_0 at the active element. Quite appropriate is the amplitude which maximizes the difference between the input and the output power (see Section 6.8.2). From the corresponding value of the describing function and Eq. (7.4), the feedback transfer function at the specified oscillation frequency is obtained. With this result the passive network can be designed. Then, with Eq. (7.4), the circuit can be analyzed once again in order to check if the oscillation condition has additional solutions at other frequencies. This must be avoided, since in this case unwanted parasitic oscillations at other frequencies are possible. If necessary, the circuit design has to be modified until a single frequency oscillation is guaranteed.

A particularly important and difficult issue in the design process is the accurate description of the active element and the describing function, either by measurement or by simulation or both. Often the description of the active element is based on a small signal characterization, which is easier to accomplish but may lead to a non-optimum output power performance. In such a situation, it is common practice to improve the performance of the oscillator empirically.

The oscillation condition is a necessary but not a sufficient condition for stationary oscillations. A further condition is that the mode of oscillation is stable, which means that, after a disturbance, the oscillation amplitude automatically returns to its former steady state value. The stability condition will be discussed later on, using results of the noise analysis.

7.3 NOISE ANALYSIS

The aim of the following noise analysis is the general calculation of the amplitude and phase noise of oscillators. For the amplitude and phase noise of linear and non-linear active networks the results can directly be taken from chapter 6. For convenience, the most important results will be repeated here.

If we denote the amplitudes of the input and output signals by X_0 and Y_0, respectively, and the corresponding amplitude and phase fluctuations by $\Delta X/X_0$ and $\Delta Y/Y_0$ or $\Delta\Phi$ and $\Delta\Psi$, then we can write

$$
\begin{bmatrix} \dfrac{\Delta Y}{Y_0} \\ \Delta\Psi \end{bmatrix} = \begin{bmatrix} K_{\alpha\alpha} & K_{\alpha\phi} \\ K_{\phi\alpha} & K_{\phi\phi} \end{bmatrix} \begin{bmatrix} \dfrac{\Delta X}{X_0} \\ \Delta\Phi \end{bmatrix} + \begin{bmatrix} \dfrac{\Delta Y_n}{Y_0} \\ \Delta\Psi_n \end{bmatrix} . \tag{7.5}
$$

The quantities $\Delta Y_n/Y_0$ and $\Delta\Psi_n$ describe the contribution of the internal noise of the active network to the amplitude and phase noise of the output signal. The matrix elements of $[K]$ denote the changes of the fluctuations of the input signal when passing through the non-linear network. For the calculation of the oscillator noise it is normally sufficient to take into account the amplitude compression only. With the amplitude compression factor k_α (see section 6.7.1) we get from Eq. (7.5):

$$\left[\begin{array}{c} \dfrac{\Delta Y}{Y_0} \\ \Delta\Psi \end{array}\right] = \left[\begin{array}{cc} 1-k_\alpha & 0 \\ 0 & 1 \end{array}\right] \left[\begin{array}{c} \dfrac{\Delta X}{X_0} \\ \Delta\Phi \end{array}\right] + \left[\begin{array}{c} \dfrac{\Delta Y_n}{Y_0} \\ \Delta\Psi_n \end{array}\right] . \tag{7.6}$$

As already discussed in Section 6.6, the contributions of the internal noise of the active element depend on the drive level. For a weak drive level i.e. quasi-linear operation, we can determine the spectra of the amplitude and phase noise approximately from the noise figure F and the input power P_{in} of the active network according to the relation

$$W_{\alpha n} = W_{\phi n} = \frac{Fk\,T_o}{2\,P_{in}} . \tag{7.7}$$

The spectra are approximately independent of frequency, i.e. white, and the amplitude and phase fluctuations are not correlated.

For a strongly non-linear mode of operation, in many oscillator circuits the internal noise can be described approximately by a low frequency noise process and a frequency up-conversion. We then may write

$$W_{\alpha n} = |m_\alpha|^2\,W_m \tag{7.8}$$

and

$$W_{\phi n} = |m_\phi|^2\,W_m , \tag{7.9}$$

with the modulation factors m_α and m_ϕ. The low frequency spectrum W_m typically has a more or less pronounced $1/f$-characteristic. Amplitude and phase noise are fully correlated according to this model.

The quantitative determination of the spectra $W_{\alpha n}$ and $W_{\phi n}$ for a non-linear network is complicated and requires accurate models for the non-linear elements. If the spectra can not be determined by measurements, the following empirical approach may be used, which results from a slight modification of the relations for the linear case:

$$W_{\alpha n} = \left(1+\frac{f_{co}}{f}\right) \cdot \left\{ \begin{array}{l} W_0 \\ (1-k_\alpha)^2\,W_0 \end{array} \right. \tag{7.10}$$

$$W_{\phi n} = \left(1+\frac{f_{co}}{f}\right) W_0 , \tag{7.11}$$

with

$$W_0 = \frac{F_{ef}\,k\,T_0}{2\,P_{in}} . \tag{7.12}$$

The spectrum W_0 corresponds to the noise in the linear case, however, with an effective large signal noise figure F_{ef}, which usually is higher than the small signal noise figure F. Typical values for F_{ef} might range from 2 dB to 20 dB. The factor $(1 + f_{co}/f)$ accounts for the $1/f$-noise of the active component. The noise corner frequency f_{co} has already been introduced in Section 6.3 and typically is found in the range of 1 kHz to 100 MHz. The spectrum W_{an} may be multiplied by the factor $(1 - k_a)^2$ if it can be assumed that the amplitude noise mainly originates from the input section of the non-linear network and that it therefore will be subject to an amplitude compression. A possible correlation between amplitude and phase noise is normally neglected.

The relationships between the amplitude and phase fluctuations of the input and the output of the linear network with the transfer function H of the block diagram of Fig. 7.2 have already been given in chapter 6, Eq. (6.20). However, it must be noticed that the meanings of X and Y are interchanged in Fig. 7.2. With the abbreviations $H_c = H(\Omega_0)$, $H_l = H(\Omega_0 - \omega)$ and $H_u = H(\Omega_0 + \omega)$ we have:

$$
\begin{bmatrix} \dfrac{\Delta X}{X_0} \\[2ex] \Delta\Phi \end{bmatrix} = \frac{1}{2} \begin{bmatrix} \dfrac{H_u}{H_c} + \dfrac{H_l^*}{H_c^*} & j\left(\dfrac{H_u}{H_c} - \dfrac{H_l^*}{H_c^*}\right) \\[3ex] -j\left(\dfrac{H_u}{H_c} - \dfrac{H_l^*}{H_c^*}\right) & \dfrac{H_u}{H_c} + \dfrac{H_l^*}{H_c^*} \end{bmatrix} \begin{bmatrix} \dfrac{\Delta Y}{Y_0} \\[2ex] \Delta\Psi \end{bmatrix} . \quad (7.13)
$$

For convenience the intrinsic thermal noise of the passive feedback circuit is neglected because it is typically small compared with the noise of the active circuit. However, in principle, it would not raise any problem to consider also the thermal noise of the passive circuitry.

Similar to Eq. (6.16) we obtain a conversion matrix for the linear output network to the load of Fig. 7.2 with the transfer function $A(\Omega)$. For the amplitude and phase fluctuations at the output, $\Delta Z/Z_0$ and $\Delta\Theta$, we get the matrix equation

$$
\begin{bmatrix} \dfrac{\Delta Z}{Z_0} \\[2ex] \Delta\Theta \end{bmatrix} = \frac{1}{2} \begin{bmatrix} \dfrac{A_u}{A_c} + \dfrac{A_l^*}{A_c^*} & j\left(\dfrac{A_u}{A_c} - \dfrac{A_l^*}{A_c^*}\right) \\[3ex] -j\left(\dfrac{A_u}{A_c} - \dfrac{A_l^*}{A_c^*}\right) & \dfrac{A_u}{A_c} + \dfrac{A_l^*}{A_c^*} \end{bmatrix} \begin{bmatrix} \dfrac{\Delta Y}{Y_0} \\[2ex] \Delta\Psi \end{bmatrix} , \quad (7.14)
$$

with the abbreviations $A_c = A(\Omega_0)$, $A_l = A(\Omega_0 - \omega)$ and $A_u = A(\Omega_0 + \omega)$.

With the results described above, the noise of the complete oscillator can be calculated in a general form, once the feedback loop is closed (see Fig. 7.2).

Here, the feedback loop is closed via the phasors of the amplitude and phase fluctuations. Using alternative algebra, which, however, delivers exactly identical results, one may use the upper and lower sideband phasors, in order to close the loop.

Combining Eqs. (7.6) and (7.13), we get

$$
\left[\begin{array}{c} \dfrac{\Delta Y}{Y_0} \\[3ex] \Delta \Psi \end{array} \right] = \dfrac{1}{2} \left[\begin{array}{cc} (1 - k_\alpha) \left(\dfrac{H_u}{H_c} + \dfrac{H_l^*}{H_c^*} \right) & j(1 - k_\alpha) \left(\dfrac{H_u}{H_c} - \dfrac{H_l^*}{H_c^*} \right) \\[4ex] -j \left(\dfrac{H_u}{H_c} - \dfrac{H_l^*}{H_c^*} \right) & \left(\dfrac{H_u}{H_c} + \dfrac{H_l^*}{H_c^*} \right) \end{array} \right]
$$

$$
\cdot \left[\begin{array}{c} \dfrac{\Delta Y}{Y_0} \\[3ex] \Delta \Psi \end{array} \right] + \left[\begin{array}{c} \dfrac{\Delta Y_n}{Y_0} \\[3ex] \Delta \Psi_n \end{array} \right] . \quad (7.15)
$$

With the abbreviations

$$
H_\Sigma = \frac{1}{2} \left(\frac{H_u}{H_c} + \frac{H_l^*}{H_c^*} \right)
$$

and

$$
H_\Delta = \frac{1}{2} \left(\frac{H_u}{H_c} - \frac{H_l^*}{H_c^*} \right) \quad (7.16)
$$

we get

$$
\left[\begin{array}{cc} 1 - (1 - k_\alpha) H_\Sigma & -j(1 - k_\alpha) H_\Delta \\[3ex] j H_\Delta & 1 - H_\Sigma \end{array} \right] \cdot \left[\begin{array}{c} \dfrac{\Delta Y}{Y_0} \\[3ex] \Delta \Psi \end{array} \right] = \left[\begin{array}{c} \dfrac{\Delta Y_n}{Y_0} \\[3ex] \Delta \Psi_n \end{array} \right] . \quad (7.17)
$$

Solving for $\Delta Y / Y_0$ and $\Delta \Psi$ we obtain:

$$
\frac{\Delta Y}{Y_0} = \frac{(1 - H_\Sigma)(\Delta Y_n / Y_0) + j(1 - k_\alpha) H_\Delta \Delta \Psi_n}{(1 - H_\Sigma)\left[1 - (1 - k_\alpha) H_\Sigma \right] - (1 - k_\alpha) H_\Delta^2} , \quad (7.18)
$$

$$
\Delta \Psi = \frac{\left[1 - (1 - k_\alpha) H_\Sigma \right] \Delta \Psi_n - j H_\Delta (\Delta Y_n / Y_0)}{(1 - H_\Sigma)\left[1 - (1 - k_\alpha) H_\Sigma \right] - (1 - k_\alpha) H_\Delta^2} . \quad (7.19)
$$

The amplitude and phase fluctuations each depend on the noise contributions $\Delta Y_n / Y_0$ as well as $\Delta \Psi_n$ of the active network, because in the linear circuit a mutual conversion of both kinds of fluctuations is possible. As was already mentioned, this conversion vanishes if the passive resonance circuit has a symmetry in the sense that $H_u / H_c = (H_l / H_c)^*$. With $H_\Sigma = H_u / H_c$ and $H_\Delta = 0$ the Eqs. (7.18) and (7.19) simplify to

$$
\frac{\Delta Y}{Y_0} = \frac{1}{1 - (1 - k_\alpha) H_u / H_c} \frac{\Delta Y_n}{Y_0} , \quad (7.20)
$$

$$
\Delta \Psi = \frac{1}{1 - H_u / H_c} \Delta \Psi_n . \quad (7.21)
$$

With Eqs. (7.13) and (7.14) it is possible to calculate from $\Delta Y/Y_0$ and $\Delta \Psi$ the amplitude and phase fluctuations at the input of the active network as well as at the load output of the oscillator.

By forming the squared magnitude, the above relations can be transformed into equations for the corresponding spectra. Apart from the spectra W_α and W_ϕ, the spectrum for the frequency fluctuations W_f is sometimes also given. The frequency fluctuations, i.e. the deviations of the instantaneous frequency from the mean value $\Omega_c/2\pi$ are obtained by a differentiation with respect to time of the phase fluctuations and a division by 2π. The phasor of the frequency fluctuations is denoted by ΔF. Then, in complex notation we can write

$$\Delta F = \frac{j\,\omega}{2\,\pi}\,\Delta \Phi \ . \tag{7.22}$$

Therefore, the relation between the spectra is given by

$$W_f = \left(\frac{\omega}{2\,\pi}\right)^2 \cdot W_\Phi = f^2\,W_\Phi \ . \tag{7.23}$$

7.4 STABILITY CONDITION

In Section 7.2, it was already mentioned that the oscillation condition (7.4) is not a sufficient condition for a stable oscillation at a certain frequency. The stability is only guaranteed if the oscillation amplitude is insensitive to disturbances. This requires that the oscillation amplitude returns to its former value after a disturbance stops. Only such a behavior corresponds to a stable mode of operation. If the amplitude deviation remains constant or even increases over time, then the oscillation mode is unstable.

The stability of the oscillation amplitude can be checked with the help of Eq. (7.18). This equation is not restricted to noise signals but relates in a general manner disturbances of the amplitude and phase in the active network to the resultant amplitude changes of the oscillator. The relation is linear and Eq. (7.18) can be considered as the transfer function of a linear system. Therefore, well-known methods can be employed in order to test for stability of the system. If we denote the denominator of the equation (7.18) as De, then De is a complex function of the offset frequency ω or $De = De(j\omega)$. For a stability test we will replace the expression $j\omega$ by the complex frequency $p = \sigma + j\omega$. Subsequently, we will search for the solutions p_i of the equation

$$De(p) = 0 \ . \tag{7.24}$$

The transfer function (7.18) belongs to a stable system, if for all solutions p_i of Eq. (7.24):

$$\mathrm{Re}\{p_i\} < 0 \ . \tag{7.25}$$

In this case, the oscillation amplitude of the oscillator is stable.

7.5   EXAMPLES

7.5.1   Two-port oscillator with transmission resonator

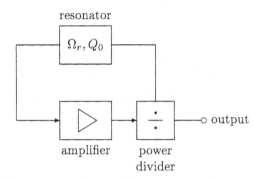

Fig. 7.3   Two-port oscillator with a transmission resonator.

As an example of a two-port oscillator we will analyze the circuit in Fig. 7.3.
The output signal of a large signal amplifier is applied via a signal divider to
the input of a transmission resonator. The transmitted and filtered signal is
fed back to the input of the amplifier, thus closing the feedback loop.

It is expedient to describe the circuit by scattering wave parameters. We
will assume for convenience that all components are matched around the oscillation frequency so that reflections do not occur. Furthermore, all signal
delays and attenuations caused by interconnecting lines are neglected. They
are thought to be included in the amplifier, the signal divider, or resonator.

The non-linear properties of the large signal amplifier have already been
treated in Section 6.8.2. The power gain G_p decreases monotonously with increasing input power P_s and approaches zero asymptotically. Quantitatively,
this dependence of the amplifier gain G_p as a function of the input power P_s
can be described by the approximate empirical expression:

$$G_p(P_s) = \frac{P_{sat}}{P_s} \left[1 - \exp\left(-\frac{G_{s0} P_s}{P_{sat}} \right) \right] \ . \qquad (7.26)$$

Here, G_{s0} is the small signal gain of the amplifier and P_{sat} is the saturated
output power of the amplifier. We can thus write for the describing function
D of the nonlinear amplifier circuit:

$$D(X_0) = \sqrt{ \frac{P_{sat}}{X_0^2} \left[1 - \exp\left(-\frac{G_{s0} X_0^2}{P_{sat}} \right) \right] } \ . \qquad (7.27)$$

The phasor amplitude X_0 in the above equation now belongs to a wave quantity.

The scattering parameters S_{21} and S_{12} of the transmission resonator can be written in a general form as

$$S_{21} = S_{12} = \frac{2\sqrt{\beta_1 \beta_2}}{1 + \beta_1 + \beta_2 + j\,Q_0 \left(\dfrac{\Omega}{\Omega_r} - \dfrac{\Omega_r}{\Omega} \right)} \qquad (7.28)$$

with Q_0 as the unloaded quality factor and Ω_r as the resonance frequency of the resonator. For $\Omega = \Omega_r$ the scattering parameters S_{12} and S_{21} are real and have their maximum magnitude. The positive and real quantities β_1 and β_2 are the coupling factors of the input and the output of the resonator. In the following, we will assume a symmetrical coupling, i.e. $\beta_1 = \beta_2 = \beta$. Furthermore, we will use for the resonator the approximation that $(\Omega/\Omega_r - \Omega_r/\Omega) = 2\omega/\Omega_r$ for $\omega \ll \Omega$ and with $\omega = \Omega - \Omega_r$. Then, Eq. (7.28) simplifies to the expression

$$S_{12} = S_{21} = \frac{2\beta}{1 + 2\beta + j\,2\,Q_0\,\omega/\Omega_r} \ . \qquad (7.29)$$

We assume that the signal divider has a frequency independent transfer function of $1/\sqrt{2}$ for both output ports. Thus we obtain for the two transfer functions of Fig. 7.2

$$H(\Omega) \;=\; H(\omega + \Omega_r) = \frac{\sqrt{2}\,\beta}{1 + 2\beta + j\,2\,Q_0\,\omega/\Omega_r} \ , \qquad (7.30)$$

$$A(\Omega) \;=\; \frac{1}{\sqrt{2}} \ . \qquad (7.31)$$

The oscillation condition (7.4) together with the equations (7.27) and (7.30) reads:

$$\sqrt{\frac{P_{sat}}{X_0^2}\left[1 - \exp\left(-\frac{G_{s0}X_0^2}{P_{sat}}\right)\right]} \cdot \frac{\sqrt{2}\,\beta}{1 + 2\beta + j\,2\,Q_0\,\omega/\Omega_r} = 1 \ . \qquad (7.32)$$

Since the imaginary part of the oscillation condition must be zero, we conclude that the oscillation frequency $\Omega_0 = \omega + \Omega_r$ is equal to the resonance frequency Ω_r of the oscillator, i.e. $\omega = 0$. Squaring the real part of the oscillation condition yields:

$$\frac{P_{sat}}{X_0^2}\left[1 - \exp\left(-\frac{G_{s0}X_0^2}{P_{sat}}\right)\right] \cdot \frac{2\beta^2}{(1 + 2\beta)^2} = 1 \ . \qquad (7.33)$$

For given circuit parameters, this equation allows one to determine the oscillation amplitude X_0 and thus also the output power P_s of the amplifier. If the amplitude of the amplifier input signal X_0 has been fixed, then the necessary coupling factor β of the resonator can be determined with Eq. (7.33).

Problem

7.1 An amplifier has the saturated output power $P_{sat} = 20\,\text{dBm}$ and the small signal gain $G_{s0} = 15\,\text{dB}$. Which signal input power P_{sopt} leads to a maximum difference between the output and input power? What is the value of the coupling factor β in order to meet the oscillation condition Eq. (7.33) for this optimum drive level? What is the value of the corresponding output power?

For the amplifier characteristic according to Eq. (7.26) the amplitude compression factor k_α has already been determined in Section 6.8.2:

$$k_\alpha = 1 - \frac{G_{s0}}{G_p(P_s)}\,\exp\left(-\frac{G_{s0}P_s}{P_{sat}}\right)\,. \tag{7.34}$$

Furthermore, for the calculation of the noise we need the quantities H_Σ and H_Δ from Eq. (7.16). With Eq. (7.30) we obtain

$$H_\Sigma = \frac{1 + 2\beta}{1 + 2\beta + j\,2\,Q_0\,\omega/\Omega_r} \tag{7.35}$$

$$H_\Delta = 0\,. \tag{7.36}$$

With Eqs. (7.20) and (7.21) the amplitude and phase fluctuations at the output of the amplifier are given by

$$\frac{\Delta Y}{Y_0} = \frac{1 + 2\beta + j\,2\,Q_0\,\omega/\Omega_r}{k_\alpha(1 + 2\beta) + j\,2\,Q_0\,\omega/\Omega_r}\cdot\frac{\Delta Y_n}{Y_0}\,, \tag{7.37}$$

$$\Delta\Psi = \frac{1 + 2\beta + j\,2\,Q_0\,\omega/\Omega_r}{j\,2\,Q_0\,\omega/\Omega_r}\cdot\Delta\Psi_n\,. \tag{7.38}$$

Because the transfer function of the signal divider is assumed to be frequency independent, the output coupling network does not change the noise. Therefore, the amplitude and phase fluctuations of the output signal of the oscillator are identical to those at the output of the amplifier. With the assumptions

$$W_{an} = (1 - k_\alpha)^2\,W_0 \tag{7.39}$$

and

$$W_{\phi n} = W_0 \tag{7.40}$$

for the internal noise of the amplifier we obtain for the spectra of the oscillator noise:

$$W_\alpha(\omega) = (1 - k_\alpha)^2\,\frac{(1 + 2\beta)^2 + (2\,Q_0\,\omega/\Omega_r)^2}{k_\alpha^2\,(1 + 2\beta)^2 + (2\,Q_0\,\omega/\Omega_r)^2}\,W_0\,, \tag{7.41}$$

$$W_\phi(\omega) = \left[1 + \left(\frac{1 + 2\beta}{2\,Q_0\,\omega/\Omega_r}\right)^2\right]W_0\,. \tag{7.42}$$

From the Eqs. (7.41) and (7.42) two corner frequencies may be derived, denoted by ω_{t1} and ω_{t2} (see Fig. 7.4):

$$\omega_{t1} = k_\alpha (1 + 2\beta) \frac{\Omega_r}{2 Q_0} , \qquad (7.43)$$

$$\omega_{t2} = (1 + 2\beta) \frac{\Omega_r}{2 Q_0} = \frac{\omega_{t1}}{k_\alpha} . \qquad (7.44)$$

Because $k_\alpha < 1$ we have $\omega_{t1} < \omega_{t2}$. Therefore, for offset frequencies below ω_{t1} we observe a frequency independent amplitude noise:

$$W_\alpha(\omega) = \left(\frac{1}{k_\alpha} - 1 \right)^2 W_0 , \qquad \omega \ll \omega_{t1} . \qquad (7.45)$$

Between ω_{t1} and ω_{t2} the amplitude noise decreases with increasing offset frequency by up to 20 dB/decade, before it again approaches a constant level above ω_{t2}:

$$W_\alpha(\omega) = (1 - k_\alpha)^2 W_0 , \qquad \omega \gg \omega_{t2} . \qquad (7.46)$$

The phase noise decreases with 20 dB/decade up to the corner frequency ω_{t2}. For even higher offset frequencies, the spectrum W_ϕ approaches the constant value W_0. Thus, we qualitatively observe a noise behavior of the two-port oscillator, as shown in Fig. 7.4a.

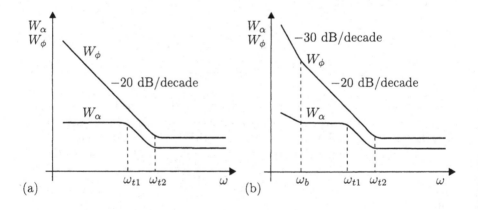

Fig. 7.4 Qualitative spectra of the amplitude and phase noise of a two-port oscillator (a) without $1/f$-noise and (b) with $1/f$-noise.

If a $1/f$-noise contribution is included in the spectra $W_{\alpha n}$ and $W_{\phi n}$, then the form of the spectra changes as shown in Fig. 7.4b, assuming that the $1/f$-noise corner frequency $\omega_b < \omega_{t1}$. A quantitative noise calculation is to be performed in the next problem.

Problem

7.2 Calculate the noise spectra W_α and W_Φ for the data of the two-port oscillator of problem 7.1. The amplifier is assumed to have an effective noise figure of $F_{ef}=20$ dB and a $1/f$-noise corner frequency f_b of 0.1 MHz. The resonance frequency of the resonator is assumed to be 10 GHz, and the unloaded quality or Q-factor 1000.

Of crucial importance for the phase noise is the resonator quality factor Q_0. For small offset frequencies, a doubling of the quality factor leads to a reduction of the phase noise by 6 dB. Therefore, for a practical realization of an oscillator circuit, a high resonator quality factor is of prime importance.

An improvement of the phase noise can also be achieved by a reduction of the coupling factor β, i.e. by a weaker coupling of the resonator. However, the range of possible coupling factors is rather small because, according to Eq. (7.29), a smaller coupling of the resonator also leads to a higher transmission loss of the resonator.

The amplitude noise is only slightly affected by the properties of the resonator. The two limits according to Eqs. (7.45) and (7.46) only depend on the properties of the amplifier. A change of the resonator parameters just leads to a shift of the two corner frequencies ω_{t1} and ω_{t2}.

The spectra of the two-port oscillator also depend on the position of the output coupling circuit as is shown in the next problem.

Problem

7.3 The circuit of a two-port oscillator as shown in Fig. 7.5 is modified in such a way that the output coupling network is located at the output of the resonator or the input of the amplifier, respectively. Calculate the spectra of the amplitude and phase noise with the same numerical values as in the problems 7.1 and 7.2.

The stability of the two-port oscillator can be analyzed by means of the denominator of Eq. (7.37). With the complex offset frequency p, a zero of the denominator occurs at the complex offset frequency p_1:

$$p_1 = -k_\alpha \left(1 + 2\,\beta\right) \frac{\Omega_0}{2\,Q_0} \quad . \tag{7.47}$$

Because $k_\alpha > 0$ and $\beta > 0$ the zero lies in the left half of the complex p−plane. The solutions of the oscillation condition (7.32) thus correspond to stable oscillation modes.

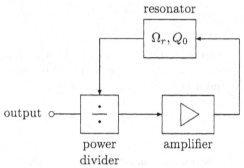

Fig. 7.5   Two-port oscillator with output coupling network at the amplifier input.

7.5.2   One-port oscillator with a series resonator

As a second example we will examine the circuit in Fig. 7.6. The active element is a one-port, which in the equivalent circuit is described by a real resistance R. Between the amplitudes of the current and the voltage there is a non-linear relationship. Therefore, the resistance is not constant.

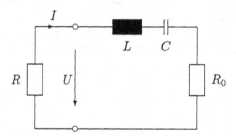

Fig. 7.6   One-port oscillator with a series resonator.

Figure 7.7 shows typical curves for the dependence of the voltage amplitude and the resistance on the current amplitude. For small amplitudes the active two-port has a nearly constant negative terminal resistance. Therefore, high-frequency power can be delivered to the load. With increasing current or voltage amplitudes the magnitude of the negative resistance will decrease, similar to the decrease of the gain in a saturated amplifier. There are, however, also some differences between a one-port with a negative terminal resistance and an amplifier. While the output signal of an amplifier increases monotonously with the input signal, for a one-port also a decrease of, e.g. the voltage amplitude with an increasing current amplitude may be observed. Therefore, the resistance can drop to zero already for a finite current amplitude, while for an amplifier the zero gain is only reached asymptotically.

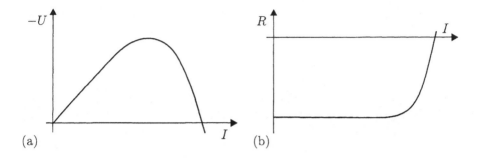

Fig. 7.7   (a) Voltage and (b) resistance of the active two-port as a function of the current.

With oriented arrows for the current and voltage phasors as defined in Fig. 7.6 we can write

$$U = -RI \ . \tag{7.48}$$

If we consider the current phasor I as the input signal and the voltage phasor U as the output signal of the non-linear network, then the negative resistance $-R$ corresponds to the describing function D of Fig. 7.2. By means of Eq. (6.79) we can determine the amplitude compression factor k_α from the current dependence of R according to Fig. 7.7. With $I_0 = |I|$ we obtain

$$k_\alpha = -\frac{I_0}{R(I_0)} \cdot \frac{dR}{dI_0} \ . \tag{7.49}$$

Due to $dR/dI_0 > 0$, $k_\alpha > 0$ for $R < 0$. For amplifiers, k_α is also smaller than one. This restriction does not apply to active one-ports. Because of Eq. (7.49), k_α may even become arbitrarily large in the vicinity of the zero crossing of the $R-$function. Even without an exact knowledge of the function $R(I)$ the value of the compression factor $k_\alpha > 0$ leading to maximum output power can be estimated, as will be shown in the next problem.

Problem

7.4   For an active non-linear two-port the amplitudes of the current and the voltage have been chosen such that the power transferred from the active one-port to the load becomes maximum. What is the value of the compression factor $k_\alpha > 0$ in this case?

The linear network of the oscillator or its equivalent circuit consists of a resonator with an inductance L and a capacitance C in series and additionally the load resistance R_0. For the linear network the voltage U is considered as the input signal and the current I as the output signal. Therefore, the transfer function $H(\Omega)$ of Fig. 7.2 is identical to the admittance of the series connection of L, C and R_0:

$$H(\Omega) = \frac{1}{R_0 + j\,\Omega\,L + 1/j\,\Omega\,C} \ . \tag{7.50}$$

If we consider the voltage at R_0 as the output signal, then the transfer function $A(\Omega)$ of the output coupling network is given by

$$A(\Omega) = \frac{R_0}{R_0 + j\,\Omega\,L + 1/j\,\Omega\,C} \ . \tag{7.51}$$

The oscillation condition (7.4) for this equivalent circuit yields

$$\frac{-R(I)}{R_0 + j\left(\Omega L - \dfrac{1}{\Omega C}\right)} = 1 \ . \tag{7.52}$$

From the real part of Eq. (7.52) a condition for the oscillation amplitude I_0 is obtained:

$$-R(I_0) = R_0 \ . \tag{7.53}$$

An oscillation amplitude I_0 will build up with a corresponding negative resistance value of the active element which in its steady state is equal to the load resistance R_0. The imaginary part of the oscillation condition Eq. (7.52) shows that the oscillation frequency Ω_0 is equal to the resonance frequency Ω_r of the series resonance circuit:

$$\Omega_0 = \frac{1}{\sqrt{LC}} = \Omega_r \ . \tag{7.54}$$

With Eq. (7.54), Eq. (7.50) yields

$$H(\Omega) = \frac{1}{R_0 + j\,\Omega_r\,L\left(\dfrac{\Omega}{\Omega_r} - \dfrac{\Omega_r}{\Omega}\right)} \tag{7.55}$$

and with $\Omega = \Omega_r + \omega$ and $\omega \ll \Omega$ we get

$$H(\Omega) = \frac{1}{R_0 + j\,2\omega\,L} \ . \tag{7.56}$$

For the quantities H_Σ and H_Δ we then obtain

$$H_\Sigma = \frac{H_u}{H_0} = \frac{R_0}{R_0 + j\,2\omega\,L} \ , \tag{7.57}$$

$$H_\Delta = 0 \ . \tag{7.58}$$

The corresponding elements of the output coupling network $A(\Omega)$ are

$$A(\Omega) = \frac{R_0}{R_0 + j\,2\omega\,L} \tag{7.59}$$

and

$$A_\Sigma = \frac{A_u}{A_0} = \frac{R_0}{R_0 + j\,2\omega\,L} \;, \tag{7.60}$$

$$A_\Delta = 0 \tag{7.61}$$

By insertion of Eqs. (7.57) and (7.58) into Eqs. (7.20) and (7.21) we obtain the amplitude and phase fluctuations of the voltage U:

$$\frac{\Delta U}{U_0} = \frac{1}{1 - (1 - k_\alpha)\dfrac{R_0}{R_0 + j\,2\omega L}} \cdot \frac{\Delta U_n}{U_0} \;, \tag{7.62}$$

$$\Delta\Psi = \frac{1}{1 - \dfrac{R_0}{R_0 + j\,2\omega L}} \cdot \Delta\Psi_n \;. \tag{7.63}$$

The fluctuations at the load resistance $\Delta U_R / U_{R0}$ and $\Delta\Theta$ are related to the fluctuations $\Delta U/U_0$ and $\Delta\Psi$ according to Eqs. (7.14), (7.59), (7.60) and (7.61):

$$\begin{bmatrix} \dfrac{\Delta U_R}{U_{R0}} \\ \Delta\Theta \end{bmatrix} = \begin{bmatrix} \dfrac{R_0}{R_0 + j\,2\,\omega\,L} & 0 \\ 0 & \dfrac{R_0}{R_0 + j\,2\,\omega\,L} \end{bmatrix} \begin{bmatrix} \dfrac{\Delta U}{U_0} \\ \Delta\Psi \end{bmatrix} \;. \tag{7.64}$$

Inserting $\Delta U/U_0$ and $\Delta\Psi$ from the Eqs. (7.62) and (7.63) into Eq. (7.64) we obtain

$$\frac{\Delta U_R}{U_{R0}} = \frac{R_0}{k_\alpha R_0 + j\,2\,\omega L} \cdot \frac{\Delta U_n}{U_0} \;, \tag{7.65}$$

$$\Delta\Theta = \frac{R_0}{j\,2\,\omega L} \cdot \Delta\Psi_n \;. \tag{7.66}$$

For the spectra of the amplitude and phase noise of the active one-port we can use a description according to the Eqs. (7.10) to (7.12). If we neglect the $1/f$-part of the noise ($f_b = 0$) and if we set $W_{an} = W_{\phi n} = W_0$, then we obtain as a final result for the amplitude and phase noise of the oscillator:

$$W_\alpha(\omega) = \frac{R_0^2}{(k_\alpha R_0)^2 + (2\,\omega L)^2} W_0 \;, \tag{7.67}$$

$$W_\phi(\omega) = \frac{R_0^2}{(2\,\omega L)^2} W_0 \;. \tag{7.68}$$

Note that the amplitude and phase noise differently depend on the offset frequency ω (see Fig. 7.5). The phase noise constantly decreases with growing distance from the carrier by 20 dB/decade. In practice, this decay is ultimately limited by the thermal noise, e.g. the thermal noise of the load resistor, which has not been included in the above formulas. For the amplitude noise we can identify an angular corner frequency ω_t from the Eq. (7.65):

$$\omega_t = \frac{k_\alpha R_0}{2\,L} \ . \tag{7.69}$$

For $\omega \ll \omega_t$ the Eq. (7.67) leads to

$$W_\alpha(\omega) = \frac{1}{k_\alpha^2}\, W_0 \ , \quad \omega \ll \omega_t \ . \tag{7.70}$$

For offset frequencies below the corner frequency ω_t, the amplitude noise is constant. Far above the corner frequency, the second term in the denominator of Eq. (7.67) dominates:

$$W_\alpha(\omega) = \frac{R_0^2}{(2\,\omega L)^2}\, W_0 \ , \quad \omega \gg \omega_t \ . \tag{7.71}$$

For high offset frequencies the amplitude noise also decreases by 20dB/decade and has, in this model, the same spectral density as the phase noise. Qualitatively, the noise spectra versus frequency are depicted in Fig. 7.8a.

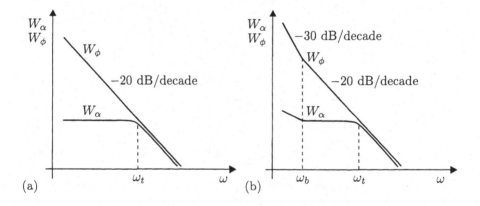

Fig. 7.8 Qualitative spectra of the amplitude and phase noise of a one-port oscillator (a) without $1/f$-noise and (b) with $1/f$-noise.

Problem

7.5   A one-port oscillator has the element values $L = 1\,\mu\text{H}$, $C = 0.1\,\text{pF}$, $R_0 = 50\,\Omega$, $k_\alpha = 2$, $P_{in} = 1\,\text{mW}$, $F_{ef} = 12\,\text{dB}$. Calculate the oscillation frequency and the spectra of the amplitude and phase noise.

The shape of the spectra changes, if with $f_b \neq 0$ also a contribution of the $1/f$-noise is considered. Qualitatively, for this case the noise spectra versus frequency are depicted in Fig. 7.8b.

Finally, we will inspect the circuit for stability. After the introduction of a complex frequency p, the denominator $De(p)$ of Eq. (7.65) reads

$$De(p) = k_\alpha R_0 + 2\,pL \ . \tag{7.72}$$

From $De(p_1){=}0$ it follows

$$p_1 = -\frac{k_\alpha R_0}{2\,L} = -\omega_t < 0 \ . \tag{7.73}$$

The zero of the denominator $De(p)$ is negative real. Therefore, the solution of the oscillation condition (7.52) corresponds to a stable oscillation mode of the oscillator.

Problem

7.6   In the oscillator circuit of Fig. 7.6 the series resonance circuit is replaced by the parallel resonance circuit of Fig. 7.9. Prove that this oscillator circuit is not stable. What modification of the U-I-characteristic of Fig. 7.7 is necessary in order to obtain stable oscillations?

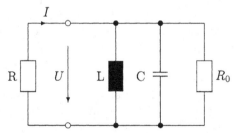

Fig. 7.9   Oscillator circuit with a parallel resonator.

Some oscillator circuits may be considered as one-port or two-port oscillators, simply by defining the interface between the active and the passive part of the circuitry accordingly. An example is given in Fig. 7.10, where the active element is an operational amplifier.

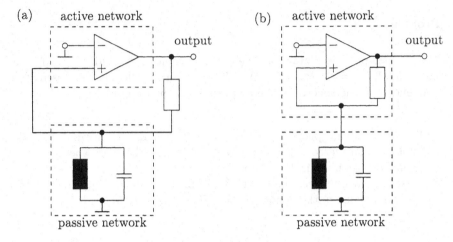

Fig. 7.10   An example of an oscillator, which may equally well be considered as (a) a two-port or (b) a one-port oscillator.

Generally, the interface should be chosen such that the active part includes all active components and only passive components with weak frequency dependence. The passive part should contain only passive components including the resonant part of the passive circuitry. Then, the convenient approximation is justified that the active part can be considered as frequency independent with respect to the small offset frequency range close to the carrier frequency. The passive part does not depend on the oscillation amplitude. Under these assumptions, the interface between these two main parts of the oscillator circuit is arbitrary and any choice will lead to the same result.

A negative input resistance or a negative real part of an input impedance is needed for the operation of a one-port oscillator. Such negative resistances can be realized with transistors. Figure 7.11 shows the idealized small signal equivalent circuit of a bipolar transistor with a voltage controlled current source, controlled by the voltage u_{be}, and with the trans-conductance g_m. As will be shown in the problem 7.7, the real part of the small signal input admittance Y_{ei} becomes negative, if the load admittance at the base, Y_B, is an inductive reactance and the collector admittance Y_C is real. The circuit can oscillate, if a resonator circuit is connected to the emitter terminal, of which the real part of the admittance is smaller than the real part of the input admittance $\mathrm{Re}\{Y_{ei}\}$. A similar argument applies to the small signal equivalent circuit of a field effect transistor, as shown in Fig. 7.12. The calculation of the input admittance Y_{si} at the source can be performed in a similar way to that described in problem 7.7.

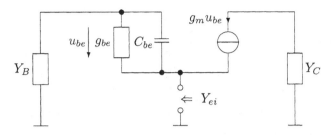

Fig. 7.11 Small signal equivalent circuit of a negative resistance with a bipolar transistor.

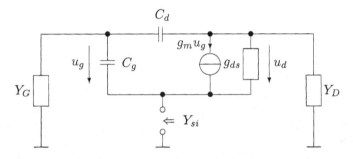

Fig. 7.12 Small signal equivalent circuit of a negative resistance with a field effect transistor.

Problem

7.7 Calculate the input admittances Y_{ei} of Fig. 7.11 and show that it can become negative for an inductive reactance at the base.

7.5.3 Voltage controlled oscillator (VCO)

A voltage controlled oscillator (VCO) is tunable with respect to the oscillation frequency via a varactor or capacitance diode, which changes the resonance frequency of the passive part of the circuitry. A one-port equivalent circuit of a VCO is shown in Fig. 7.13. The active element is a bipolar high-frequency transistor. For a given oscillation frequency the noise theory of the VCO does not differ from the noise theory of a fixed tuned oscillator, which has been presented above. The varactor diode can be treated as a passive lossy capacitor, of which the instantaneous capacitance is determined by the bias voltage of the varactor. The losses of the varactor give rise to thermal noise with a noise temperature, which is equivalent to the physical junction tem-

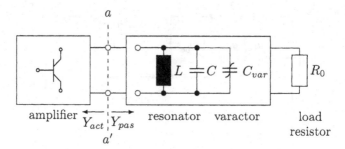

Fig. 7.13   A one-port equivalent circuit of a VCO.

perature of the varactor. A necessary condition for such a noise behavior is a low noise bias supply of the varactor. The bias and tuning voltage should be well filtered, stabilized and the bias voltage source impedance should be low. If these precautions are taken, then the excess noise of the varactor due to a frequency modulation by the bias circuit can be negligible as compared with the thermal noise of the capacitance diode and the other passive circuitry.

Across the tuning range of the VCO the small signal real part of the active admittance Y_{act}, as seen from the plane a-a' in Fig. 7.13, must be higher than the real part of the admittance of the passive branch Y_{pas} of the one-port oscillator. According to a rule of thumb it should be at least about twice as high, i.e. across the intended tuning range. The oscillation frequency is determined by the condition that the imaginary parts of the admittances are of opposite sign. Figure 7.14 shows the simulated small signal admittances versus frequency of a VCO with a structure as shown in Fig. 7.13, which is tunable from 5 to 9 GHz.

7.6   NOISE IN PHASE-LOCKED LOOP CIRCUITS

Figure 7.15 shows the principal structure of a phase locked loop circuit. The oscillator to be stabilized must be tunable, i.e. it must be a VCO, a voltage controlled oscillator. The VCO signal with the frequency f_v is divided in frequency by a factor of N by means of a programmable divider and is then compared with the phase of a reference signal with the frequency f_{ref}. The phase comparison is performed with a so called phase detector, which generates an output voltage that is proportional to the phase difference of the divided VCO signal and the reference signal. The reference signal often is derived from a quartz crystal oscillator signal, possibly after a frequency division. In the locked mode of the phase control loop, the phase of the VCO signal is rigidly locked to the phase of the reference signal in such a way that the phase difference is constant and typically small. Then the divided VCO angular frequency $\Omega_v = 2\pi f_v$ is equal to the reference angular frequency

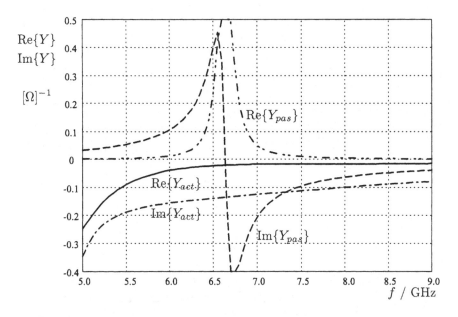

Fig. 7.14 Small signal admittances Y_{act} of the active and Y_{pas} of the passive circuit of the VCO.

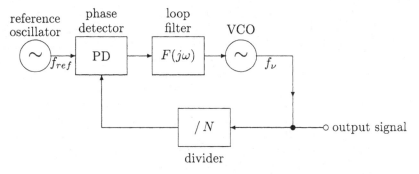

Fig. 7.15 Block diagram of a phase locked loop circuit.

$\Omega_{ref} = 2\pi f_{ref}$, i.e.

$$\Omega_v = N \cdot \Omega_{ref} \tag{7.74}$$

or

$$f_v = N \cdot f_{ref} \ . \tag{7.75}$$

The frequency stability of the VCO output signal only depends on the stability of the reference signal and the properties of the control loop. This feature is the essential reason for the great importance of phase locked loops for the realization of stable signal sources at high frequencies. The long term and partly also the short term frequency stability of a crystal oscillator can thus be transferred e.g. to microwave frequencies, where crystal oscillators are not available. In addition, the frequency of the VCO signal can be changed in steps, simply by a variation of the division ratio N in integer steps. Thus, a huge number of different frequencies can be generated in equal frequency steps, all with the same relative long term stability as the crystal oscillator signal.

Besides the long term frequency stability, the short term stability or phase or frequency noise of an oscillator can also be improved. Figure 7.16 shows two signal flow diagrams for the phase fluctuations corresponding to Fig. 7.15. In this section about phase locked loops, only phase fluctuations will be con-

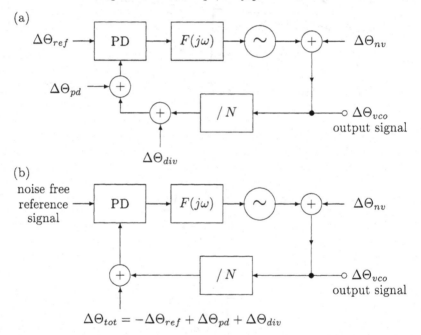

Fig. 7.16   Signal flow diagrams for the phase fluctuations within a phase locked loop circuit.

sidered, amplitude fluctuations will be neglected, because their influence normally is of minor importance. The phase fluctuations of the free-running VCO will be denoted by the phasor $\Delta\theta_{nv}$, those of the reference oscillator by the phasor $\Delta\theta_{ref}$ and the phase noise of the stabilized VCO output signal by the phasor $\Delta\theta_{vco}$ (Fig. 7.16a). Furthermore, the equivalent phase noise produced by a noisy phase discriminator including the loop filter and amplifier as referred to the divide-by-N output port will be described by the phasor $\Delta\theta_{pd}$.

Finally, the equivalent phase noise of the divide-by-N circuit, also referred to the output port of the divider, will be described by the phasor $\Delta\theta_{div}$. We can introduce a resultant phasor $\Delta\theta_{tot}$, referred to the plane of the reference oscillator with

$$\Delta\theta_{tot} = -\Delta\theta_{ref} + \Delta\theta_{pd} + \Delta\theta_{div} \ , \qquad (7.76)$$

where $\Delta\theta_{tot}$ describes the total equivalent phase noise at the input of the phase detector (Fig. 7.16b). The noise phasors $\Delta\theta_{ref}$ and $\Delta\theta_{pd}$ and $\Delta\theta_{div}$ typically are all mutually uncorrelated. Therefore, we can write for the corresponding phase spectra

$$W_{tot} = W_{ref} + W_{pd} + W_{div} \ . \qquad (7.77)$$

The frequency divider also reduces the amplitude of the phase fluctuations by a factor of N (compare with Section 6.8.3). The sensitivity of the phase detector is described by the constant K_{pd}. The loop filter has a transfer function $F(j\omega)$, which normally is frequency dependent. The VCO may have the constant tuning sensitivity K_{vco}, which is assumed to be independent of the offset frequency ω. The transfer function of the control voltage at the tuning input of the VCO to the phase of the VCO is given by $K_{vco}/j\omega$. This is because the frequency of the oscillator as controlled by the tuning voltage is the time derivative of the phase. We therefore obtain the following relationship from the flow diagram of Fig. 7.16:

$$\Delta\theta_{vco} = \cfrac{1}{1 + \cfrac{1}{N}K_{pd}\,F(j\,\omega)\,\cfrac{K_{vco}}{j\,\omega}}\,\Delta\theta_{nv} + \cfrac{K_{pd}\,F(j\,\omega)\,\cfrac{K_{vco}}{j\,\omega}}{1 + \cfrac{1}{N}K_{pd}\,F(j\,\omega)\,\cfrac{K_{vco}}{j\,\omega}}\,\Delta\theta_{tot} \ .$$

$$(7.78)$$

The control behavior strongly depends on the open loop gain $V(j\omega)$:

$$V(j\omega) = K_{pd}\,F(j\,\omega)\,\frac{K_{vco}}{j\,\omega} \ . \qquad (7.79)$$

A loop filter may be designed by means of an operational amplifier according to the schematic diagram in Fig. 7.17 and with the transfer function $F(j\omega)$:

$$F(j\omega) = \frac{U_2}{U_1} = -\frac{j\omega\tau_2 + 1}{j\omega\tau_1} \quad \text{with} \quad \tau_1 = R_1 C \quad \text{and} \quad \tau_2 = R_2 C \ . \quad (7.80)$$

The magnitude of the transfer function $|F(j\omega)|$ of the loop filter as a function of frequency is schematically shown in Fig. 7.18. Typically the open loop gain will show a low-pass behavior. Let us define an angular corner frequency ω_{con} of the loop filter such that $|V(j\omega_{con})| = 1$. For $|V(j\omega)| \gg 1$, i.e. $\omega \ll \omega_{con}$, the phase noise of the oscillator output signal $\Delta\theta_{vco}$ is determined by the phase fluctuations of the reference signal plus the equivalent noise of the frequency divider and the phase discriminator plus the control circuit, i.e. $\Delta\theta_{tot}$:

$$\Delta\theta_{vco} = N \cdot \Delta\theta_{tot} \quad \text{for} \quad \omega \ll \omega_{con} \ . \qquad (7.81)$$

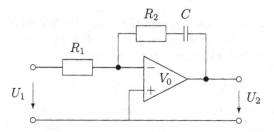

Fig. 7.17 Equivalent circuit of an active loop filter of first order with the amplification factor $V_0 \approx \infty$.

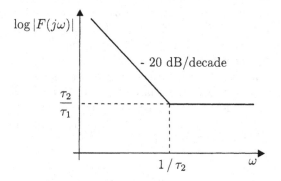

Fig. 7.18 Transfer function $F(j\omega)$ of the loop filter versus frequency.

For $|V(j\omega)| \ll 1$ and $\omega \gg \omega_{con}$ the phase noise of the oscillator output is equal to the phase noise of the free running VCO, $\Delta\theta_{nv}$.

$$\Delta\theta_{vco} = \Delta\theta_{nv} \qquad \text{for} \qquad \omega \gg \omega_{con} \ . \qquad (7.82)$$

In summary, a phase locked loop circuit offers the possibility to reduce the phase noise of a tunable oscillator almost to the level of the reference oscillator, but only for offset frequencies ω that are small with respect to the angular corner frequency ω_{con} of the open loop gain.

The reference oscillator will often be a quartz crystal oscillator with rather low phase noise and a low $1/f$-noise corner frequency. Further noise will add, e.g. noise from the phase detector and the frequency divider. In the following text, we will denote the total phase noise $\Delta\theta_{tot}$ by the name *reference oscillator phase noise* or *reference phase noise* for simplicity. Typically, this reference phase noise will show a rather flat and frequency independent behavior, although in absolute terms the noise level will increase considerably due to the times N multiplication factor, as predicted by Eq. (7.81). On the other hand, the internal phase noise of the free-running VCO will typically drop with increasing offset frequencies by 20 dB/decade. Therefore, we will normally observe an intersection of the multiplied reference phase noise curve and the free-running VCO phase noise curve. Let us call the angular frequency belonging to this intercept frequency ω_{int}. A loop filter angular corner frequency ω_{con}, which is approximately equal to ω_{int}, may be a good choice in many circumstances and will give an overall phase noise behavior versus offset frequency as shown qualitatively in Fig. 7.19.

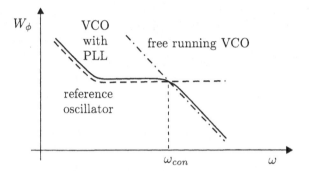

Fig. 7.19 Qualitative phase noise spectrum of a PLL stabilized VCO with $\omega_{con} \approx \omega_{int}$.

If we choose the cut-off frequency of the open loop gain ω_{con} smaller than the intercept frequency ω_{int}, i.e. $\omega_{con} < \omega_{int}$, then we will observe a phase noise behavior versus offset frequency as shown in Fig. 7.20. We observe an increase of the phase noise of the stabilized oscillator around the intercept frequency ω_{int}, because the phase noise of the free running VCO is now dominating around the offset frequency ω_{con}. If we choose the cut-off fre-

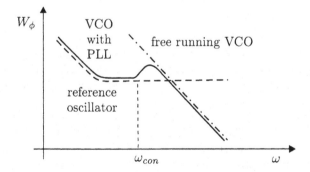

Fig. 7.20 Qualitative phase noise spectrum of a PLL stabilized VCO with $\omega_{con} < \omega_{int}$.

quency of the open loop gain ω_{con} greater than the intercept frequency ω_{int}, i.e. $\omega_{con} > \omega_{int}$, then we will observe a phase noise function versus offset frequency as shown in Fig. 7.21. We observe an increase of the phase noise of the stabilized oscillator above the intercept frequency ω_{int}, because the phase noise of the reference oscillator is now dominating around the offset frequency ω_{con}. In summary, within the loop bandwidth, a VCO stabilized by a phase

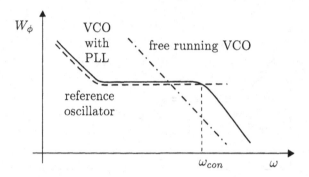

Fig. 7.21   Qualitative phase noise spectrum of a PLL stabilized VCO with $\omega_{con} > \omega_{int}$.

locked loop behaves much like a frequency multiplier concerning the input and output phase noise. Because the input phase noise spectrum is multiplied by the division factor N^2, care should be taken to keep the reference frequency as high as possible. If this does not comply with the desired frequency step, a solution with more than one loop has to be chosen.

7.7   MEASUREMENT OF THE OSCILLATOR NOISE

7.7.1   Amplitude noise

An apparently obvious method for the measurement of the amplitude noise, namely the application of a spectrum analyzer, does not work because the phase noise is dominant and covers the amplitude noise. Therefore, we need a selective method to measure the amplitude noise.

In the simplest case, the amplitude noise can be measured with a setup as shown in Fig. 7.22. The amplitude noise detector consists of a rectifier diode, which can be a Schottky diode, a low pass filter and an amplifier. The output voltage $u(t)$ is proportional to the time dependent amplitude of the input signal $x(t)$. With $x(t) = [X_0 + \Delta x(t)] \cos[\Omega_0 t + \Delta\Phi(t)]$ the relation

$$u(t) = K_{ad}[X_0 + \Delta x(t)] \tag{7.83}$$

may be formulated for the output signal $u(t)$ of the amplitude detector. Here, K_{ad} is a proportionality constant characterizing the sensitivity of the ampli-

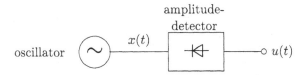

Fig. 7.22 Simple circuit for the measurement of amplitude noise.

tude detector. The output signal consists of a constant part $U_0 = K_{ad}X_0$ and a time dependent part $\Delta u(t) = K_{ad}\Delta x(t)$. We obtain the spectrum $W_\alpha(\omega)$ of the normalized amplitude fluctuations $\Delta x(t)/X_0$ from the d.c. voltage U_0 and the spectrum $W_u(\omega)$ of the signal part $\Delta u(t)$:

$$W_\alpha(\omega) = \frac{1}{U_0^2} W_u(\omega) \ . \tag{7.84}$$

In this form, the measurement method is only partly useful in practice. The amplitude noise of most oscillators is so weak, that it will give rise to a detector output signal of only the same order of magnitude as the internal detector noise. In order to achieve a better sensitivity, it is necessary to determine the internal detector noise first and then subtract it from the measured total noise spectrum W_u. Because the internal noise depends upon the carrier drive level, it would be necessary to drive the detector with a noise-free carrier signal of the same amplitude and frequency. This, however, is not possible in practice. The problem can be solved with a measurement circuit as shown in Fig. 7.23. The oscillator signal is split into two equal signals by means of a signal divider

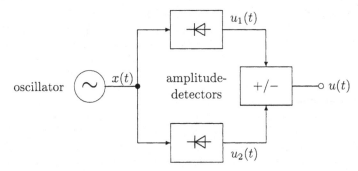

Fig. 7.23 Measurement of the amplitude noise with two amplitude detectors.

and is then applied to a couple of paired amplitude detectors. We then obtain for the output signals or voltages

$$u_1(t) = K_{ad}[X_0 + \Delta x(t)] + u_{n1}(t) \ , \tag{7.85}$$

$$u_2(t) = K_{ad}[X_0 + \Delta x(t)] + u_{n2}(t) \ , \tag{7.86}$$

where the signals u_{n1} and u_{n2} describe the internal noise of the detectors. The output voltage can either be the sum or the difference of the voltages u_1 or u_2:

$$u_\Sigma(t) = u_1(t) + u_2(t) = 2K_{ad}[X_0 + \Delta x(t)] + u_{n1}(t) + u_{n2}(t) \quad, (7.87)$$
$$u_\Delta(t) = u_1(t) - u_2(t) = u_{n1}(t) - u_{n2}(t) \quad, \tag{7.88}$$

The voltage u_Σ includes a d.c. component $U_0 = 2K_{ad}X_0$, while the voltage u_Δ only consists of the internal noise of the detectors. Because the noise contributions from the two detectors are uncorrelated, it follows for the spectra of the time dependent parts:

$$W_{u\Sigma}(\omega) = (2\,K_{ad}\,X_0)^2 W_\alpha(\omega) + W_{n1} + W_{n2} \quad, \tag{7.89}$$
$$W_{u\Delta}(\omega) = W_{n1} + W_{n2} \quad. \tag{7.90}$$

Then, we obtain the spectrum $W_\alpha(\omega)$ of the normalized amplitude fluctuations $\Delta x(t)/X_0$ by the algebraic operation

$$W_\alpha(\omega) = \frac{1}{U_0^2} \left[W_{u\Sigma}(\omega) - W_{u\Delta}(\omega) \right] \quad. \tag{7.91}$$

In this manner, the influence of the internal detector noise can be eliminated to a large extent.

Alternatively, the influence of the detector noise may also be eliminated by calculating the cross-correlation of the signals $u_1(t)$ and $u_2(t)$, instead of taking the sum and difference of the signals.

As an example, Fig. 7.24 schematically shows the measured amplitude noise of an oscillator with a GaAs field effect transistor as the active element. The oscillation frequency is approximately 10 GHz.

Because the amplitude noise is less important for practical applications than the phase noise, it is seldom measured and in most cases not specified in data sheets.

7.7.2   Phase noise

The most convenient method in order to measure the phase noise of an oscillator is by means of a high quality spectrum analyzer as has already been mentioned in Section 6.9. There are, however, some restrictions on the use of a spectrum analyzer, as has also been discussed in Section 6.9. A successful measurement is only possible if the amplitude noise is sufficiently small and if the internal phase noise of the spectrum analyzer is lower than the phase noise to be measured. Here, we will examine other possibilities for the measurement of phase noise, which do not need a spectrum analyzer.

The first method needs a second oscillator, e.g. at the same frequency. The measurement principle shows similarities with the measurement of the phase noise of two-ports, described in Section 6.9. Figure 7.25 shows a block diagram of the measurement circuit.

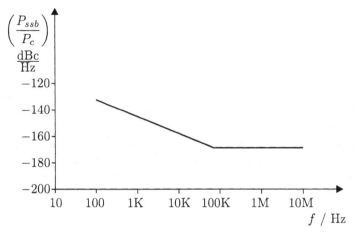

Fig. 7.24   Schematic diagram of the measured amplitude noise of a GaAs field effect transistor oscillator with an oscillation frequency of approximately 10 GHz.

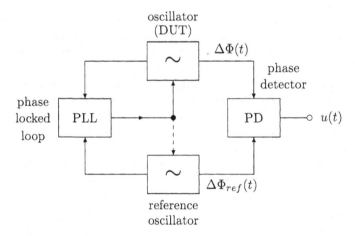

Fig. 7.25   Measurement of the phase noise with a reference oscillator.

The phase fluctuations of the output signals of the device under test and the reference oscillator are compared by a linear phase detector. With $\Delta\Phi(t)$ as the phase fluctuations of the measurement oscillator and $\Delta\Phi_{ref}(t)$ as the phase fluctuations of the reference oscillator, we can write for the signal $u(t)$ at the output of the phase detector

$$u(t) = K_{pd} \left[\Delta\Phi(t) - \Delta\Phi_{ref}(t)\right] \ , \tag{7.92}$$

or for the corresponding spectra, respectively,

$$W_u(\omega) = K_{pd}^2 \left[W_\phi(\omega) + W_{ref}(\omega)\right] \ . \tag{7.93}$$

The constant K_{pd} again describes the sensitivity of the phase detector. The condition for the validity of the Eqs. (7.92) and (7.93) is that both sources oscillate at exactly the same frequency. Furthermore, the phase difference of these two oscillators must either be constant or show only small and slow variations in time. The stringent requirements regarding the frequency equality of the two oscillators normally can not be met with free running oscillators but rather requires a phase locked loop system for stabilization. Therefore, either the measurement oscillator or the reference oscillator needs to be tunable in frequency. It is important that the corner frequency of the phased locked loop system is smaller than the smallest offset frequency to be measured. Otherwise the phase fluctuations of both oscillators are equal and due to $\Delta\Phi(t) = \Delta\Phi_{ref}(t)$ the output signal of the phase detector becomes zero according to Eq. (7.92). The input signals of the phase detector may also be derived from the oscillator signals by frequency division, if the oscillation frequencies are too high as direct inputs to a highly linear digital phase detector.

The spectrum $W_u(\omega)$ can be analyzed with a low frequency spectrum analyzer after a proper low noise amplification. A high-frequency spectrum analyzer may be used for an absolute calibration. According to Eq. (7.92) the spectrum $W_u(\omega)$ is the sum of the phase noise spectra of the measurement oscillator and the reference oscillator. Therefore, it is obvious to use a reference oscillator with a very low phase noise so that its contribution to W_n can be neglected. Otherwise the measurement procedures discussed in section 6.9 can also be applied to oscillator measurements. Thus, the reference oscillator may be replaced by a second measurement oscillator, closely matched to the first one, which requires a correction of the measurement result by 3 dB, or one needs to evaluate the combined noise of the three possible pairs of three different oscillators (see problem 6.8).

The calibration of the different discriminator circuits can be performed by modulating the oscillator under test in frequency or in phase at a given offset frequency ω. For this purpose, also a sinusoidal calibration signal may be introduced by phase or frequency modulating the tunable oscillator. The modulation index and thus the calibration factor can be determined via the generated carrier sidebands. The sidebands can be observed on a high-frequency spectrum analyzer. In case that the oscillator under test is not tunable, a modulation of the reference oscillator might be necessary. Then, a tunable reference oscillator is needed.

The use of a reference oscillator and a frequency stabilizing circuit is somewhat disadvantageous because of the high effort involved required for its realization. The next method does not need a reference oscillator because the oscillator signal is directly applied to a frequency discriminator, where the phase or frequency fluctuations of the oscillator under test are converted to a low frequency noise signal.

The working principle of most frequency discriminators relies on the conversion of phase or frequency fluctuations of the discriminator input signal into amplitude fluctuations, which then can be measured quite easily with an amplitude discriminator. Figure 7.26 shows a block diagram of the basic setup of such frequency discriminators. The output signal of the oscillator under test is divided by a signal divider into two parts and is applied to two linear networks with strongly different transfer functions $H(j\Omega)$ and $G(j\Omega)$. In the following section, $H(j\Omega)$ is assumed to be the transfer function of a transmission type resonator with a high quality factor and $G(j\Omega)$ is assumed to be the transfer function of a frequency independent phase shift network. The output signals of the two networks H and G are added and yield a sum

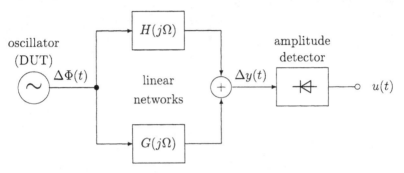

Fig. 7.26   Block diagram of a frequency discriminator for the measurement of phase noise.

signal phasor Y with the amplitude Y_0. The amplitude detector converts the amplitude fluctuations $\Delta y(t)$ of this sum signal to the output signal $u(t)$.

Up to the amplitude detector input, the discriminator is a linear system with the transfer function $H(j\Omega) + G(j\Omega)$. The normalized amplitude fluctuations can therefore be calculated with Eq. (7.14) in complex form:

$$\frac{\Delta Y}{Y_0} = \frac{j}{2}\left[\frac{H(\Omega_r + \omega) + G(\Omega_r + \omega)}{H(\Omega_r) + G(\Omega_r)} - \frac{H^*(\Omega_r - \omega) + G^*(\Omega_r - \omega)}{H^*(\Omega_r) + G^*(\Omega_r)}\right]\Delta\Phi \ .$$

$$(7.94)$$

In this equation, the amplitude noise of the sinusoidal carrier signal under test has been neglected.

The relation between the amplitude fluctuations $\Delta y(t)$ of the sum signal and the output signal $u(t)$ of the detector can again be described by means of Eq. (7.83). The output signal consists of a DC part $U_0 = K_{ad}Y_0$ and a fluctuating part $\Delta u(t) = K_{ad} \cdot \Delta y(t)$. We are only interested in the time dependent part $\Delta u(t)$ or the discriminated phase noise and obtain the following expression for the spectrum $W_u(\omega)$:

$$W_u(\omega) = \left(\frac{K_{ad}Y_0}{2}\right)^2$$

$$\cdot \left|\frac{H(\Omega_r + \omega) + G(\Omega_r + \omega)}{H(\Omega_r) + G(\Omega_r)} - \frac{H^*(\Omega_r - \omega) + G^*(\Omega_r - \omega)}{H^*(\Omega_r) + G^*(\Omega_r)}\right|^2 W_\phi(\omega) \ . \ (7.95)$$

As an example for a frequency discriminator we will examine the circuit of Fig. 7.27 in more detail. Here, the frequency dependent network $H(\Omega)$ consists of a circulator and a reflection type resonator, which is tuned to the oscillation frequency of the oscillator under test. This circuit structure has a particularly high discrimination efficiency. If the signals are assumed to be wave quantities,

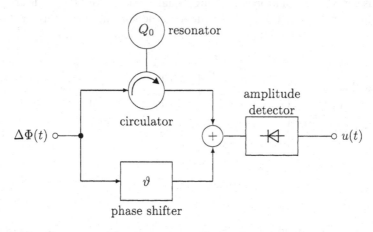

Fig. 7.27   Frequency discriminator with a reflection type resonator.

then the transfer function $H(j\Omega)$ is identical to the reflection coefficient of the resonator:

$$H(\Omega) = \frac{\beta - 1 - j\,Q_0 \left(\dfrac{\Omega}{\Omega_r} - \dfrac{\Omega_r}{\Omega}\right)}{\beta + 1 + j\,Q_0 \left(\dfrac{\Omega}{\Omega_r} - \dfrac{\Omega_r}{\Omega}\right)} \quad . \tag{7.96}$$

In the last equation, $\beta > 1$ is the coupling factor to the resonator and Q_0 is the quality factor of the unloaded resonator. With $\Omega = \Omega_r + \omega$ and $\omega \ll \Omega_r$ Eq. (7.96) simplifies to

$$H(\Omega_r + \omega) = \frac{\beta - 1 - j2\,Q_0\,\omega/\Omega_r}{\beta + 1 + j2\,Q_0\,\omega/\Omega_r} \quad . \tag{7.97}$$

The other linear network is a phase shifter with a frequency independent transfer function

$$G(j\Omega) = e^{j\vartheta} \quad . \tag{7.98}$$

It follows from the Eqs. (7.95), (7.97) and (7.98)

$$W_u(\omega) = \left(\frac{K_{ad}\,Y_0}{2}\right)^2$$

$$\cdot \left| \frac{\dfrac{\beta - 1 - j2\,Q_0\,\omega/\Omega_r}{\beta + 1 + j2\,Q_0\,\omega/\Omega_r} + e^{j\vartheta}}{\dfrac{\beta - 1}{\beta + 1} + e^{j\vartheta}} - \frac{\dfrac{\beta - 1 - j2\,Q_0\,\omega/\Omega_r}{\beta + 1 + j2\,Q_0\,\omega/\Omega_r} + e^{-j\vartheta}}{\dfrac{\beta - 1}{\beta + 1} + e^{-j\vartheta}} \right| W_\phi(\omega) \ ,$$

$$(7.99)$$

or, after some manipulations,

$$W_u(\omega) = (K_{ad}\,Y_0)^2 \left[\frac{\sin\vartheta}{1 + \left(\dfrac{\beta - 1}{\beta + 1}\right)^2 + 2\,\dfrac{\beta - 1}{\beta + 1}\cos\vartheta} \right]^2$$

$$\cdot \left| \frac{4\beta\,Q_0\,\omega/\Omega_r}{(\beta + 1)(\beta + 1 + j2\,Q_0\,\omega/\Omega_r)} \right|^2 W_\phi(\omega) \ . \qquad (7.100)$$

By forming the derivative with respect to ϑ we can find the optimum setting of the phase shifter, as a function of the coupling factor β. The conversion efficiency of the discriminator becomes maximum for

$$\vartheta = \vartheta_0 = \arccos\left[\frac{-2\,\dfrac{\beta - 1}{\beta + 1}}{1 + \left(\dfrac{\beta - 1}{\beta + 1}\right)^2} \right] \ . \qquad (7.101)$$

A variation of the coupling factor β between 0 and ∞ causes a change of the optimum phase shift value ϑ_0 between $0°$ and $180°$. For a very weak coupling ($\beta \approx 0$) as well as for a very strong coupling ($\beta \gg 1$), the discrimination efficiency becomes very small. In between there exists an optimum coupling value of $\beta = 1$, i.e. $H(j\Omega_r) = 0$. This coupling is also called *critical coupling*. The corresponding phase shift value is $\vartheta_0 = 90°$.

If we limit the discussion to offset frequencies that are small compared with the 3-dB bandwidth of the resonator, then $2\,Q_0\,\omega/\Omega_r \ll 1$ is valid. For a constant phase angle $\vartheta = 90°$ we obtain from Eq. (7.100)

$$W_u(\omega) = \left(K_{ad}\,Y_0\,\frac{2\beta\,Q_0}{1 + \beta^2}\,\frac{\omega}{\Omega_r} \right)^2 W_\phi(\omega) \ . \qquad (7.102)$$

It also follows from Eq. (7.102) that the critically coupled resonator ($\beta = 1$) has the best discrimination efficiency. Once the coupling β and the phase angle ϑ have optimally been chosen, the discrimination efficiency entirely depends on the unloaded Q-factor of the resonator. Therefore, for the measurement of low-noise oscillators the Q-factor of the resonator should be sufficiently high.

If a circulator in the circuit of Fig. 7.27 is not available or not suitable, one may employ a 3 dB coupler instead, however at the expense of 6 dB additional loss in conversion efficiency.

The ratio of W_u and W_ϕ is proportional to the square of the offset frequency ω, as may be seen from Eq. (7.102). This shows that the circuit basically acts as a frequency discriminator and not as a phase discriminator. If in Eq. (7.102) the spectrum W_ϕ is replaced by the frequency noise spectrum $W_f = (\omega/2\pi)^2\, W_\phi$, then the conversion efficiency is independent of the offset frequency.

The circuit of Fig. 7.27 may be simplified by using a transmission type resonator, as shown in Fig. 7.28. In this case, the transmission line bypass with the phase shifter ϑ is not needed. However, for this configuration the carrier

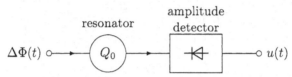

Fig. 7.28   Frequency discriminator with a transmission type resonator.

frequency must not coincide with the resonance frequency of the resonator but should rather be positioned on the slope of the resonance curve. There is an optimum frequency position which has to be found in the next problem. The discrimination efficiency is lower than that of the circuit of Fig. 7.27. The details are left to problem 7.8.

Problem

7.8   Calculate the discrimination efficiency for the circuit of Fig. 7.28. At what offset frequency relative to the resonance frequency of the resonator does the maximum discrimination efficiency occur? What is the degradation in efficiency relative to the circuit of Fig. 7.27, if we assume the same unloaded Q-factor for both circuits?

A frequency discriminator can also be realized with a delay line as shown in Fig. 7.29. The discriminator circuit is similar to the one in Fig. 7.27. It is assumed that the delay line has a length l and an attenuation factor $\alpha = \alpha' \cdot l$. Furthermore, we will assume that the total length of the is an integer multiple n of the wavelength at the oscillation frequency. As will be shown in problem 7.9, the relation between the spectra W_ϕ and W_u is given by the expression

$$W_u(\omega) = (K_{ad}\, Y_0)^2\left[\frac{\sin\dfrac{\omega l}{2v}}{\cosh\alpha' l}\right]^2 W_\phi(\omega) \,, \qquad (7.103)$$

where $v = \Omega_0\, l/(n2\pi)$ is the phase velocity of the transmission or delay line.

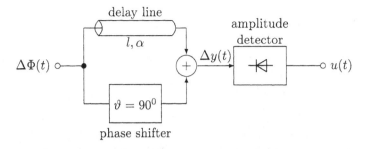

Fig. 7.29 Frequency discriminator with a delay line.

Problem

7.9 Derive Eq. (7.103). Which length of the delay line leads to the highest discrimination efficiency?

Up to now we have assumed that the oscillator signal to be measured is free of amplitude fluctuations. If this approximation is not valid, then the measurement circuit of Fig. 7.27 can be modified by substituting the amplitude detector by a balanced mixer (see Fig. 7.30).

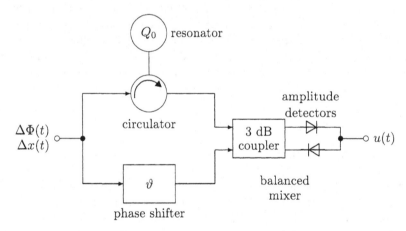

Fig. 7.30 Frequency discriminator with a balanced mixer.

The balanced mixer allows one to reduce the impact of the amplitude noise significantly, as is proven in problem 7.10.

Problem

7.10   Show that for the balanced discriminator circuit of Fig. 7.30 the amplitude noise $\Delta x(t)$ of the carrier signal does not contribute to the output signal $u(t)$ for a perfectly balanced mixer.

The calibration of the different discriminator circuits can be performed by modulating the oscillator under test in frequency or in phase at a given offset frequency ω. As has been discussed before, the modulation index and thus the calibration factor can be determined via the generated carrier sidebands. The sidebands can be measured by a high-frequency spectrum analyzer. In the case that the oscillator under test is not tunable, a frequency modulation of the resonator might be an alternative. Because only a very small modulation index is needed, the resonator may loosely be coupled to e.g. a varactor diode, without deteriorating the quality factor of the resonator significantly. The calibration factor of the resonator, which is modulated in its resonance frequency, can be determined by a substitution measurement with a tunable oscillator of the same frequency and output power. The comparison is done on the basis of equal low frequency signals at the output of the discriminator. The calibrations should be evaluated as a function of the offset frequency ω.

High-frequency systems, for instance radar systems or parts of radar systems, often have a structure similar to the circuit shown in the figures 7.26 to 7.30. Therefore, it is possible that a circuit in the system acts as a frequency or phase discriminator and converts the phase noise of oscillator or carrier signals into low frequency noise signals. These noise signals are sometimes considerably stronger than the internal noise of preamplifiers or mixers. A reduction of the carrier phase noise may then be necessary.

7.7.3   Injection locking

For the method of injection locking, the signal of an auxiliary oscillator or an injecting oscillator or a reference oscillator is directly injected or coupled into the main oscillator. One may use an amplifier and a circulator or a coupler to isolate the auxiliary oscillator from the main oscillator and the output as shown in Fig. 7.31. If the fixed frequency Ω_i of the auxiliary oscillator and the frequency Ω_0 of the free running main oscillator are sufficiently close together, then the main oscillator will also oscillate at exactly the frequency Ω_i, i.e. the frequency of the reference oscillator. We can observe that the main oscillator is synchronized by the reference oscillator. For a quantitative analysis of this effect we must extend the signal flow diagram of Fig. 7.2 as shown in Fig. 7.32. The injecting oscillator signal Z_i is coupled to the oscillation circuit via the coupling network $E(\Omega)$ and induces the signal Y_i into the oscillation loop.

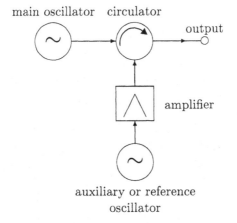

Fig. 7.31 Circuit for injection locking of an oscillator.

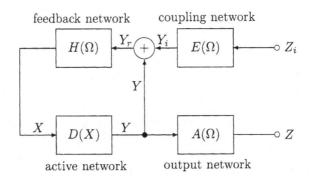

Fig. 7.32 Signal flow diagram for an injection locked oscillator.

This modification alters the oscillation condition. With the definition

$$\frac{Y_i}{Y} = q_i \, e^{j \, \vartheta} \tag{7.104}$$

we obtain for the oscillation condition

$$D(X) \, H(\Omega_i) \left(1 + q_i \, e^{j\vartheta}\right) = 1 \; . \tag{7.105}$$

Despite the formal similarity to Eq. (7.4), a substantial difference exists between both oscillation conditions. For the case of the free running oscillator, the oscillation condition determines the oscillation frequency and the amplitude. For the case of injection locking, the frequency of the main oscillator is equal to the frequency Ω_i of the auxiliary oscillator and thus is known. But now the angle ϑ appears as an unknown quantity, which describes the phase difference between the injecting signal Y_i and the oscillation signal Y.

By means of Eq. (7.105) some general properties of an injection locked oscillator can be derived. For this purpose, we will assume in the following section that the oscillation condition has a solution with the amplitude X_0 and the angular frequency Ω_0 for $q_i = 0$, i.e. for the free running case. For $q_i \neq 0$, generally both the amplitude and the frequency of the oscillation will change. However, usually the changes are small. Therefore, we assume that due to $X \approx X_0$ and $\Omega_i \approx \Omega_0$ the describing function $D(X)$ will only change its magnitude and the transfer function $H(\Omega)$ will only change its phase.

Under these conditions we will first consider the case $\Omega_i = \Omega_0$, i.e. the injection locked oscillator has the same frequency as the reference oscillator. Then, $D(X) H(\Omega_i) = D(X) H(\Omega_0)$ is real and due to Eq. (7.105) also the factor $1 + q_i e^{j\vartheta}$ must be real. Thus, the only solutions for the angle ϑ are 0 and $\pm\pi$. For $\vartheta = 0$, we have $1 + q_i e^{j\vartheta} > 1$ and the magnitude of $D(X)$ must decrease as compared to the free running case. For the amplifier, this will lead to an increased drive level and also to an increased oscillation amplitude. For $\vartheta = \pm\pi$, the relations will reverse. However, this operating point is unstable and, therefore, will not appear in practice. If the frequency of the reference oscillator is changed, then the product $D(X) H(\Omega_i)$ becomes complex, and therefore $1 + q_i e^{j\vartheta}$ must also become complex. From this we conclude $\vartheta \neq 0$. Because of $q_i < 1$, the phase angle of the expression $1 + q_i e^{j\vartheta}$ is limited. The largest angle occurs for $\vartheta = \pm 90°$ and is equal to $\arctan q_i$. This operating point corresponds to the largest phase change of the function $H(\Omega_i)$ and, therefore, to the largest deviation of the frequency Ω_i from Ω_0. If the reference frequency deviates even further from Ω_0, then Eq. (7.105) no longer has a solution. Thus, only a limited frequency range $\Omega_0 \pm \Delta\Omega_m$ exists, within which an injection locking or a phase synchronization of the main oscillator is possible. By varying Ω_i within this range, the phase ϑ will change from $-90°$ via 0° at $\Omega_i = \Omega_0$ to $+90°$.

The synchronization range is proportional to q_i and thus to $\sqrt{P_i/P_0}$, where P_i is the injected power and P_0 is the output power of the main oscillator. Furthermore, the frequency dependence of $H(\Omega)$ is important. The less the phase of $H(\Omega)$ changes as a function of Ω, the wider the synchronization range $2\Delta\Omega_m$ becomes. For a resonator the phase slope is proportional to the quality factor Q. Therefore, we expect a relationship of the type

$$\Delta\Omega_m \sim \frac{1}{Q}\sqrt{\frac{P_i}{P_0}} \ . \tag{7.106}$$

For a more rigorous evaluation of the synchronization range, we will consider the one-port oscillator with the series resonance circuit of Section 7.5. Figure 7.33 shows the circuit which has been supplemented by the voltage source U_i, which corresponds to the injection signal Y_i of Fig. 7.32. We thus have $U_i/U = q_i \exp(j\vartheta)$, and therefore the oscillation condition with respect to the real and imaginary part is given by:

$$R_0 + R(I)(1 + q_i \cos\vartheta) \ = \ 0 \ , \tag{7.107}$$

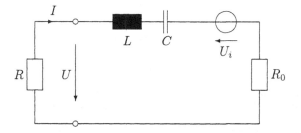

Fig. 7.33   Phase synchronization of a one-port oscillator.

$$2 \left(\Omega_i - \Omega_0 \right) L + R(I) \, q_i \sin \vartheta \;\; = \;\; 0 \; . \qquad (7.108)$$

At the limits of the synchronization range we obtain from $\Omega_i - \Omega_0 = \pm \Delta \Omega_m$ and $\vartheta = \pm 90°$:

$$R_0 + R(I) \;\; = \;\; 0 \; , \qquad (7.109)$$

$$2 \cdot \Delta \Omega_m \cdot L + R(I) \cdot q_i \;\; = \;\; 0 \; . \qquad (7.110)$$

The output power of the oscillator is given by

$$P_0 = \frac{1}{2} R_0 I^2 = \frac{1}{2} R_0 \left| \frac{U}{R(I)} \right|^2 \; , \qquad (7.111)$$

and with Eq. (7.109) we get

$$P_0 = \frac{|U|^2}{2 \, R_0} \; . \qquad (7.112)$$

If we define as injection power P_i the available power of the voltage source U_i with the resistance R_0, then

$$P_i = \frac{|U_i|^2}{8 \, R_0} \; . \qquad (7.113)$$

From this we get

$$q_i = \frac{|U_i|}{U} = 2 \sqrt{\frac{P_i}{P_0}} \; , \qquad (7.114)$$

and with the Eqs. (7.109) and (7.110):

$$\Delta \Omega_m = \frac{R_0}{L} \sqrt{\frac{P_i}{P_0}} \; . \qquad (7.115)$$

For a resonator it is customary to define an external quality factor Q_{ext} by

$$Q_{ext} = \frac{\Omega_0 \, L}{R_0} \; . \qquad (7.116)$$

If we insert this into Eq. (7.115) we obtain

$$\Delta\Omega_m = \frac{\Omega_0}{Q_{ext}} \sqrt{\frac{P_i}{P_0}} \; . \tag{7.117}$$

This is a very useful equation for the determination of the external quality factor of an oscillator by means of a measurement of the synchronization range.

Problem

7.11   In the circuit of a two-port oscillator with a transmission resonator

resonator

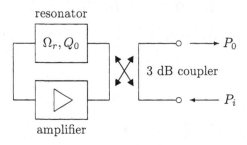

amplifier

as shown in Fig. 7.3 the signal divider is replaced by an ideal 3 dB coupler. The injection signal with the available power P_i is fed into one of the ports (see figure). By analogy with Eq. (7.117) derive for this oscillator circuit a relationship between the parameters $\Delta\Omega_m, P_i, P_0$ and the properties of the resonator.

A general determination of the amplitude and phase noise for the injection locked oscillator with the signal flow diagram of Fig. 7.32 is tedious. Therefore, we will only treat the special case that $\Omega_i = \Omega_o$, i.e. the frequency of the free running main oscillator is equal to the frequency of the reference oscillator. Then, we have $\vartheta = 0$. Furthermore, the amplitude noise will be neglected and only the phase noise will be discussed.

If the phase fluctuations of the signal Y are denoted by $\Delta\Psi$ and those of the signal Y_r by $\Delta\Psi_r$, then Eqs. (7.6), (7.14) and (7.15) yield

$$\Delta\Psi = H_\Sigma \, \Delta\Psi_r + \Delta\Psi_n \; . \tag{7.118}$$

Due to $\vartheta = 0$ and thus $Y_r = (1 + q_i)Y$ the phase fluctuations $\Delta\Psi_r$ can be determined from the fluctuations $\Delta\Psi$ and the fluctuations $\Delta\Psi_i$ of the injecting signal. Figure 7.34 illustrates the situation for the case that only the injecting signal Y_i has phase fluctuations $\Delta\Psi_i$. Since the sum signal Y_r has an amplitude which is larger by $(1 + q_i)/q_i$, the corresponding phase disturbance is reduced by approximately the same factor. A similar reasoning applies to

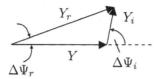

Fig. 7.34   Superposition of a noisy injecting signal.

the phase fluctuations $\Delta\Psi$ of the signal Y. In summary, we obtain the result

$$\Delta\Psi_r = \frac{1}{1+q_i}\left(\Delta\Psi + q_i\Delta\Psi_i\right) . \tag{7.119}$$

From the Eqs. (7.118) and (7.119) we get

$$\Delta\Psi = \frac{q_i\, H_\Sigma\, \Delta\Psi_i + (1+q_i)\Delta\Psi_n}{1+q_i-H_\Sigma} . \tag{7.120}$$

For a better insight into the last equation we will again consider the one-port oscillator with a series resonant circuit. With Eq. (7.57) and taking into account the output network, we obtain for the phase fluctuations $\Delta\Theta$ at the load resistance

$$\Delta\Theta = R_0\, \frac{\dfrac{R_0}{R_0+j\,2\,\omega\,L}\,\dfrac{q_i}{1+q_i}\,\Delta\Psi_i + \Delta\Psi_n}{R_0\,\dfrac{q_i}{1+q_i}+j\,2\,\omega\,L} . \tag{7.121}$$

For $q_i \to 0$ the Eq. (7.121) coincides with the Eq. (7.66).

If the phase noise of the injection signal can be neglected, then Eq. (7.121) simplifies to

$$\Delta\Theta = \frac{R_0}{R_0\,\dfrac{q_i}{1+q_i}+j\,2\,\omega\,L}\,\Delta\Psi_n . \tag{7.122}$$

In contrast to the free running oscillator, we observe for the synchronized oscillator a corner frequency ω_c, which is given by the expression

$$\omega_c = \frac{R_o}{2\,L}\,\frac{q_i}{q_i+1} . \tag{7.123}$$

Above the corner frequency, the phase noise approaches the value of the free running oscillator. For offset frequencies below the corner frequency, i.e. $\omega < \omega_c$, the noise level approaches the constant value

$$\Delta\Theta = (1+1/q_i)\,\Delta\Psi_n , \tag{7.124}$$

if the injection signal is assumed to be noise-free. If the injection signal has phase noise itself, then for frequencies $\omega < \omega_c$ we obtain from Eq. (7.121) the expression

$$\Delta\Theta = \Delta\Psi_i + \left(1 + \frac{1}{q_i}\right)\Delta\Psi_n . \qquad (7.125)$$

Thus, the injected phase noise can simply be added to the case without noise from the injection signal, as given by Eq. (7.124). Qualitatively, we may observe noise spectra as shown in Fig. 7.35, if we assume that the phase noise of the injection oscillator is much lower than that of the main oscillator. For small offset frequencies, the total phase noise of the synchronized main oscillator approaches the noise of the injection oscillator, for high offset frequencies it approaches the noise of the free running oscillator. In between there is a range of nearly constant phase noise. Thus, the situation is very similar to a phase locked loop circuit.

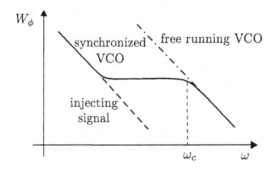

Fig. 7.35   Noise spectra of an injection locked oscillator.

Injection locking has not received technical importance for the stabilization of oscillators. However, the effect of injection locking can occur unintentionally, if two ore more oscillators operate at frequencies that are close to each other. Particularly critical is the situation when a swept source temporarily gets close to the frequency of a fixed frequency oscillator. If for instance the isolation between the two oscillators is not sufficient, they may synchronize each other and oscillate at the same frequency for a short time. From a system point of view, such a situation may be quite disastrous and normally must be avoided, e.g. by means of a better isolation of the sources.

7.8 DISTURBING EFFECTS OF OSCILLATOR NOISE

In this section, we will discuss some examples for the deteriorating effects of oscillator noise, i.e. mainly of oscillator phase noise, in high-frequency circuits and systems.

7.8.1 Heterodyne reception

Receivers for high-frequency signals are usually realized using the heterodyne principle. By means of a mixer the signal spectrum of input frequencies is mixed with the sinusoidal signal of a so-called local oscillator, which is often tunable in frequency, and thus converted to a lower intermediate frequency band (i.f. band). The wanted i.f. signal is filtered by a fixed i.f. band-pass filter and separated from the unwanted signals. Figure 7.36 illustrates the situation with the help of two sinusoidal input signals at angular frequencies Ω_{s1} and Ω_{s2} but significantly different amplitudes. By the frequency conversion with

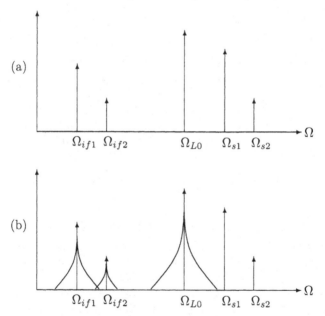

Fig. 7.36 Heterodyne reception with a noise-free (a) and a noisy (b) local oscillator.

the local oscillator of frequency Ω_{LO} two intermediate frequency signals with angular frequencies $\Omega_{if1} = \Omega_{s1} - \Omega_{LO}$ and $\Omega_{if2} = \Omega_{s2} - \Omega_{LO}$ will appear at the mixer i.f. output. Any amplitude noise of the local oscillator can be made ineffective by the use of a well balanced mixer. However, the phase noise of the local oscillator signal will lead to i.f. spectra as shown in Fig. 7.36. The noise sidebands are transferred to the i.f. signals through the mixing process. Then, it might happen that noise sidebands of the strong i.f. signal cover the weaker i.f. signal, so that a detection of the weak i.f. signal is no longer possible. Here, the phase noise of the local oscillator ultimately leads to a reduction of the input sensitivity and the dynamic range of the receiver.

7.8.2   Sensitivity of a spectrum analyzer

A block diagram of a standard spectrum analyzer system with a frequency range of approximately 100 kHz to 4 GHz is shown in Fig. 7.37. Typical relevant frequencies are given for clarity. Let us assume that the input signal

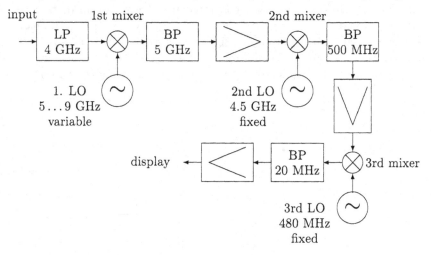

Fig. 7.37   Block diagram of a 4 GHz spectrum analyzer.

is an ideal sinusoidal signal of 2.2 GHz, i.e. an input carrier signal without noise sidebands. Then, nevertheless, we will observe a line spectrum with noise sidebands on the display, as shown schematically in Fig. 7.38. The noise sidebands originate from the phase noise sidebands of the internal VCO and the fixed local oscillators of the spectrum analyzer. Their phase or frequency fluctuations are transferred to the intermediate frequency range by linear frequency conversions in such a way that the phase deviations remain invariant and thus also the noise-to-carrier ratio. This result has already been discussed in Section 6.8.3. Therefore, the phase noise of the internal oscillators of the spectrum analyzer will determine the limiting sensitivity, if the phase noise of carrier signals are measured.

Typically the VCO will contribute the most significant part of the internal phase noise of the spectrum analyzer. A YIG-tuned oscillator instead of a VCO will be found in high quality instruments, if a particularly good phase noise behavior is required. A small YIG (Yttrium Iron Garnet) sphere is used

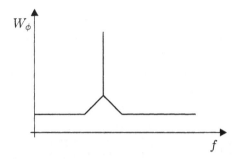

Fig. 7.38 Idealized spectrum of a sinusoidal carrier signal displayed on a spectrum analyzer.

as a high quality resonator, which determines the resonant frequency of the tunable oscillator. The resonant frequency is controlled by a magnetic bias field.

7.8.3 Distance measurements

Because electromagnetic waves propagate with the speed of light c, they can be utilized for the measurement of the distance l to a reflecting object. If a signal

$$x_1(t) = X_1 \cos[\phi(t)] = X_1 \cos[\Omega_0 t] \qquad (7.126)$$

is radiated by the transmitting antenna, then we obtain a time delayed signal of the form

$$x_2(t) = X_2 \cos[\phi(t - \tau)] = X_1 \cos[\Omega_0(t - \tau)] \qquad (7.127)$$

at the receiving antenna. Here,

$$\tau = \frac{2l}{c} \qquad (7.128)$$

is the propagation time of the signal forth to the object and back to the antenna. The distance information is given by the phase difference between the transmitted and received signal:

$$l = \frac{c}{2\Omega_o}[\phi(t) - \phi(t - \tau)] \ , \qquad (7.129)$$

because for a noise-free system we can write

$$\phi(t) - \phi(t - \tau) = \Omega_0 \tau \ . \qquad (7.130)$$

If the transmitted signal has phase fluctuations $\Delta\phi$, then Eq. (7.130) should be substituted by

$$\phi(t) - \phi(t - \tau) = \Omega_0 \tau + \Delta\phi(t) - \Delta\phi(t - \tau) \ . \qquad (7.131)$$

We conclude from the last equation that for $\Delta\phi(t) \neq 0$ the measured distance l will also show fluctuations which can be determined by means of Eq. (7.129). Quantitatively the measurement error can be described by the standard deviation σ_l. Due to Eq. (7.129) we have

$$\sigma_l = \frac{c}{2\,\Omega_0}\,\sigma_\phi \ , \tag{7.132}$$

where, according to Eq. (7.131), σ_ϕ is the standard deviation of the phase difference. Generally the standard deviation σ_X of a stochastic variable X can be determined from different mean values:

$$\sigma_X = \sqrt{\overline{(X - \overline{X})^2}} = \sqrt{\overline{X^2} - (\overline{X})^2} \ . \tag{7.133}$$

Equations (7.131) and (7.133) yield

$$\sigma_\phi^2 = \overline{[\Delta\phi(t)]^2} + \overline{[\Delta\phi(t - \tau)]^2} - 2\,\overline{\Delta\phi(t)\,\Delta\phi(t - \tau)} \ . \tag{7.134}$$

If the products of the phase fluctuation functions are expressed by their autocorrelation functions R_ϕ, we obtain for the standard deviation σ_ϕ

$$\sigma_\phi^2 = 2\,[R_\phi(0) - R_\phi(\tau)] \ . \tag{7.135}$$

The autocorrelation function R_ϕ can be expressed by the spectrum W_ϕ via a Fourier transformation:

$$\sigma_\phi^2 = 2 \int_{-\infty}^{+\infty} W_\phi(f)\,[1 - \exp(j\,2\,\pi f\tau)]\,df \tag{7.136}$$

$$= 2 \int_{-\infty}^{+\infty} W_\phi(f)\,[1 - \cos(2\,\pi f\tau)]\,df \ . \tag{7.137}$$

The error in the measurement of the distance results as a weighted integral of the spectrum W_ϕ. The stronger the phase noise of the signal source, the greater is the statistical error in the distance measurement.

7.8.4   Velocity measurements

Using the principle described in the preceding section, it is also possible to determine the velocity v of a reflecting object that moves parallel to the direction of the electromagnetic wave. This is a practical application of the doppler effect. For a transmitted signal according to Eq. (7.126) we obtain for the received signal:

$$x_2(t) = X_2 \cos\left[\Omega_2 \cdot (t - \tau)\right] = X_2 \cos\left[\left(1 - \frac{v}{c}\right)\Omega_0 \cdot (t - \tau)\right] \tag{7.138}$$

In the noise-free case, we have

$$v = c\left(1 - \frac{\Omega_2}{\Omega_1}\right) \ , \tag{7.139}$$

with

$$\Omega_1 = \Omega_0 = \text{const.} \tag{7.140}$$

and

$$\Omega_2 = \left(1 - \frac{v}{c}\right)\Omega_0 \ . \tag{7.141}$$

For a signal source with phase noise the frequencies Ω_1 and Ω_2 are given by

$$\begin{aligned}
\Omega_1 &= \Omega_0 + \frac{d}{dt}[\Delta\phi(t)] = \Omega_0 + \Delta\Omega(t) \ , \tag{7.142}\\
\Omega_2 &= \left(1 - \frac{v}{c}\right)\left(\Omega_0 + \frac{d}{dt}[\Delta\phi(t-\tau)]\right) \\
&= \left(1 - \frac{v}{c}\right)[\Omega_0 + \Delta\Omega(t-\tau)] \ . \tag{7.143}
\end{aligned}$$

Due to the phase noise the measured velocity v_m also shows statistical fluctuations:

$$\begin{aligned}
v_m &= c\left[1 - \frac{\left(1 - \dfrac{v}{c}\right)[\Omega_0 + \Delta\Omega(t-\tau)]}{\Omega_0 + \Delta\Omega(t)}\right] \\
&\approx c\,\frac{\Delta\Omega(t) - \Delta\Omega(t-\tau) + \dfrac{v}{c}[\Omega_0 + \Delta\Omega(t-\tau)]}{\Omega_0} \\
&\approx \frac{c}{\Omega_0}\left[\Delta\Omega(t) - \Delta\Omega(t-\tau) + \frac{v}{c}\Omega_0\right] \ . \tag{7.144}
\end{aligned}$$

For the last approximations, use has been made of the relations $\Delta\Omega \ll \Omega_0$ and $v \ll c$. This leads to a measurement error which depends on the frequency fluctuations of the signal source:

$$v_m - v = \frac{c}{\Omega_0}[\Delta\Omega(t) - \Delta\Omega(t-\tau)] \ . \tag{7.145}$$

Problem

7.12   Calculate, by analogy with Eq. (7.137), the standard deviation σ_v of the measured velocity v_m as a function of the phase noise spectrum W_ϕ of the signal source.

Apart from the above measurement error, a doppler radar for the measurement of speed is affected by another disturbing effect. The Eq. (7.145) also holds for $v = 0$. This implies that a fixed object can give rise to a noise signal at the output of the receiver, due to the mechanism of phase or frequency discrimination of the delayed signal. Such discriminated noise signals, if caused by strong reflectors, can disturb or even cover the signals from small moving objects. In radar systems this type of noise is called clutter noise.

7.8.5   Transmission of information by a frequency or phase modulated carrier signal

For frequency or phase modulation in a communication system, the frequency or phase of a highfrequency carrier signal is controlled in a well defined manner by the information to be transmitted. In the receiver the information can be retrieved by a proper demodulation process. Since the demodulator can not distinguish between the intentional changes of the phase or frequency by the modulation and the unintentional phase noise, the phase fluctuations will lead to a disturbing noise signal at the output of the demodulator. Furthermore, in a practical system, the phase noise contributions of several oscillators may add up, due to multiple mixing processes.

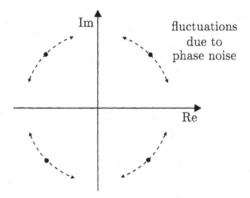

Fig. 7.39   Phase states for a QPSK-modulation in a complex plane.

These discussions not only apply to analog but also to digital communication systems. In digital systems the phase noise will increase the bit error rate. Figure 7.39 shows as an example the modulation states of a QPSK-system (**Q**uarternary **P**hase **S**hift **K**eying). For this modulation method the amplitude remains constant while the phase of the carrier signal is changed according to the digital modulation. The phase difference between adjacent states is 90°. The four states can be described by points in the complex plane, equally distributed on a circle. However, the oscillators are not ideal but show statistical fluctuations of their phase. If the resultant phase error exceeds 45°, then the actual state cannot correctly be detected and a bit error results. The stronger the phase noise, the higher the probability that a bit error occurs. The stability requirements of the source become even higher, if more sophisticated modulation methods are employed. E.g. , the 64 QAM-modulation (**Q**uadrature **A**mplitude **M**odulation) consists of a combination of 64 amplitude and phase states of the carrier. Accordingly, already smaller phase disturbances may lead to a bit error.

7.8.6   Measurement system for the microwave gas spectroscopy

Figure 7.40 shows a block diagram of a microwave gas spectroscopy system. A sinusoidal microwave signal is transmitted through a gas cell and rectified

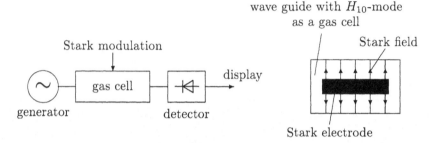

Fig. 7.40   Block diagram of a microwave gas spectroscopy system.

by a microwave detector. In the gas cell, the gas to be analyzed has a low pressure. The frequency of the microwave signal coincides with the resonance absorption frequency of the gas. In the gas cell a periodic change of the electric field, the so-called Stark field, e.g. with a frequency of 30 kHz, will also shift periodically the gas resonance absorption frequency by a very small amount. This may lead to a weak periodic amplitude modulation of the microwave signal, which is demodulated by the detector. The sensitivity of the detector may be limited e.g. by the shot noise of the diode.

The transmission behavior of the gas cell including the coupling networks may show some irregularities versus frequency, i.e. the magnitude of the transfer function may have a ripple. This can lead to a discrimination of the phase noise of the microwave source, which then is converted into amplitude noise and also measured by the detector. For this example, the important contribution of the phase noise stems from an offset frequency of 30 kHz, i.e. the Stark modulation frequency. Under unfavorable circumstances the discriminated phase noise may be stronger than the intrinsic noise at the detector.

In order to improve the sensitivity, the gas cell may be converted into a long resonator by placing short circuits at both ends of the gas cell. Such a resonator gas cell tends to discriminate phase noise even more effectively.

8

Quantization Noise

8.1 QUANTIZATION NOISE OF ANALOG-TO-DIGITAL CONVERTERS

A simple but useful model of a quantizer is shown in Fig. 8.1. In this model the quantization error is treated as an additive noise signal with a constant frequency spectrum.

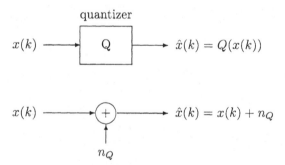

Fig. 8.1 Quantization error model with additive white noise.

The analog signal $V(\omega)$ which has been analog-to-digital (A/D) converted is assumed to be limited in frequency by a low pass filter with the corner frequency f_{lp}. Furthermore, it is assumed that the band limited analog signal has been sampled with a sampling frequency f_{sa} which is at least twice the frequency of f_{lp}. The statistical description of the quantization noise is based on the following assumptions:

1) The error sequence n_Q is a stationary random process.

2) The error sequence is uncorrelated with the sampled sequence $x(k)$.

3) The spectrum of the error sequence can be described as white noise.

4) The amplitude density distribution of the error sequence is constant across the range of the quantization error.

The A/D converter may have m quantization bits, a full range excursion of X_{pp} and a quantization step of Δ. Then we can write

$$X_{pp} = 2^m \cdot \Delta \ . \tag{8.1}$$

For small values of Δ and large values of m it is justified to assume that n_Q is a stochastic variable which is evenly distributed in the range $-\Delta/2$ to $+\Delta/2$, as shown in Fig. 8.2.

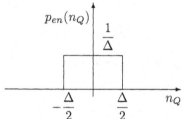

Fig. 8.2   Probability density distribution of the quantization error.

Then, the expected value of n_Q is zero and its variance σ_e^2 is

$$\sigma_e^2 = \frac{\Delta^2}{12} \ . \tag{8.2}$$

Assuming a full range excitation by a sinusoidal signal with a peak-to-peak amplitude of X_{pp} and an effective value of $\sqrt{2}/2\,X_{pp}$, the ratio R_{sn} of the signal power to the quantization noise power is given by

$$
\begin{aligned}
R_{sn} &= \frac{12\,X_{pp}^2}{8\,\Delta^2} \\
&= \frac{3 \cdot 2^{2m}}{2} \ .
\end{aligned}
\tag{8.3}
$$

In decibels the signal to noise ratio is thus given by $10 \log R_{sn} = m \cdot 6.02\,\mathrm{dB} + 1.76\,\mathrm{dB}$.

With the assumption of a constant or white spectrum and an effective frequency band of half the sampling frequency f_{sa}, the spectrum of the quantization noise W_{qu} is given by

$$W_{qu} = \frac{2 \cdot \Delta^2}{12 \cdot f_{sa}} = \frac{1}{6} \frac{\Delta^2}{f_{sa}} \ . \tag{8.4}$$

In a practical circuit the influence of the quantization noise can be reduced to an insignificant level, simply by increasing the quantization bit number m. An insignificant noise level of the quantization noise is definitely reached when it is lower than the noise level of the other physical noise sources, e.g. the amplified thermal noise.

Another possibility to reduce the level of the quantization noise is to increase the sampling frequency f_{sa} and to apply subsequently a low-pass filter with the corner frequency f_{lp} to the signal $V(\omega)$ containing the information. The low-pass filter may also be a digital filter.

The influence of the quantization noise is much more pronounced in the example of the next section, namely the example of a fractional divider phase locked loop system. Here, the quantization noise is inherently connected to the solution of the given problem and its influence can only be reduced by increasing the number of stages and, thus, also increasing the complexity of the system.

8.2   QUANTIZATION NOISE OF FRACTIONAL DIVIDER PHASE LOCKED LOOPS

For a phase locked loop (PLL) circuit consisting of a single loop the smallest achievable frequency step is equal to the reference frequency f_{ref}. Fig. 8.3 shows a block diagram of a simple PLL-circuit. In the locked mode the VCO

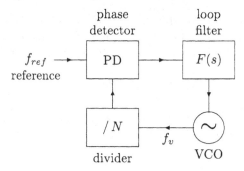

Fig. 8.3   Block diagram of a single loop PLL.

frequency f_v changes by a frequency step of Δf, if the integer division factor N is changed by an integer unit to $N \pm 1$. Thus, we can recognize a dilemma of the single loop PLL, namely that a reduction of the frequency step width can only be achieved via a reduction of the reference frequency f_{ref}. A reduction of the reference frequency, however, has a number of disadvantages, e.g. a smaller control loop bandwidth and a longer turn-on time of the loop and also a higher total multiplication factor with an associated higher phase noise level. One alternative of a single phase locked loop circuit is the use of several interlocked loops at the price of a higher complexity. Still another method applies a fractional divider concept, in order to create fractions of the reference frequency as step width. With the integer division factor P in front of the decimal point and the fractional division factor F behind the decimal point we obtain for the VCO-frequency f_v:

$$f_v = P.F \cdot f_{ref} \tag{8.5}$$

The fractional part F is generated by a variation versus time of the integer division factor N. A simple possibility for the variation of the division factor is shown in Fig. 8.4.

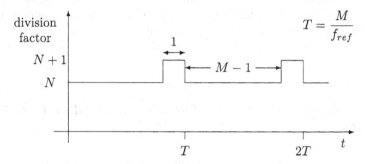

Fig. 8.4   Periodical variation of the division factor.

During M cycles of a period T, the integer division factor N is switched once from N to $N + 1$, i.e. for the duration of one period of the reference frequency f_{ref}. We obtain for the mean value of N, i.e. $\overline{N}$ the expression:

$$\overline{N} = \frac{N(M-1) + (N+1) \cdot 1}{M} = N + \frac{1}{M} \ . \tag{8.6}$$

Within one period T, it is possible to change the division factor N to the division factor $N + 1$ just n-times, with $n = 0, 1, 2, \ldots, M$. In this case, we get for the mean value of N:

$$\overline{N} = \frac{N(M-n) + (N+1) \cdot n}{M} = N + \frac{n}{M}, \quad n \in N, \quad 0 \le n \le M \ . \tag{8.7}$$

We see that the division factor can be varied in steps of $1/M$. Then, the VCO frequency of the phase locked loop can take on the values

$$f_v = \left(N + \frac{n}{M}\right) f_{ref} \ . \tag{8.8}$$

The realization of a fractional division ratio by a periodic variation of the division ratio is, however, not a practical solution for a PLL-system. Due to the periodic switching of the division ratio, a strong periodic phase modulation of the carrier signal occurs, which leads to intolerably strong disturbing spectral lines in the vicinity of the carrier signal. This disturbing spectrum changes its shape with the fractional division factor F and can not be removed by filtering, because the spectral lines often lie close to the carrier frequency. Apparently, the division factor variation should not be performed periodically but in a pseudo-random manner, such that the mean value of the division factor exactly equals the value $P.F$. Furthermore, the broad spectrum which arises from the pseudo-random switching of the division factor should be performed in such a way that the spectral density close to the carrier frequency becomes small. A circuit concept, which fulfills all these requirements, can be derived from the so-called Sigma-Delta (Σ/Δ) modulation.

8.2.1 Application of the Sigma-Delta modulation

Figure 8.5 shows a block diagram of the sigma-delta modulation. After the integrator and an analog-to-digital converter (ADC), an integer digital signal $y(k)$ is derived, which due to the analog-to-digital conversion possesses a quantization error n_Q. The digital output signal $y(k)$ is fed back to the input with a time delay. The feedback signal, at least in principle, must again be converted to an analog signal (DAC, digital-to-analog converter), because the input signal $x(t)$ was assumed to be an analog signal. In order to obtain a

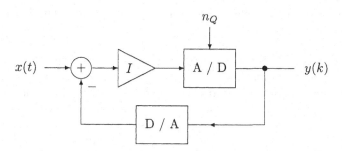

Fig. 8.5   Block diagram of the sigma-delta modulation.

fully digital circuit, already the input signal $x(t)$, which describes the fractional part F of the mean division factor, is a high resolution digital signal. This digital signal is applied to the adder, which now also operates digitally. The D/A converter shown in Fig. 8.5 can thus be omitted.

The output signal of the adder is first integrated, which is accomplished digitally (symbol I), before it is applied to the integer quantization (Fig. 8.6). Again the error signal is described by n_Q. This error signal n_Q serves for a mathematical description, it is not a signal which is added to the circuit. The mathematical description of the circuit is performed with the aid of the Z-transformation. This transformation can advantageously be applied for the description of discrete-time signals, as is the case here.

The different components of Fig. 8.6 are designated by the corresponding transfer functions of the variable $z = \exp(j\omega)$ of the Z-transformation. Here, ω is a normalized angular frequency. The delay element d is related to z via $d = z^{-1}$. The output signal $Y(z)$ of the one-stage system is obtained as

$$Y(z) \;=\; \frac{1}{1 - z^{-1}}\left(F - z^{-1}\,Y(z)\right) + n_Q$$

$$\text{or} \quad Y(z) \;=\; F + \left(1 - z^{-1}\right) n_Q\,. \tag{8.9}$$

One may recognize that the mean value of Y, i.e. $Y(z(\omega = 0))$ or $Y(1)$, is

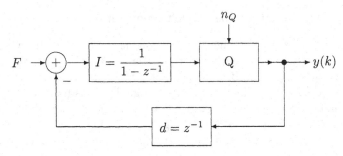

Fig. 8.6   Block diagram of a one-stage fractional digital circuit.

exactly equal to F. Furthermore, one may notice that the quantization error n_Q is differentiated once. On a logarithmic scale this corresponds to a decrease of the quantization noise spectrum by 6 dB/octave when approaching the carrier frequency. In the scale of the angular offset frequency ω the carrier frequency is at $\omega = 0$. For the case that F is a fractional number, the discrete-time signal $y(k)$ has a non-periodic quasi-stochastic, noise-like character. Therefore, the spectral power density of the quantization noise is approximately white, i.e. frequency independent. Due to the differentiation, however, we observe a decrease of the spectrum of the signal $y(k)$ towards the carrier by 6 dB/octave, as has been mentioned before. If the reciprocal value of the fractional number F is an integer number, then the corresponding spectrum is a line spectrum.

8.2.2   Multiple integration

In order to reduce the phase or frequency noise in the vicinity of the carrier, one may employ multistage circuits. As an example, Fig. 8.7 shows a three-stage circuit. This circuit has the property that the quantization noise of the first and the second stage is compensated exactly and just the quantization noise of the third stage remains. For this three-stage circuit the noise spectrum decreases by $3 \cdot 6\,\mathrm{dB} = 18\,\mathrm{dB}$/octave toward the carrier frequency.

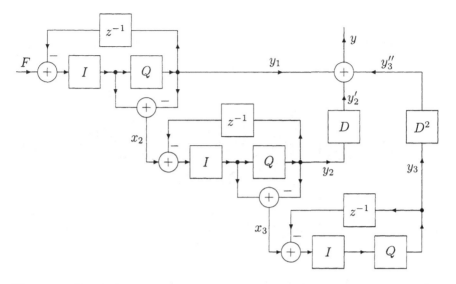

Fig. 8.7   Block diagram of a cascade three-stage fractional digital circuit.

In this multistage system the quantization error signal is fed to the input of the following stage. With the transfer function of a differentiation $D = 1 - z^{-1}$ and D^2 and D^3 as a double and a triple differentiation, respectively, we can write

$$Y = Y_1 + Y_2' + Y_3'' \tag{8.10}$$

and, furthermore,

$$Y_1 = F + n_{Q1} \cdot D \ . \tag{8.11}$$

We also find that

$$X_2 = Y_1 - n_{Q1} - Y_1 = -n_{Q1}$$

and

$$
\begin{aligned}
Y_2 &= \left(X_2 - z^{-1} \cdot Y_2 \right) I + n_{Q2} \\
Y_2' &= -n_{Q1} \cdot D + n_{Q2} \cdot D^2
\end{aligned} \tag{8.12}
$$

and similarly

$$Y_3'' = -n_{Q2} \cdot D^2 + n_{Q3} \cdot D^3 \ . \tag{8.13}$$

From the above equations we finally get the result

$$Y = F + n_{Q3} \cdot D^3 \ . \tag{8.14}$$

We see that the quantization noise of the first and second stage is compensated and the quantization noise of the third and last stage is multiplied by D^3, which leads to a decrease of the spectrum toward the carrier by the already mentioned 18dB/octave. Again the mean value of Y is exactly equal to F.

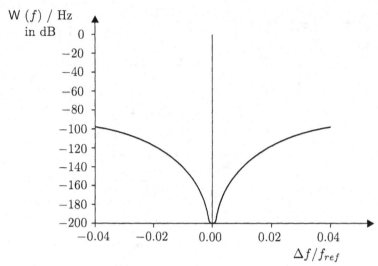

Fig. 8.8   Simulated noise spectrum of a three-stage fractional divider circuit.

Figure 8.8 shows a spectrum $W(f)$ as it results from a simulation of the three-stage circuit of Fig. 8.7 with $F = 0.05$. The output frequency is exactly equal to $f_v/P.F$, where f_v is the VCO-frequency. The deviations of the division factor ΔN are, for example, for a certain period of time: 1, −1, 0, 0, 1, −1, 0, −2, 3, −2....

In the tree-stage configuration, the deviations of the division factor ΔN stay in the range from −3 to +4. In the two-stage configuration, they remain in the range −1, 0, 1, 2 and in the one-stage configuration the deviations of the division factor are 0, 1. This is proven in problem 8.1.

Problem

8.1   Show that for a one-stage configuration the possible division factor deviations ΔN are 0 and 1, for a two-stage configuration −1, 0, 1, and 2 and for a three stage configuration −3, −2, −1, 0, 1, 2, 3, and 4.

The high noise level which appears further away from the carrier frequency, must strongly be reduced by analog filtering in the loop filter of the PLL. It is interesting to note that this filtering can only be performed by an analog filter and not with a digital filter.

Another possible configuration for the generation of the sequence of division factors is shown in Fig. 8.9. This circuit will be designated as a chain circuit in contrast to the cascade circuit of Fig. 8.7. Again a three-stage configuration is shown which, however, in addition contains multipliers with the real weighting factors κ_1, κ_2 and κ_3. For this circuit we obtain for the output signal $Y(z)$ as

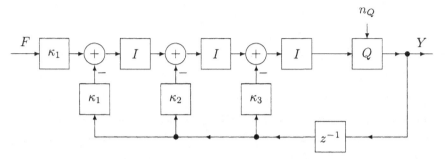

Fig. 8.9 Block diagram of a chain circuit of integrators with weighting factors in the feedback pathes.

a function of z:

$$Y(z) = \left(\left[\kappa_1 \left(F - z^{-1} \cdot Y(z) \right) I - \kappa_2 \cdot z^{-1} \cdot Y(z) \right] I - \kappa_3 \cdot z^{-1} \cdot Y(z) \right) I + n_Q$$
$$(8.15)$$

or

$$Y(z) = \frac{\kappa_1 \cdot F + n_Q \cdot D^3}{D^3 + \left(D^2 \cdot \kappa_3 + D \cdot \kappa_2 + \kappa_1 \right) z^{-1}} .$$
$$(8.16)$$

As will be shown in the next section, the cascade circuit of Fig. 8.7 and the chain circuit of Fig. 8.9 provide identical results for coefficients $\kappa_1 = \kappa_2 = \kappa_3 = 1$ in the chain circuit. This not only is true for a three-stage circuit but for any number of stages. The identity also holds for the time sequence of the output values of $y(k)$ and, therefore, for the division factor deviations. Thus, the possible deviations of ΔN range from -3 to 4 also for the three-stage chain circuit, provided that the weighting coefficients κ are all equal to one.

8.2.3 Identity of the cascade and the chain circuit

It is possible to transfer both the chain circuit of Fig. 8.9 and the cascade circuit of Fig. 8.7 into another circuit, which is identical to both circuits. For this purpose, the weighting factors κ_i with $i = 1, 2, 3 \ldots$ are set equal to one in the chain circuit:

$$\kappa_1 = \kappa_2 = \kappa_3 = 1 .$$
$$(8.17)$$

The chain circuit can be simplified in several steps, which are self-explanatory, as shown by the following figures.

The feedback signals are quantized and remain quantized after passing through the delay and the integration units. Under the assumption of infinitely wide quantizers which, however, are not realizable in practice, the signals pass the quantizer without a further quantization. For already quantized signals the quantizer has no effect. Therefore, the feedback paths can

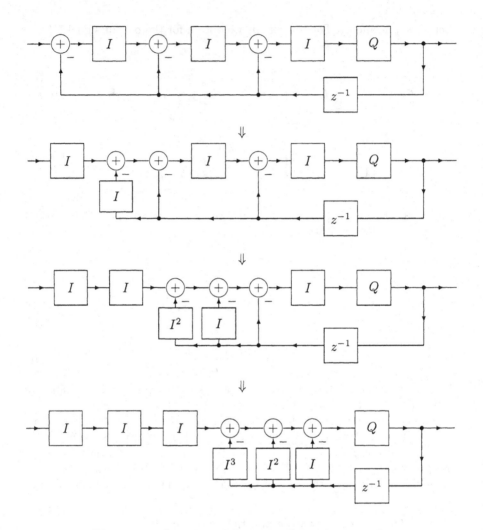

Fig. 8.10 Some simplifications of the chain circuit.

also be located directly behind the quantizer. This fact is used for the next conversion steps as shown in Fig. 8.11. A last evident conversion leads to the final circuit of Fig. 8.12 which, however, is not a practical circuit, as was mentioned before.

Next, the chain circuit will be simplified. For this purpose, the first stage is considered. The already quantized signals are transferred with the transfer factor one by the integrator of the second stage. Thus, the quantized signals of the first and second stage directly reach the differentiator at the output of the second stage. Similarly, the signals of the third stage directly reach the double differentiator. After several conversion steps one obtains the same

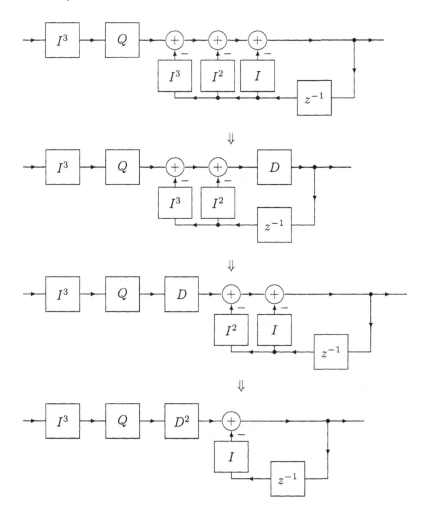

Fig. 8.11   Further conversions of the chain circuit.

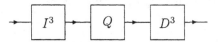

Fig. 8.12   Conversion of the chain circuit to the final circuit.

block diagram for the cascade circuit as for the chain circuit. Thus, it is shown that both logic circuits can be transformed into each other, which proves that they are identical.

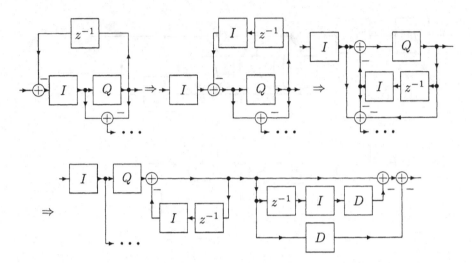

Fig. 8.13   Conversion steps for the cascade circuit.

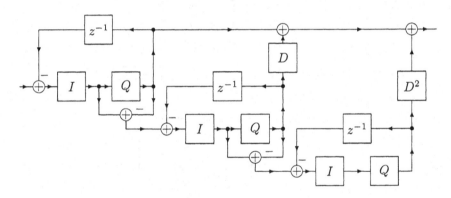

Fig. 8.14   Further conversion steps of the cascade circuit.

8.2.4 Chain circuit with weighting coefficients

The weighting factors κ_1 to κ_3 of the chain circuit can not be chosen arbitrarily. It has to be guaranteed that the denominator of the polynomial in Eq. (8.16) is stable. As is to be verified in problem 8.2, a choice of the weighting coefficients according to e.g. $\kappa_1 = 1/4$, $\kappa_2 = 1/2$ and $\kappa_3 = 1$ leads to a stable system. With respect to the z-plane the system is stable if all zeros z_i of the denominator polynomial lie within the unit circle, i.e. if

$$|z_i| < 1 . \tag{8.18}$$

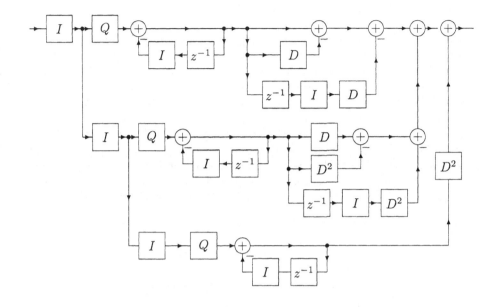

Fig. 8.15 Further conversion steps of the cascade circuit.

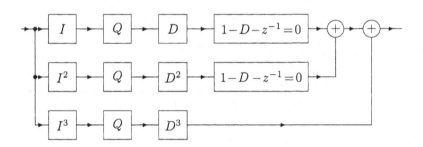

Fig. 8.16 Further conversion steps of the cascade circuit.

Fig. 8.17 Final conversion steps of the cascade circuit.

Problem

8.2 Show that a choice of the weighting coefficients according to $\kappa_1 = 1/4$, $\kappa_2 = 1/2$ and $\kappa_3 = 1$ of Fig. 8.9 leads to a stable system.

The variation of the division factor in integer steps around the mean division factor $\overline{N}$ leads to a corresponding phase modulation at the output of the divider chain. At the output of the phase discriminator of the phase locked loop circuit these phase fluctuations are transferred to corresponding voltage fluctuations, which are proportional to the time dependent variations of the division factor. In order to achieve a rapid decrease of the quantization noise close to the carrier frequency, it is essential that the phase discriminator has a very high linearity between the input phase and the output voltage. Non-linear effects in the phase discriminator may lead to a strong increase of the quantization noise close to the carrier frequency due to non-linear mixing processes. Then, the noise spectrum of Fig. 8.8 may change to a spectrum as shown in Fig. 8.18. The influence of the non-linear effects in the phase discrim-

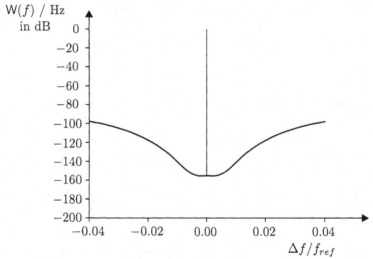

Fig. 8.18   Noise spectrum of a fractional divider circuit with increased close-to-carrier noise due to non-linear effects in the phase discriminator.

inator can be reduced by reducing the peak-to-peak division factor variation ΔN. For the example of the three-stage chain circuit with $\kappa_1 = 1/4$, $\kappa_2 = 1/2$ and $\kappa_3 = 1$ the division factors at the output of the fractional divider vary in a range between $-1 \ldots 2$, corresponding to a peak-to-peak deviation of 3, which is a significant reduction compared with the value of 7 for the three stage system with unity coefficients. Thus, the peak-to-peak variation of the division factor has been reduced by a factor of 2.3, as compared to the chain circuit with unity weighting factors.

The circuit of Fig. 8.19 shows a four-stage fractional logic core with an additional delay element in the integrator chain.

The weighting coefficients are chosen as $\kappa_1 = 3/16$, $\kappa_2 = 1/2$, $\kappa_3 = 1$ and $\kappa_4 = 1$. This leads to a stable circuit.

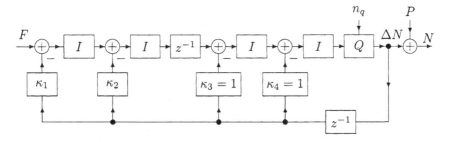

Fig. 8.19   A four-stage fractional logic core with an additional delay element in the integrator chain.

Problem

8.3   Show that a choice of the weighting coefficients of Fig. 8.19 according to $\kappa_1 = 3/16$, $\kappa_2 = 1/2$, $\kappa_3 = 1$ and $\kappa_4 = 1$ and an additional delay element in the integrator chain leads to a stable system.

The additional delay element in the logic core behind an integrator allows one to increase the maximum clock frequency of the logic circuit. This delay does not alter significantly the frequency characteristic of the transfer function compared with a four-stage circuit without this additional delay element. Stability is guaranteed by the proper choice of the weighting factors. This four-stage circuit shows a decrease towards the carrier frequency of $4 \cdot 6\mathrm{dB} = 24\,\mathrm{dB/octave}$. With an average division ratio $\overline{N} = 20.05$ the division factors at the output of the fractional divider vary in a range from $18 \leq 22$. Thus, the peak-to-peak variation of the division factor has been reduced by a factor of 3.75 compared with the four-stage chain circuit with unity weighting factors, which has a peak-to-peak deviation of 15.

The numerical calculation of first the sequence and then the spectrum is straightforward and not particularly difficult. Figure 8.20 shows a typical phase noise spectrum of a VCO stabilized by a four-stage fractional divider circuit.

Instead of determining the sequence of division factors in real time, it is also possible to determine all division factors once in advance and to store them in a fast digital memory. This procedure is possible due to another remarkable property of the sequence of division factors, namely that it is periodic with a period of the length n_{max}. The length of the period depends on the number of stages Q and on the bit length m of the fractional part F and is approximately given by

$$n_{max} \approx 2^{[\mathrm{ld}(Q)\,+\,m]} \ . \tag{8.19}$$

The bracket in the exponent in the last equation denotes a rounding to the nearest integer number. The fact that the sequence is not much longer than

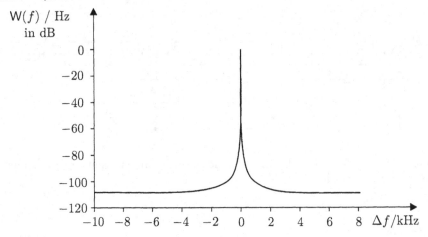

Fig. 8.20 Typical phase noise spectrum of a 8 GHz VCO stabilized by a four-stage fractional divider circuit.

the inverse of the fractional part of the division factor F makes it in some cases feasible to build fractional synthesizers on the basis of fast digital memories. One can take advantage of the relatively short period of the sequence and store exactly one sequence period in the memory. The memory is read out periodically, in order to realize, for example, one fixed frequency. Another fixed frequency makes it necessary to store and read out another stored sequence.

8.2.5 Transient behavior of a fractional logic circuit

When starting a real time fractional logic circuit, some initial starting values are assigned to the integrators and delay units and normally all these values will be zero. We may arrange all these contents and in addition the output signal in the form of a vector and we name this a vector element $[V]$ of the chain circuit. A sequence consists of a number of consecutive vector elements $[V]_1, [V]_2, [V]_3 \ldots [V]_l$, that are exactly reproduced in the following period of the sequence. Depending on the chosen weighting coefficients κ_i, however, the associated starting zero vector element $[V]_{st}$ may not be an element of the sequence. If nevertheless one starts with $[V]_{st}$, then a transitory behavior is observed with a limited number t_r of steps which is always smaller than the length of the sequence.

After t_r steps a transitory vector element will become necessarily identical to a vector element of the sequence. Once this has happened, the transient path will be left and the vector elements will follow those of the sequence with its periodicity. However, a large number of different sequences exist that do not have any vector element in common, but the same length and the same statistical properties, e.g. identical spectra. Which one of the different sequences the transient path will cross and then follow, depends on the starting vector element. However, in case of the chain circuit with unity coefficients the zero starting vector element is an element of one of the sequences. The same is true for the cascade circuit.

In order to illustrate the transient behavior, we choose the following numerical example: A three-stage chain circuit has the coefficients $\kappa_1 = 1/4$, $\kappa_2 = 1/2$, $\kappa_3 = 1$ and the input signal $F = 2^{-8} + 2^{-13}$. The length of the sequence is $l = s^{17}$ and with a zero starting vector $[V]_{st}$ the transient path length turns out to be $t_r = 30$.

8.2.6 Fractional divider without a PLL

Figure 8.21 shows a fractional divider circuit without a phase locked loop. An oscillator with the fixed frequency f_v is divided in frequency by the vari-

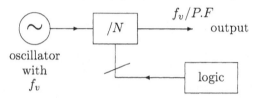

Fig. 8.21   Fractional divider circuit without a phase locked loop.

able division factor N. The sequence of the division factor versus time may e.g. be the one produced by the fractional division factor generator as shown in Fig. 8.7 or 8.9. One might expect to observe a spectrum as shown in Fig. 8.8. Passing the output signal through a narrow band-pass filter such a circuit could find use as a fine tunable oscillator signal. Unfortunately, however, the output signal has a spectrum as shown in Fig. 8.22. The phase modulation of the output signal also leads to a parasitic amplitude modulation which does not decrease towards the carrier frequency and which considerably raises the noise floor in the vicinity of the carrier frequency. Therefore the qualitative shape of the spectrum looks like the spectrum shown in Fig. 8.22.

The reason for the parasitic amplitude modulation is the constant peak signal amplitude at the divider output, which together with the phase modulation or pulse width modulation leads to an amplitude modulation and thus stochastic amplitude fluctuations with an approximately constant spectrum. Numerically, it is not particularly difficult to calculate first the sequence and then the spectrum at the output, while an analytical solution is more difficult and needs some simplifications.

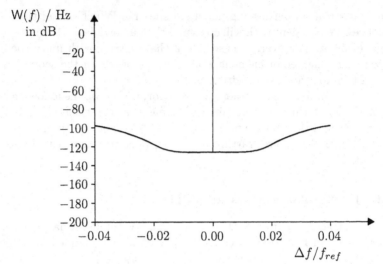

Fig. 8.22   Noise spectrum of a fractional divider circuit without a phase locked loop.

Appendix A
Solutions to the
problems of Chapter 1

Chapter 1

Problem 1.1

The normalization condition is given by Eq. (1.1). With the normal distribution given by Eq. (1.4) the following equation holds:

$$\int\limits_{-\infty}^{+\infty} \frac{1}{\sigma \cdot \sqrt{2\pi}} \exp\left[\frac{-(y-\mu)^2}{2\sigma^2}\right] dy = 1 \ . \tag{A.1}$$

This can be transformed with the substitution

$$z = \frac{y-\mu}{\sigma} \qquad \text{and} \qquad dy = \sigma \cdot dz \ . \tag{A.2}$$

Inserting Eq. (A.2) into Eq. (A.1) leads to

$$\int\limits_{-\infty}^{+\infty} p\,(y)\,dy = \frac{1}{\sqrt{2\pi}} \int\limits_{-\infty}^{+\infty} \exp\left[-\frac{1}{2}z^2\right] dz \ . \tag{A.3}$$

The integral can either be solved by using an integral table or by squaring and converting into polar coordinates. The latter method yields

$$\left[\int_{-\infty}^{+\infty} \exp\left(-\frac{1}{2}z^2 \right) dz \right]^2$$

$$= \int_{-\infty}^{+\infty} \exp\left[-\frac{1}{2}z^2 \right] dz \cdot \int_{-\infty}^{+\infty} \exp\left[-\frac{1}{2}v^2 \right] dv$$

$$= \int\int_{-\infty}^{+\infty} \exp\left(-\frac{z^2 + v^2}{2} \right) dz\, dv \ . \tag{A.4}$$

With a transformation into polar coordinates according to $z = r \cdot \cos\phi$ and $v = r \cdot \sin\phi$ and $dz \cdot dv = r \cdot dr \cdot d\phi$, we obtain

$$\left[\int_{-\infty}^{+\infty} \exp\left(-\frac{1}{2}z^2 \right) dz \right]^2 = \int_0^{2\pi} d\phi \cdot \int_0^{\infty} \exp\left(-\frac{r^2}{2} \right) r\, dr \ . \tag{A.5}$$

The integration over ϕ yields a factor of 2π and the substitution $r^2/2 = s$ leads to

$$\left[\int_{-\infty}^{+\infty} \exp\left(-\frac{1}{2}z^2 \right) dz \right]^2 = 2\pi \cdot \int_0^{\infty} \exp\left(-s \right) ds = 2\pi \ . \tag{A.6}$$

Thus, for a normal distribution,

$$\int_{-\infty}^{+\infty} \frac{1}{\sigma \cdot \sqrt{2\pi}} \exp\left[\frac{-(y-\mu)^2}{2\sigma^2} \right] dy = \frac{1}{\sqrt{2\pi}} \cdot \sqrt{2\pi} = 1 \ , \tag{A.7}$$

which means that the normalization condition is fulfilled.

Problem 1.2

The probability density of the sum variables can directly be written as a convolution of the probability densities by using Eq. (1.40). But the resulting convolution integrals are rather complicated. Therefore, the probability density will be calculated with the help of the convolution theorem and the characteristic function. Each of the rectangular distributions must meet the normalization condition Eq. (1.1). Thus we have for the density of the rectangular distribution $p_1(x)$:

$$p_1(x) = \frac{1}{x_2 - x_1} \left(u(x - x_1) - u(x - x_2) \right) \tag{A.8}$$

$$\text{with} \quad u(x) = \left\{ \begin{array}{ll} 1 & \text{for} \quad x > 0 \\ 0 & \text{for} \quad x \le 0 \end{array} \right\} \ .$$

The characteristic function of the rectangular distribution $p_1(x)$ is given by

$$C_1(u) = \frac{1}{x_2 - x_1} \int_{x_1}^{x_2} e^{jux} dx = \frac{1}{x_2 - x_1} \cdot \frac{1}{ju} \left(e^{jux_2} - e^{jux_1} \right) . \quad (A.9)$$

Similar expressions are obtained for the rectangular distributions $p_2(x)$ and $p_3(x)$. The characteristic function $C_s(u)$ of the sum variable is given by Eq. (1.41) as the product of the individual characteristic functions. Inserting the numerical values leads to the following expression for $C_s(u)$:

$$C_S(u) = \frac{1}{8 \cdot (ju)^3} \cdot \left(e^{13ju} - e^{12ju} - e^{11ju} + e^{10ju} \right.$$
$$\left. - e^{9ju} + e^{8ju} + e^{7ju} - e^{6ju} \right) . \quad (A.10)$$

An inverse Fourier transformation yields for $p_s(s)$:

$$\begin{aligned}
p_s(s) = \frac{1}{16} \Big[&(s-6)^2 u(s-6) - (s-7)^2 u(s-7) - (s-8)^2 u(s-8) \\
&+ (s-9)^2 u(s-9) - (s-10)^2 u(s-10) + (s-11)^2 u(s-11) \\
&+ (s-12)^2 u(s-12) - (s-13)^2 u(s-13) \Big] . \quad (A.11)
\end{aligned}$$

In Fig. A.1 a graphical representation of $p_s(s)$ is shown. For comparison a graph of the normal distribution is also shown in the figure. For the calculation

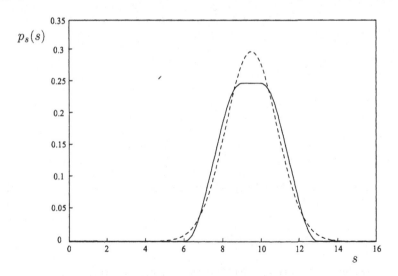

Fig. A.1

of the variance σ_s^2 the sum of the individual variances was used.

Problem 1.3

If X is a random variable with a Gaussian distribution, the characteristic function of X is

$$C_x(u) = \exp\left(ju\mu_x - \frac{1}{2}\sigma_x^2 \cdot u^2\right) . \qquad (A.12)$$

Proof: Inserting $p(x)$ as given by Eq. (1.4) into the Eq. (1.35) of the characteristic function, the exponent of the integrand is given by

$$h(x) = \frac{-1}{2\sigma_x^2}\left[(x - \mu_x)^2 - 2ju\sigma_x^2 x\right] . \qquad (A.13)$$

It follows by a quadratic complement:

$$h(x) = \frac{-1}{2\sigma_x^2}\left[(x - \mu_x - ju\sigma_x^2)^2 + \frac{1}{2\sigma_x^2}(u^2\sigma_x^4 - 2ju\sigma_x^2\mu_x)\right] . \qquad (A.14)$$

The second term is independent of x and can be placed in front of the integral. With the substitution $z = x - \mu_x - ju\sigma_x^2$ and an intermediate result from problem 1.1, we obtain for the first term:

$$\int_{-\infty}^{+\infty} \exp\left[\frac{-(x - \mu_x - ju\sigma_x^2)^2}{2\sigma_x^2}\right] dx$$

$$= \int_{-\infty}^{+\infty} \exp\left[\frac{-z^2}{2\sigma_x^2}\right] dz = \sigma_x\sqrt{2\pi} . \qquad (A.15)$$

This leads to the statement:

$$C_x(u) = \int_{-\infty}^{+\infty} \frac{1}{\sqrt{2\pi}\sigma_x} \exp\left(h(x)\right) dx$$

$$= \exp\left(ju\mu_x - \frac{\sigma_x^2 \cdot u^2}{2}\right) . \qquad (A.16)$$

For the random variable Y, which is independent of X, we correspondingly obtain:

$$C_y(u) = \exp\left(ju\mu_y - \frac{1}{2}\sigma_y^2 \cdot u^2\right) . \qquad (A.17)$$

For the characteristic function of the sum of the random variables the following equation holds:

$$C_s(u) = C_x(u) \cdot C_y(u) = \exp\left[ju\left(\mu_x + \mu_y\right) - \frac{u^2}{2}\left(\sigma_x^2 + \sigma_y^2\right)\right] . \qquad (A.18)$$

With $\mu_s = \mu_x + \mu_y$ and $\sigma_s^2 = \sigma_x^2 + \sigma_y^2$ the characteristic function of a Gaussian distributed random variable is given by

$$C_s(u) = \exp\left[j u \mu_s - \frac{\sigma_s^2 \cdot u^2}{2}\right] . \tag{A.19}$$

Thus the inverse Fourier transformation leads to a Gaussian distribution with the variance σ_s^2 and the mean value μ_s.

Problem 1.4

For the proof of Eq. (1.63) two random variables Y_1 and Y_2 are introduced. Y_1 and Y_2 are assumed to be statistically independent of each other and in addition to be normally distributed, that is $E\{Y\} = 0$ and $\sigma^2 = 1$. For a random variable with these properties Eq. (1.63) could easily be shown. But for X_1 and X_2 the statistical independence was not assumed. In order to get an expression for $C(u_1, u_2)$, nevertheless, the variables X_1 and X_2 with

$$\begin{aligned} X_1 &= a_{11}Y_1 + a_{12}Y_2 \\ X_2 &= a_{21}Y_1 + a_{22}Y_2 \end{aligned} \tag{A.20}$$

are expressed by the statistically independent variables Y_1 and Y_2. Because of

$$\begin{aligned} E\{X_1\} &= a_{11}E\{Y_1\} + a_{12}E\{Y_2\} \\ E\{X_2\} &= a_{21}E\{Y_1\} + a_{22}E\{Y_2\} \end{aligned} \tag{A.21}$$

the random variables X_1 and X_2 have zero expectation values. According to a theorem of probability theory, a random variable resulting from a linear combination of normal distributed random variables is normally distributed again. But the quality of statistical independence and the value of the variance are not maintained. Therefore the assumptions made for X_1 and X_2 in conjunction with Eq. (1.63) are valid. For the following steps it is convenient to use matrices:

$$\mathbf{A} = \begin{pmatrix} a_{11} & a_{12} \\ a_{21} & a_{22} \end{pmatrix} \quad \mathbf{Y} = \begin{pmatrix} Y_1 \\ Y_2 \end{pmatrix} \quad \mathbf{X} = \begin{pmatrix} X_1 \\ X_2 \end{pmatrix} . \tag{A.22}$$

Thus we have

$$\mathbf{X} = \mathbf{A} \cdot \mathbf{Y} . \tag{A.23}$$

On rather general conditions the inverse matrix $\mathbf{A}^{-1}$ exists and Eq. (A.23) can be written as

$$\mathbf{Y} = \mathbf{A}^{-1} \cdot \mathbf{X} . \tag{A.24}$$

With the transposed matrix $\mathbf{Y}^T$ the quadratic matrix

$$\mathbf{Y} \cdot \mathbf{Y}^T = \begin{pmatrix} Y_1 Y_1 & Y_1 Y_2 \\ Y_2 Y_1 & Y_2 Y_2 \end{pmatrix} \tag{A.25}$$

can be formed. Replacing the matrix elements $Y_i Y_k$ by the respective expectation values $E\{Y_i Y_k\}$ leads to the covariance matrix $\rho_y = E\{\mathbf{YY^T}\}$. The main diagonal of this matrix consists of the variances $E\{Y_i Y_k\}|_{i=k}$ and the matrix also contains all covariances $E\{Y_i Y_k\}|_{i \neq k}$. $E\{\mathbf{YY}^T\} = \rho_y$ is identical to the unit matrix $\mathbf{I}$ due to the assumptions made for Y_i. The covariance matrix $\rho_\mathbf{x}$ of the variable X_i is given by

$$\begin{aligned} \rho_\mathbf{x} &= E\{\mathbf{XX}^T\} \\ &= E\{\mathbf{AY}(\mathbf{AY})^T\} = E\{\mathbf{AYY}^T\mathbf{A}^T\} = \mathbf{A}\,E\{\mathbf{YY}^T\}\mathbf{A}^T \\ &= \mathbf{AIA}^T = \mathbf{AA}^T \ . \end{aligned} \tag{A.26}$$

For the statistically independent variables Y_1 and Y_2 the characteristic function of the bivariate Gaussian distribution can be written as the product of the characteristic functions of the variables Y_1 and Y_2:

$$\begin{aligned} C(v_1, v_2) &= \int\!\!\!\int_{-\infty}^{+\infty} \exp\left(jv_1 y_1 + jv_2 y_2\right) \cdot p(y_1, y_2)\, dy_1 dy_2 \\ &= \int\!\!\!\int_{-\infty}^{+\infty} \exp\left(j\mathbf{v}^T \mathbf{y}\right) \cdot p(y_1, y_2)\, dy_1 dy_2 \\ &= \exp\left(-\frac{1}{2}\mathbf{v}^T\mathbf{v}\right) \ . \end{aligned} \tag{A.27}$$

By switching to the variables X_1 and X_2 the expression $\exp\left(j\mathbf{v}^T\mathbf{y}\right)$ becomes $\exp\left(j\mathbf{v}^T\mathbf{A}^{-1}\mathbf{x}\right)$. In order to return to the form of Eq. (1.59) the following substitutions are made:

$$\mathbf{v}^T\mathbf{A}^{-1} = \mathbf{u}^T; \quad \mathbf{v}^T = \mathbf{u}^T\mathbf{A}; \quad \mathbf{v} = \mathbf{A}^T\mathbf{u} \ . \tag{A.28}$$

Then we have

$$\begin{aligned} C(u_1, u_2) &= \int\!\!\!\int_{-\infty}^{+\infty} \exp\left(j\mathbf{u}^T\mathbf{x}\right) \cdot p(x_1, x_2)\, dx_1 dx_2 \\ &= \exp\left(-\frac{1}{2}\mathbf{u}^T\mathbf{AA}^T\mathbf{u}\right) \\ &= \exp\left(-\frac{1}{2}\mathbf{u}^T\rho_\mathbf{x}\mathbf{u}\right) \ . \end{aligned} \tag{A.29}$$

Since the covariance matrix $\rho_{\mathbf{x}}$ is a symmetric matrix, this equation is identical to Eq. (1.63) and can easily be extended to multivariate Gaussian distributions.

Problem 1.5

As mentioned earlier in problem 1.4, Eq. (1.63) can easily be extended to e.g. four variables. Thus

$$C\left(u_1, u_2, u_3, u_4\right) = \exp\left(-\frac{1}{2}\sum_{i=1}^{4}\sum_{k=1}^{4}\rho_{ik}u_i u_k\right) \, . \tag{A.30}$$

In order to obtain the specified moment, an equation similar to Eq. (1.57) has to be solved:

$$\frac{1}{j^4}\frac{\partial^4 C\left(u_1, u_2, u_3, u_4\right)}{\partial u_1 \partial u_2 \partial u_3 \partial u_4}\bigg|_{u_1=\ldots=u_4=0}$$

$$= \rho_{12}\rho_{34} + \rho_{13}\rho_{24} + \rho_{14}\rho_{23} \, . \tag{A.31}$$

Thus the moment of fourth order is

$$\mathrm{E}\{X(t_1) \cdot X(t_2) \cdot X(t_3) \cdot X(t_4)\} \tag{A.32}$$

$$= \rho(t_2 - t_1) \cdot \rho(t_4 - t_3) + \rho(t_3 - t_1) \cdot \rho(t_4 - t_2) + \rho(t_4 - t_1) \cdot \rho(t_3 - t_2) \, .$$

For the special case $t_1 = t_2 = t$ and $t_3 = t_4 = t + \theta$ this relation is reduced to

$$\mathrm{E}\{X^2(t) \cdot X^2(t + \theta)\} = \rho^2(0) + 2\rho^2(\theta) \, . \tag{A.33}$$

Problem 1.6

According to the definition of the correlation the noise signals of two frequency bands at different frequencies filtered from arbitrary broadband noise are totally uncorrelated. The noise signals filtered from white noise are even uncorrelated, if they are transferred to the same frequency band by frequency translations.

Problem 1.7

Rectangularly shaped white noise has the following power density spectrum:

$$W_b(f) = \left\{\begin{array}{ll} W_0 & \text{for } f_1 \leq f \leq f_2 \text{ and } -f_2 \leq f \leq -f_1 \\ 0 & \text{else} \end{array}\right\} \tag{A.34}$$

with $W_0 > 0$ and real.

By application of an inverse Fourier transformation the corresponding autocorrelation function $\rho(\theta)$ is obtained:

$$
\begin{aligned}
\rho(\theta) &= \int_{-\infty}^{+\infty} W_b(f) \cdot \exp\left(j2\pi f\theta\right) df \\
&= \frac{W_0}{j2\pi\theta} \left(\exp\left(j2\pi f_2\theta\right) - \exp\left(j2\pi f_1\theta\right) + \exp\left(-j2\pi f_1\theta\right) - \exp\left(-j2\pi f_2\theta\right) \right) \\
&= \frac{W_0}{\pi\theta} \left(\sin\left(2\pi f_2\theta\right) - \sin\left(2\pi f_1\theta\right) \right) \\
&= W_0\Delta f \cdot 2 \cdot \cos\left(2\pi\theta f_0\right) \cdot \mathrm{si}\left(\pi\Delta f\theta\right) \quad\quad\quad\quad\quad (A.35)
\end{aligned}
$$

$$
\text{with} \quad f_0 = \frac{f_1 + f_2}{2} \quad \text{and} \quad \Delta f = f_2 - f_1 \; .
$$

In the figure below the autocorrelation function and its envelope are drawn.

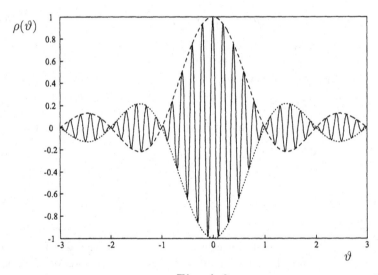

Fig. A.2

Problem 1.8

With Eq. (1.80) the power spectrum of the output noise is given by

$$
W_a(\omega) = \frac{W_0}{1 + \omega^2 R^2 C^2} \; , \quad\quad\quad\quad\quad (A.36)
$$

with the following transfer function $V(\omega)$ of the low pass filter:

$$
V(\omega) = \frac{U_a}{U_e} = \frac{1}{1 + j\omega RC} \; . \quad\quad\quad\quad\quad (A.37)
$$

In order to apply the Fourier transform correspondence

$$\sigma^2 \frac{2 \cdot k}{k^2 + \omega^2} \quad \bullet\!\!-\!\!\circ \quad \sigma^2 \cdot \exp\left(-k \cdot |\,\theta\,|\right) \ , \tag{A.38}$$

the result for $W_a(\omega)$ is rewritten as

$$W_a(\omega) = \frac{W_0}{1 + \omega^2 R^2 C^2} = \frac{W_0}{2RC} \frac{\dfrac{2}{R \cdot C}}{\dfrac{1}{R^2 C^2} + \omega^2} = \sigma^2 \frac{2 \cdot k}{k^2 + \omega^2} \ . \tag{A.39}$$

Thus the autocorrelation function of the output noise is given by

$$\rho_a(\theta) = \frac{W_0}{2RC} \exp\left(-\frac{|\,\theta\,|}{R \cdot C}\right) \ . \tag{A.40}$$

Appendix B
Solutions to the
Problems of Chapter 2

Chapter 2

Problem 2.1

1) The overall resistance R_i of the parallel connection of R_3 and the two resistors R_1 and R_2 in series can be calculated as follows:

$$R_i = (R_1 + R_2)\|R_3 = \frac{R_3(R_1 + R_2)}{R_1 + R_2 + R_3} \quad . \tag{B.1}$$

Thus the following noise equivalent circuit with the spectrum W_u of the noise equivalent voltage source can directly be specified:

$$W_u = 4kT_0 \cdot R_i = 4kT_0 \cdot \frac{R_3(R_1 + R_2)}{R_1 + R_2 + R_3} \quad . \tag{B.2}$$

2) Determining the noise equivalent source first, each of the resistors is replaced by a noise equivalent circuit.

The spectra of the noise equivalent voltage sources are given as follows:

$$W_{u1} = 4kT_1 R_1$$

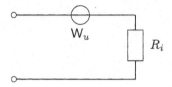

Fig. B.1   Noise equivalent circuit.

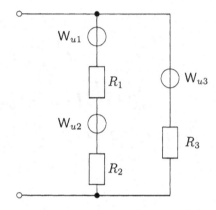

Fig. B.2   Noise equivalent circuit.

$$W_{u2} = 4kT_2R_2 \qquad (B.3)$$
$$W_{u3} = 4kT_3R_3 \ .$$

In order to calculate the resulting noise equivalent source, each source has to be transformed towards the input of the circuit. For that purpose, all other sources are short circuited. Furthermore, note that the spectra are related according to the squared magnitude of the corresponding voltage transfer functions:

$$W'_{u1} = W_{u1}\left(\frac{R_3}{R_1 + R_2 + R_3}\right)^2$$
$$W'_{u2} = W_{u2}\left(\frac{R_3}{R_1 + R_2 + R_3}\right)^2 \qquad (B.4)$$
$$W'_{u3} = W_{u3}\left(\frac{R_1 + R_2}{R_1 + R_2 + R_3}\right)^2 \ .$$

This leads to the following noise equivalent circuit:

In order to calculate the resulting noise equivalent voltage source W_u, the three spectra have to be summed up:

$$W_u = W'_{u1} + W'_{u2} + W'_{u3}$$

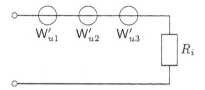

Fig. B.3 Noise equivalent circuit.

$$= \frac{4k\left(T_1 R_1 R_3^2 + T_2 R_2 R_3^2 + T_3 R_3 (R_1 + R_2)^2\right)}{(R_1 + R_2 + R_3)^2} \ . \tag{B.5}$$

So far, the calculations also apply to different temperatures T_j. The special case $T_1 = T_2 = T_3 = T_0$ leads to the following equation:

$$W_u = 4kT_0 \cdot \frac{R_3(R_1 + R_2)}{R_1 + R_2 + R_3} \ . \tag{B.6}$$

Thus the same overall noise equivalent circuit results for 1) and 2).

Problem 2.2

The ratio of the real power P_j dissipated in each impedance $Z = 1/Y$ and the total real power P_t has to be calculated. First the real power dissipated by a complex impedance is calculated in general as a function of the voltage phasor U and the current phasor I

$$
\begin{aligned}
P &= \mathrm{Re}\{U \cdot I^*\} = \mathrm{Re}\left\{U \cdot \frac{U^*}{Z^*}\right\} = \mathrm{Re}\left\{|U|^2 \cdot \frac{1}{Z^*}\right\} = |U|^2 \cdot \mathrm{Re}\{\frac{1}{Z}\} \\
&= \mathrm{Re}\{|U|^2 \cdot Y^*\} = |U|^2 \cdot \mathrm{Re}\{Y^*\} = |U|^2 \cdot \mathrm{Re}\{Y\} \ .
\end{aligned}
\tag{B.7}
$$

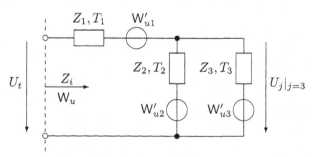

Fig. B.4 Noise equivalent circuit.

With U_t and U_j as shown in Fig. B.4, we obtain for the coefficients β_j:

$$\beta_j = \frac{P_j}{P_t} = \frac{|U_j|^2 \cdot \mathrm{Re}\{Y_j\}}{|U_t|^2 \cdot \mathrm{Re}\{Y_i\}} = \beta_j' \cdot \frac{\mathrm{Re}\{Y_j\}}{\mathrm{Re}\{Y_i\}} \ , \tag{B.8}$$

where P_t is the total power dissipated in the circuit. By using Eq. (2.19) the equivalent noise temperature T_n can be calculated. For the circuit in Fig. 2.7 the result is as follows:

$$\beta_1' = \frac{|U_1|^2}{|U_t|^2} = \left| \frac{Z_1}{Z_i} \right|^2 = \left| \frac{Z_1(Z_2 + Z_3)}{Z_1(Z_2 + Z_3) + Z_2 Z_3} \right|^2 \tag{B.9}$$

$$\beta_2' = \frac{|U_2|^2}{|U_t|^2} = \left| \frac{Z_2 || Z_3}{Z_i} \right|^2 = \left| \frac{Z_2 Z_3}{Z_1(Z_2 + Z_3) + Z_2 Z_3} \right|^2 \tag{B.10}$$

$$\beta_3' = \beta_2' \tag{B.11}$$

$$
\begin{aligned}
T_n &= \beta_1 T_1 + \beta_2 T_2 + \beta_3 T_3 \\
&= \beta_1' \frac{\text{Re}\{Y_1\}}{\text{Re}\{Y_i\}} T_1 + \beta_2' \frac{\text{Re}\{Y_2\}}{\text{Re}\{Y_i\}} T_2 + \beta_3' \frac{\text{Re}\{Y_3\}}{\text{Re}\{Y_i\}} T_3 \ . \tag{B.12}
\end{aligned}
$$

Problem 2.3

The input temperature is the equivalent temperature of the noise equivalent source of the overall circuit. T_n can be calculated by using Eq.(2.19):

$$T_n = \beta_1 T_1 + \beta_2 T_2 + \beta_3 T_3 \ . \tag{B.13}$$

According to the dissipation theorem the coefficients β_j can be derived from the real power, dissipated in the lossy elements of the circuit. The attenuator with a fixed attenuation of 6 dB dissipates a fraction of 0.75 of the injected power. The variable attenuator dissipates the fraction $(1 - \alpha_2)$ of the remaining power. The remaining real power is dissipated in the impedance Z_0. Thus the input temperature is given by

$$T_n = \frac{3}{4} \cdot 77\text{K} + \frac{1}{4} \cdot (1 - \alpha_2) \cdot 300\text{K} + \frac{1}{4} \cdot \alpha_2 \cdot 1200\text{K} \ . \tag{B.14}$$

Problem 2.4

With ρ as the reflection coefficient of the absorber, a fraction $(1 - |\rho|^2)$ of the radiated power of the antenna is converted into heat in the absorber. The reflected part of the radiated power is absorbed by the background. The dissipation theorem leads to the following equation for the noise temperature:

$$T_n = (1 - |\rho|^2) \cdot T_A + |\rho|^2 \cdot T_{ex} \ . \tag{B.15}$$

The part of the power reentering the antenna is assumed to be negligibly small.

Problem 2.5

For a noise-free two-port, described by impedance parameters, we have

$$
\begin{aligned}
U_1 &= Z_{11} \cdot I_1 + Z_{12} \cdot I_2 \\
U_2 &= Z_{21} \cdot I_1 + Z_{22} \cdot I_2 \; .
\end{aligned}
\tag{B.16}
$$

Adding the noise source U_n leads to

$$
\begin{aligned}
U_1 &= Z_{11} \cdot I_1 + Z_{12} \cdot I_2 \\
U_2 - U_n &= Z_{21} \cdot I_1 + Z_{22} \cdot I_2 \; .
\end{aligned}
\tag{B.17}
$$

The directions for the new noise sources to be calculated are chosen according to Fig. B.5.

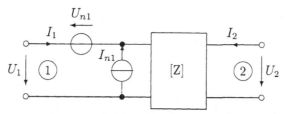

Fig. B.5   Noise equivalent circuit.

Then the circuit can be described by

$$
\begin{aligned}
U_1 + U_{n1} &= Z_{11} \cdot (I_1 + I_{n1}) + Z_{12} \cdot I_2 \\
U_2 &= Z_{21} \cdot (I_1 + I_{n1}) + Z_{22} \cdot I_2 \; .
\end{aligned}
\tag{B.18}
$$

Solving Eqs. (B.17) and (B.18) for U_1 and U_2, respectively, equating and reorganizing leads to

$$
\begin{aligned}
I_{n1} &= \frac{1}{Z_{21}} \cdot U_n \\
U_{n1} &= \frac{Z_{11}}{Z_{21}} \cdot U_n \; .
\end{aligned}
\tag{B.19}
$$

Next, the correlation between the sources has to be determined. Using the symbolic description for the cross-spectrum of the new noise current and voltage sources we have

$$
\begin{aligned}
W_{21} &= U_{n1}^* \cdot I_{n1} \\
&= \left(\frac{Z_{11}}{Z_{21}} \cdot U_n \right)^* \cdot \left(\frac{1}{Z_{21}} \cdot U_n \right) \\
&= \frac{Z_{11}^*}{|Z_{21}|^2} \cdot |U_n|^2 \\
&= \frac{Z_{11}^*}{|Z_{21}|^2} \cdot W_n \; .
\end{aligned}
\tag{B.20}
$$

The magnitude of the related normalized cross-spectrum is equal to one, because U_{n1} and I_{n1} are completely correlated:

$$
\begin{aligned}
\frac{|W_{21}|}{\sqrt{W_{u1} \cdot W_{i1}}} &= \frac{|W_{21}|}{\sqrt{|U_{n1}|^2 \cdot |I_{n1}|^2}} \\
&= \frac{|W_{21}|}{\sqrt{|U_{n1}^* \cdot I_{n1}|^2}} \\
&= \frac{|Z_{11}| \cdot W_n \cdot |Z_{21}|^2}{|Z_{21}|^2 \cdot |Z_{11}| \cdot W_n} = 1 \ .
\end{aligned}
\tag{B.21}
$$

Problem 2.6

As depicted in Fig. B.6, the current sources I_g and I_{n1}, which are connected in parallel, can be combined to one current source $I_{c1} = I_{n1} + I_g$.

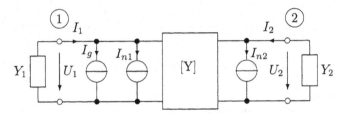

Fig. B.6   Noise equivalent circuit.

In an admittance representation we get the two-port equations:

$$
\begin{aligned}
I_1 &= Y_{11} \cdot U_1 + Y_{12} \cdot U_2 + I_{c1} \\
I_2 &= Y_{21} \cdot U_1 + Y_{22} \cdot U_2 + I_{n2}
\end{aligned}
\tag{B.22}
$$

The following equations apply for the currents and voltages at the input and output of the given circuit:

$$
I_1 = -Y_1 \cdot U_1 \quad \text{and} \quad I_2 = -Y_2 \cdot U_2 \ .
\tag{B.23}
$$

Equation (B.22) can be solved for U_2:

$$
U_2 = \frac{Y_{21}}{\det[Y']} \cdot I_{c1} - \frac{Y_{11}'}{\det[Y']} \cdot I_{n2} \quad \text{with} \quad Y_{11}' = Y_{11} + Y_1 \ .
\tag{B.24}
$$

Here, $\det[Y']$ is the determinant of the matrix

$$
[Y'] = \begin{bmatrix} Y_{11} + Y_1 & Y_{12} \\ Y_{21} & Y_{22} + Y_2 \end{bmatrix} \ .
\tag{B.25}
$$

According to the rules of the symbolic calculus of Chapter 1 the squared magnitude of Eq. (B.25) can be calculated by multiplying with the complex conjugate:

$$|U_2|^2 = \frac{1}{|\det[Y']|^2} \cdot \left(|Y_{21}|^2 \cdot |I_{c1}|^2 + |Y'_{11}|^2 \cdot |I_{n2}|^2 - Y'_{11}Y^*_{21}I_{n2}I_{c1}{}^*\right.$$
$$\left. - Y'_{11}{}^*Y_{21}I^*_{n2}I_{c1}\right) \quad \text{(B.26)}$$

Thus the noise power at the load admittance Y_2 is known. A conversion to spectra leads to

$$W_{u2} = \frac{1}{|\det[Y']|^2} \cdot \left(|Y_{21}|^2 \cdot W_{c1} + |Y'_{11}|^2 \cdot W_{n2} - Y'_{11}Y^*_{21}W_{c1n2}\right.$$
$$\left. - Y'_{11}{}^*Y_{21}W_{n2c1}\right) \quad \text{(B.27)}$$

With Eq. (1.33) we can write

$$W_{u2} = \frac{1}{|\det[Y']|^2} \cdot \left(|Y_{21}|^2 \cdot W_{c1} + |Y'_{11}|^2 \cdot W_{n2}\right.$$
$$\left. - 2 \cdot \text{Re}\{Y'_{11}Y^*_{21}W_{c1n2}\}\right) \quad \text{(B.28)}$$

Next, the noise spectrum W_{c1} is considered. The calculation of the squared absolute values of $I_{c1} = I_{n1} + I_g$ and a conversion to the spectra leads to

$$W_{c1} = W_{n1} + W_g + W_{gn1} + W_{n1g} = W_{n1} + W_g \ , \quad \text{(B.29)}$$

because for the given example the two sources I_{n1} and I_g are completely uncorrelated. Also the thermal noise of the complex admittance Y_1 is completely uncorrelated with the equivalent noise source I_{n2} at the output. The final result for the spectrum W_{u2} at the load admittance is

$$W_{u2} = \frac{1}{|\det[Y']|^2} \cdot \left(|Y_{21}|^2 \cdot (W_{n1} + 2kT \cdot \text{Re}\{Y_1\}) + |Y'_{11}|^2 \cdot W_{n2}\right.$$
$$\left. - 2 \cdot \text{Re}\{Y'_{11}Y^*_{21}W_{n1n2}\}\right) \quad \text{(B.30)}$$

Problem 2.7

We have the following equations:

$$I^*_{n1}I_{n2}\frac{Y_{12}}{Y_{22}} + I_{n1}I^*_{n2}\left(\frac{Y_{12}}{Y_{22}}\right)^* = \left|\frac{Y_{12}}{Y_{22}}\right|^2 \cdot 2kT \cdot \text{Re}\{Y_{22}\}$$
$$+ 2kT \cdot \text{Re}\left\{\frac{Y_{12}Y_{21}}{Y_{22}}\right\} \quad \text{(B.31)}$$

and

$$I_{n1}^* I_{n2} \left(\frac{Y_{21}}{Y_{11}}\right)^* + I_{n1} I_{n2}^* \frac{Y_{21}}{Y_{11}} = \left|\frac{Y_{21}}{Y_{11}}\right|^2 \cdot 2kT \cdot \text{Re}\{Y_{11}\}$$

$$+ 2kT \cdot \text{Re}\left\{\frac{Y_{12}Y_{21}}{Y_{11}}\right\} . \qquad \text{(B.32)}$$

This system of equations has to be solved for $I_{n1}^* I_{n2}$. For this purpose the first equation is multiplied by Y_{21}/Y_{11} and the second equation is multiplied by $(Y_{12}/Y_{22})^*$. Taking the difference of the two equations leads to

$$I_{n1}^* I_{n2} \left\{ \frac{Y_{21}Y_{12}}{Y_{11}Y_{22}} - \left(\frac{Y_{12}Y_{21}}{Y_{11}Y_{22}}\right)^* \right\}$$

$$= kT \left\{ \frac{Y_{21}}{Y_{11}} \left[\left|\frac{Y_{12}}{Y_{22}}\right|^2 (Y_{22} + Y_{22}^*) + \frac{Y_{12}Y_{21}}{Y_{22}} + \left(\frac{Y_{12}Y_{21}}{Y_{22}}\right)^* \right] \right.$$

$$\left. - \left(\frac{Y_{12}}{Y_{22}}\right)^* \left[\left|\frac{Y_{21}}{Y_{11}}\right|^2 (Y_{11} + Y_{11}^*) + \frac{Y_{12}Y_{21}}{Y_{11}} + \left(\frac{Y_{12}Y_{21}}{Y_{11}}\right)^* \right] \right\}$$

$$= kT \left\{ \frac{Y_{21}Y_{12}}{Y_{11}Y_{22}} \left[\frac{Y_{12}^*}{Y_{22}^*} (Y_{22} + Y_{22}^*) + Y_{21} \right] + \frac{Y_{21}Y_{12}^* Y_{21}^*}{Y_{11}Y_{22}^*} \right.$$

$$\left. - \frac{Y_{12}^* Y_{21}^*}{Y_{22}^* Y_{11}^*} \left[\frac{Y_{21}}{Y_{11}} (Y_{11} + Y_{11}^*) + Y_{12}^* \right] - \frac{Y_{12}^* Y_{12}Y_{21}}{Y_{22}^* Y_{11}} \right\}$$

$$= kT \left\{ \frac{Y_{21}Y_{12}}{Y_{11}Y_{22}} \left[\frac{Y_{12}^*}{Y_{22}^*} Y_{22} + Y_{12}^* + Y_{21} - \frac{Y_{12}^*}{Y_{22}^*} Y_{22} \right] \right.$$

$$\left. - \frac{Y_{12}^* Y_{21}^*}{Y_{22}^* Y_{11}^*} \left[Y_{21} + \frac{Y_{21}}{Y_{11}} Y_{11}^* + Y_{12}^* - \frac{Y_{21}}{Y_{11}} Y_{11}^* \right] \right\} . \qquad \text{(B.33)}$$

Thus we have

$$I_{n1}^* I_{n2} \left\{ \frac{Y_{21}Y_{12}}{Y_{11}Y_{22}} - \frac{Y_{12}^* Y_{21}^*}{Y_{11}^* Y_{22}^*} \right\} = kT\, (Y_{12}^* + Y_{21}) \left\{ \frac{Y_{21}Y_{12}}{Y_{11}Y_{22}} - \frac{Y_{12}^* Y_{21}^*}{Y_{22}^* Y_{11}^*} \right\} . \qquad \text{(B.34)}$$

Finally, we obtain

$$I_{n1}^* I_{n2} = kT\, (Y_{12}^* + Y_{21}) , \qquad \text{(B.35)}$$

the solution of Eq. (2.47).

Problem 2.8

This problem can be understood as a continuation of problem 2.6, if the open circuit at the output is included in the matrix $[Y']$ by setting $Y_2 = 0$. The direction of the current I_g has no relevance in this case because the sources I_g and I_{n1} are uncorrelated. The solution of problem 2.6

$$W_{u2} = \frac{1}{|\det[Y']|^2} \left(|Y_{21}|^2 \cdot (W_{In1} + 2kT \cdot \text{Re}\{Y_1\}) + |Y_{11}'|^2 W_{In2} \right.$$

$$\left. - 2\, \text{Re}\{Y_{11}' Y_{21}^* W_{In1In2}\} \right) \qquad \text{(B.36)}$$

is modified by inserting the equations (2.40) and (2.47):

$$W_{u2} = \frac{2kT}{|\det[Y']|^2}\left(|Y_{21}|^2 \cdot \mathrm{Re}\{Y_{11}'\} + |Y_{11}'|^2 \cdot \mathrm{Re}\{Y_{22}\}\right.$$
$$\left. - \mathrm{Re}\{Y_{11}'Y_{21}^*(Y_{12}^* + Y_{21})\}\right) \tag{B.37}$$

$$= \frac{2kT}{|\det[Y']|^2}\left(|Y_{11}'|^2 \cdot \mathrm{Re}\{Y_{22}\} - \mathrm{Re}\{Y_{11}'Y_{21}^*Y_{12}^*\}\right) . \tag{B.38}$$

With $\mathrm{Re}\{Y_{11}'Y_{21}^*Y_{12}^*\} = \mathrm{Re}\{Y_{11}'^*Y_{21}Y_{12}\}$ a short calculation leads to

$$W_{u2} = \frac{2kT}{|\det[Y']|^2} \cdot \mathrm{Re}\{Y_{11}'^* \cdot \det[Y']\}$$
$$= \frac{kT(Y_{11}'^* \cdot \det[Y'] + Y_{11}' \cdot \det[Y'^*])}{\det[Y'] \cdot \det[Y'^*]}$$
$$= \frac{kTY_{11}'^*}{\det[Y'^*]} + \frac{kTY_{11}'}{\det[Y']} . \tag{B.39}$$

Introducing the input admittance from the load side Y_{in},

$$Y_{in} = Y_{22} - \frac{Y_{12}Y_{21}}{Y_{11}'} = \frac{\det[Y']}{Y_{11}'} \tag{B.40}$$

one finally obtains:

$$W_{u2} = \frac{kT}{Y_{in}^*} + \frac{kT}{Y_{in}} = 2kT \cdot \mathrm{Re}\left\{\frac{1}{Y_{in}}\right\} .$$

This result was expected because $1/Y_{in}$ is the source impedance related to port 2. Thus it is possible to describe the resultant one-port by a thermally noisy resistor with the temperature T.

Problem 2.9

In order to facilitate the following calculation, the two-port equations (2.31) are transferred into a normalized form. For this purpose normalized currents and voltages are introduced:

$$i_i = I_i \cdot \sqrt{Z_0} \qquad u_i = \frac{U_i}{\sqrt{Z_0}}$$
$$\qquad\qquad\qquad\qquad\qquad i = 1,2 . \tag{B.41}$$
$$i_{ni} = I_{ni}\sqrt{Z_0} \qquad u_{ni} = \frac{U_{ni}}{\sqrt{Z_0}}$$

The elements of the admittance matrix are also normalized to the real reference conductance $Y_0 = 1/Z_0$:

$$y_{ik} = Y_{ik} \cdot Z_0 \qquad i = 1,2 \qquad k = 1,2 . \tag{B.42}$$

For the normalized representation of the two-port, Eq. (2.31), a matrix form results:

$$[i] = [y][u] + [i_n] .$$ (B.43)

Equation (B.43) can be rewritten with Eq. (2.50) in the following way:

$$[A] - [B] = [y]([A] + [B]) + [i_n] .$$ (B.44)

This equation can be converted according to Eq. (2.49):

$$\begin{aligned} [A] - [B] &= [y]([A] + [B]) + [i_n] \\ [B] + [y][B] &= [A] - [y][A] - [i_n] . \end{aligned}$$ (B.45)

Introducing the unity matrix $[I]$ yields

$$[B] = ([y] + [I])^{-1}([I] - [y])[A] - ([y] + [I])^{-1}[i_n] .$$ (B.46)

A comparison with Eq. (2.49) leads to

$$\begin{aligned} [S] &= ([y] + [I])^{-1} \cdot ([I] - [y]) \\ [X] &= -([y] + [I])^{-1}[i_n] . \end{aligned}$$ (B.47)

Thus the known noise current sources from Fig. 2.12 can be transformed into noise waves using these equations.

Problem 2.10

The parameters as shown in Fig. B.7 are used for the following calculations:

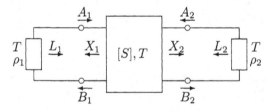

Fig. B.7   Noise equivalent circuit on the basis of noise waves.

The noise waves of the two-port are denoted by X_1, X_2, those of the load impedances by L_1, L_2. The load impedances are at the same temperature as the two-port. Equation (2.56) thus holds for the noise powers $|X_1|^2$ and $|X_2|^2$. The noise powers of the load impedances are defined according to the dissipation theorem:

$$|L_1|^2 = kT(1 - |\rho_1|^2) \quad \text{and} \quad |L_2|^2 = kT(1 - |\rho_2|^2) .$$ (B.48)

Here ρ_1, ρ_2 are the reflection coefficients of the load impedances. The noise waves of the two-port are not correlated with the noise waves of the load impedances. The noise waves of the circuit are given by

$$
\begin{aligned}
B_1 &= X_1 + S_{11}A_1 + S_{12}A_2 & B_2 &= X_2 + S_{22}A_2 + S_{21}A_1 \\
A_1 &= L_1 + \rho_1 B_1 & A_2 &= L_2 + \rho_2 B_2 \ .
\end{aligned}
\tag{B.49}
$$

First, an open circuit is assumed at port 2, and port 1 is terminated by a matched load:

$$
\rho_1 = 0, \quad \rho_2 = 1, \quad L_2 = 0 \quad \Rightarrow \quad A_1 = L_1 \ .
\tag{B.50}
$$

The following relation results for B_1 after a short calculation:

$$
B_1 = X_1 + S_{11}L_1 + \frac{S_{12}}{1 - S_{22}} \cdot (X_2 + S_{21}L_1) \ .
\tag{B.51}
$$

Under the condition of thermodynamic equilibrium, Eq. (2.54) applies to the noise waves of the circuit. With

$$
|B_1|^2 = |L_1|^2 = kT
\tag{B.52}
$$

and Eq. (2.56) some manipulations yield

$$
\begin{aligned}
&\frac{S_{12}}{1 - S_{22}} X_1^* X_2 + \frac{S_{12}^*}{1 - S_{22}^*} X_1 X_2^* \\
&= kT \left[|S_{12}|^2 - \left| \frac{S_{12}}{1 - S_{22}} \right|^2 (1 - |S_{22}|^2) - \frac{S_{11}^* S_{12} S_{21}}{1 - S_{22}} - \frac{S_{11} S_{12}^* S_{21}^*}{1 - S_{22}^*} \right] .
\end{aligned}
\tag{B.53}
$$

Next, port 2 is shorted and port 1 is terminated by a matched load:

$$
\rho_1 = 0, \quad \rho_2 = -1, \quad L_2 = 0 \quad \Rightarrow \quad A_1 = L_1
\tag{B.54}
$$

A similar calculation as above leads to a second equation for $X_1^* X_2$ and $X_1 X_2^*$:

$$
\begin{aligned}
&\frac{S_{12}}{1 + S_{22}} X_1^* X_2 + \frac{S_{12}^*}{1 + S_{22}^*} X_1 X_2^* \\
&= kT \left[-|S_{12}|^2 + \left| \frac{S_{12}}{1 + S_{22}} \right|^2 (1 - |S_{22}|^2) - \frac{S_{11}^* S_{12} S_{21}}{1 + S_{22}} - \frac{S_{11} S_{12}^* S_{21}^*}{1 + S_{22}^*} \right] .
\end{aligned}
\tag{B.55}
$$

Both equations establish a system of linear equations for $X_1^* X_2$ and $X_1 X_2^*$. After some calculations the equations can be solved for $X_1^* X_2$:

$$
X_1^* X_2 = -kT(S_{11}^* S_{21} + S_{12}^* S_{22}) \ .
\tag{B.56}
$$

This is the result for the cross-correlation already given by Eq. (2.58) .

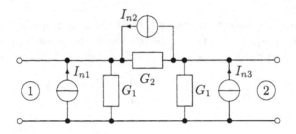

Fig. B.8   Noise sources of the two-port.

Problem 2.11

The noise of each of the three real admittances is described by an equivalent noise current source as shown in the following picture:

The source I_{n2} can be replaced by two sources, one connected in parallel to I_{n1}, the other one connected in parallel to I_{n3}. This leads to the following circuit:

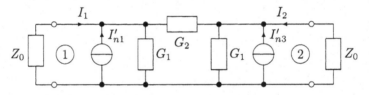

Fig. B.9   Noise equivalent circuit of the two-port.

The purpose of this problem is not to get the correlation $I_{n1}'^{*} I_{n2}'$ (it could easily be calculated with Eq. (2.47) as $I_{n1}'^{*} I_{n2}' = -2kTG_2$), but to calculate the correlation $I_1^{*} I_2$ of the currents flowing through the two terminating impedances with the impedance values Z_0. Generally, we can write for the currents I_1 and I_2:

$$
\begin{aligned}
I_1 &= V_1 I_{n1}' + V_2 I_{n2}' \\
I_2 &= V_1 I_{n2}' + V_2 I_{n1}' \ .
\end{aligned}
\tag{B.57}
$$

Here V_1 and V_2 are real transfer functions. Forming the correlation $I_1^{*} I_2$ and using the following equations:

$$
\begin{aligned}
I_{n1}'^{*} I_{n2}' &= 2kT \cdot \mathrm{Re}\{Y_{12}\} = -2kT \cdot G_2 = I_{n1}' I_{n2}'^{*} \\
|I_{ni}'|^2 &= 2kT \cdot \mathrm{Re}\{Y_{ii}\} = 2kT \cdot (G_1 + G_2) \quad i = 1, 2
\end{aligned}
\tag{B.58}
$$

yields after a short calculation:

$$
I_1^{*} I_2 = -2kT \left(-2V_1 V_2 G_1 + (V_1 - V_2)^2 G_2 \right) \ .
\tag{B.59}
$$

The π-attenuator shall be matched on both sides. Then the input impedance on both sides is equal to Z_0. Therefore, half of the currents I'_{n1} and I'_{n2} flows into the attenuator, the other half flows into the terminating impedance. The current flowing into the attenuator is transferred by the current transfer function

$$S = \frac{Z_0 G_1 - 1}{Z_0 G_1 + 1} \ . \tag{B.60}$$

The transfer functions V_1 and V_2 are given by

$$V_1 = -\frac{1}{2} \qquad V_2 = \frac{1}{2}S \ . \tag{B.61}$$

For a simultaneous match at both ports, G_2 in dependence of G_1 is given by

$$G_2 = \frac{1 - (G_1 Z_0)^2}{2 G_1 Z_0^2} \ . \tag{B.62}$$

With these relations for V_1, V_2 and G_2 and $V_1 = -1/2$ the bracket term in Eq. (B.59) can be written as follows:

$$-2V_1 V_2 G_1 + (V_1 - V_2)^2 G_2 = V_2 G_1 + G_2 (V_1 - V_2)^2$$
$$= \frac{1}{2}G_1 \frac{Z_0 G_1 - 1}{Z_0 G_1 + 1} + \frac{1 - (G_1 Z_0)^2}{2 G_1 Z_0^2} \left(\frac{1}{2} + \frac{1}{2}\frac{Z_0 G_1 - 1}{Z_0 G_1 + 1} \right)^2$$
$$= \frac{1}{2}G_1 \frac{Z_0 G_1 - 1}{Z_0 G_1 + 1} + \frac{1}{2}G_1 \frac{(1 - G_1 Z_0)(1 + G_1 Z_0)}{(G_1 Z_0)^2} \cdot \left(\frac{G_1 Z_0}{G_1 Z_0 + 1} \right)^2$$
$$= \frac{1}{2}G_1 \frac{Z_0 G_1 - 1}{Z_0 G_1 + 1} + \frac{1}{2}G_1 \frac{1 - G_1 Z_0}{1 + G_1 Z_0}$$
$$= 0 \ . \tag{B.63}$$

Thus it holds that the correlation $I_1^* I_2$ is equal to zero.

Problem 2.12

The noise generated by R and Z_0 will be analyzed separately and afterwards summed up according to the principle of superposition. First, the influence of the noise generated by R at port 2 and port 3 will be calculated. For this purpose an equivalent noise current source for the resistor R is implemented (Fig. B.10a). Fig. B.10b shows an equivalent circuit with two identical current sources. The wire in the symmetry plane carries no current. Thus it can be connected directly to the ground potential and the ports 2 and 3 are operating in the odd mode. Therefore, the entire symmetry plane of the transmission line structure can be connected to ground. The resulting short circuit at the end of the line is transformed into an open circuit at the input by means of a $\lambda/4$ transformation, so that for port 2 and port 3, respectively, the equivalent circuits in Fig. B.10c and B.10d are obtained.

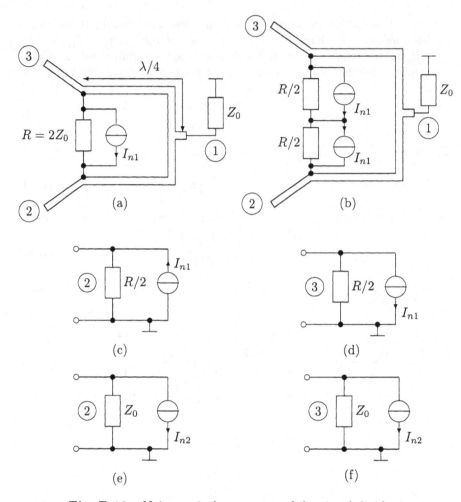

Fig. B.10   Noise equivalent sources of the signal divider.

Similarly, the noise of Z_0 can be expressed by an equivalent noise current source I'_{n2}, with $|I'_{n2}|^2 = 2kT/Z_0$. The signal of the source I'_{n2} is transferred to the ports 2 and 3, being attenuated by $\sqrt{2}$, because of the 0-degree and 3 dB-signal divider. The noise of Z_0 can also be described by two equivalent sources (Fig. B.10e,f)

$$I_{n2} = \frac{1}{\sqrt{2}}I'_{n2} \tag{B.64}$$

at port 2 and port 3. Using the isolation condition $R = 2Z_0$, the squared magnitude of the noise equivalent sources are related by

$$|I_{n1}|^2 = |I_{n2}|^2 . \tag{B.65}$$

Furthermore, the sources are uncorrelated:

$$I_{n1}^* I_{n2} = I_{n2}^* I_{n1} = 0 \ . \tag{B.66}$$

The equivalent sources resulting from the superposition at port 2 and port 3 are also uncorrelated, because

$$X_1^* X_2 = (I_{n1} + I_{n2})^* (I_{n2} - I_{n1}) = 0 \ . \tag{B.67}$$

Problem 2.13

Replacing the equivalent noise voltage source of the generator resistance by an equivalent noise current source leads to the same relations as given by problem 2.6, if the conductance Y_1 is replaced by Y_g. The noise figure is given by

$$F = 1 + \frac{\Delta W_2}{W_{20}} \ . \tag{B.68}$$

ΔW_2 is the noise at the output as induced by the two-port. This spectrum is given by the result of problem 2.6, if Y_g is considered to be noise-free, leading to $W_{n1}' = W_{n1}$. Thus

$$\Delta W_2 = \frac{1}{|\det[Y']|^2} \left(|Y_{21}|^2 \cdot W_{n1} + |Y_{11} + Y_g|^2 \cdot W_{n2} \right.$$
$$\left. - 2\mathrm{Re}\{Y_{21}^* \cdot (Y_{11} + Y_g) \cdot W_{n12}\} \right) \ . \tag{B.69}$$

W_{20} is the noise at the output caused solely by Y_g. This spectrum is given by the result of problem 2.6, if the two-port is assumed to be noise-free, i.e. W_{n1}, W_{n2} and W_{n12} are chosen to be zero. This leads to

$$W_{20} = \frac{1}{|\det[Y']|^2} \cdot |Y_{21}|^2 \cdot 2kT_0 \mathrm{Re}\{Y_g\} \ . \tag{B.70}$$

Inserting the result into Eq. (B.68) leads to Eq. (2.85).

Problem 2.14

First, the gain of the two-port will be described by scattering parameters. According to Eq. 2.72 the gain is given by

$$G_p = \frac{P_2}{P_g} \ . \tag{B.71}$$

With the designations introduced in Fig. B.11 the real power P_2 dissipated by the load Z_1 is

$$P_2 = |b_2|^2 - |a_2|^2 = |b_2|^2 \cdot (1 - |r_l|^2) \ ,$$

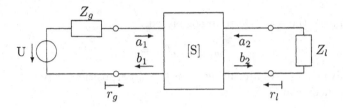

Fig. B.11

where $r_l = a_2/b_2$ is the reflection coefficient of the load. In order to calculate the available generator power P_g, the incident wave a_1 of the two-port is examined. It is the sum of the part a_g generated by the source, and a part $r_g b_1$, reflected by the generator, where r_g is the reflection coefficient of the internal generator impedance:

$$a_1 = a_g + r_g b_1 \; . \tag{B.72}$$

Then the available generator power is given by:

$$P_g = |a_1|^2 - |b_1|^2 = \frac{|a_g|^2}{1 - |r_g|^2} \; . \tag{B.73}$$

Thus the gain is obtained as

$$G_{p21} = \frac{|b_2|^2}{|a_g|^2}(1 - |r_l|^2)(1 - |r_g|^2) \; . \tag{B.74}$$

With Eq. (B.72), the definition of r_l, and

$$[b] = [S][a] \tag{B.75}$$

a short calculation leads to an equation for the ratio b_2/a_g. By inserting this ratio into the equation for the gain we obtain

$$G_{p21} = |S_{21}|^2 \cdot \frac{(1 - |r_l|^2)(1 - |r_g|^2)}{|(1 - S_{11}r_g)(1 - S_{22}r_l) - S_{12}S_{21}r_g r_l|^2} \; . \tag{B.76}$$

In order to show that the gain of a reciprocal two-port is independent of the direction, the voltage source in Fig B.11 is placed at port 2 and the power transfer from port 2 to port 1 is examined (Z_g and Z_l remain in place). For the determination of G_{p12} the same calculation as for G_{p21} has to be performed. A comparison with the calculation above shows that the index 2 has to be replaced by the index 1 (and vice versa) and the index g has to be replaced by the index l (and vice versa). This leads to

$$G_{p12} = |S_{12}|^2 \cdot \frac{(1 - |r_l|^2)(1 - |r_g|^2)}{|(1 - S_{11}r_g)(1 - S_{22}r_l) - S_{12}S_{21}r_g r_l|^2} \; . \tag{B.77}$$

For reciprocal two-ports $S_{12} = S_{21}$, and thus the gain is independent of the direction.

Problem 2.15

The cascade connection of the two attenuators can be considered as one single two-port with two temperature regions. Equation (2.70) is used for the determination of the noise figure of the two-port. ΔP_2 is composed of the part ΔP_{T1} generated in temperature region 1 and the corresponding part ΔP_{T2} for the region 2. We thus have

$$\Delta P_2 = \Delta P_{T1} + \Delta P_{T2} \tag{B.78}$$

with

$$\begin{aligned} \Delta P_{T1} &= kT_1\beta_1\Delta f \\ \Delta P_{T2} &= kT_2\beta_2\Delta f \ . \end{aligned} \tag{B.79}$$

For the part P_{20} from the generator we obtain with $\kappa_{tot} = \kappa_1 \cdot \kappa_2$:

$$P_{20} = \kappa_1\kappa_2 kT_0\Delta f \ . \tag{B.80}$$

The coefficients β_i are equal to the power fractions absorbed in the respective temperature regions $i = 1, 2$, if the power is fed in from the output side:

$$\begin{aligned} \beta_2 &= (1 - \kappa_2) \\ \beta_1 &= \kappa_2(1 - \kappa_1) \ . \end{aligned} \tag{B.81}$$

Inserting these expressions into Eq. (2.70) leads to:

$$\begin{aligned} F &= 1 + \frac{T_1\beta_1 + T_2\beta_2}{\kappa_1\kappa_2 T_0} \\ &= 1 + \frac{T_1(1 - \kappa_1)}{\kappa_1 T_0} + \frac{T_2(1 - \kappa_2)}{\kappa_1\kappa_2 T_0} \\ &= F_1 + \frac{F_2 - 1}{\kappa_1} \ . \end{aligned} \tag{B.82}$$

With the numerical values $\kappa_1 = 0.5$ and $\kappa_2 = 0.25$ we obtain for the noise figure:

$$F = 1 + \frac{T_1}{T_0} + 6\frac{T_2}{T_0} \ . \tag{B.83}$$

Problem 2.16

If the whole circuit (R_1, R_2, Z_g, Z_l) is taken as one resistive network with two temperature regions, namely, with the temperatures T_0 and T_1, the spectrum W_{20} at Z_l as caused exclusively by Z_g is given by

$$W_{20} = 4kT_0 Z_g \left| \frac{R_2 || Z_l}{R_1 + Z_g + R_2 || Z_l} \right|^2 . \tag{B.84}$$

Z_l is assumed as noise-free. The inclusion of the noise contributions of R_1, R_2 and Z_g leads to the spectrum W_2:

$$W_2 = 4kT_1 R_1 \left| \frac{R_2 || Z_l}{R_1 + Z_g + R_2 || Z_l} \right|^2$$

$$+ 4kT_1 R_2 \left| \frac{Z_l || (Z_g + R_1)}{R_2 + Z_l || (Z_g + R_1)} \right|^2$$

$$+ 4kT_0 Z_g \left| \frac{R_2 || Z_l}{R_1 + Z_g + R_2 || Z_l} \right|^2 . \tag{B.85}$$

Calculating the ratio of these terms leads to the following expression for the noise figure:

$$F = 1 + \frac{T_1}{T_0} \cdot \left(\frac{(R_1 + Z_g)(R_2 + R_1 + Z_g)}{Z_g R_2} - 1 \right) . \tag{B.86}$$

Thus it is evident, that the noise figure is independent of the value of the load resistance. To show the validity of Eq. (2.92), the available gain must directly be calculated. G_{av} is defined as the ratio of the available output power P_{2av} to the available generator power P_g. We have

$$P_g = \frac{|U_g|^2}{4Z_g} \tag{B.87}$$

and

$$P_{2av} = \frac{|U_{20}|^2}{4Z_l} = \frac{|U_g|^2 \left(\frac{R_2}{R_1 + R_2 + Z_g} \right)^2}{4 \left(R_2 || (R_1 + Z_g) \right)} . \tag{B.88}$$

For the ratio of both terms a short calculation leads to

$$\frac{1}{G_{av}} = \frac{P_g}{P_{2av}} = \frac{(R_1 + Z_g)(R_2 + R_1 + Z_g)}{Z_g R_2} = F|_{T_1 = T_0} . \tag{B.89}$$

This result is equal to equation (2.92).

Problem 2.17

According to Eq. (2.99) it is more advantageous to place the first amplifier in front of the second amplifier, if the condition

$$F_1 + \frac{F_2 - 1}{G_{1av}} < F_2 + \frac{F_1 - 1}{G_{2av}} \tag{B.90}$$

is fulfilled. A short conversion leads to an equation, which is more useful in practice:

$$\frac{F_1 - 1}{1 - \dfrac{1}{G_{1av}}} < \frac{F_2 - 1}{1 - \dfrac{1}{G_{2av}}} \ . \tag{B.91}$$

The terms compared here are called "noise measure".

Problem 2.18

In order to calculate the circles with a constant noise figure in the complex generator impedance plane, equation (2.106) is modified as follows:

$$F = \text{const} = 1 + \frac{W_u + |Z|^2 W_i + 2\text{Re}\{Z \cdot W_{ui}\}}{2kT_0 \cdot \text{Re}\{Z\}} \ . \tag{B.92}$$

Setting $Z = R + jX$ leads to

$$F = 1 + \frac{W_u + (R^2 + X^2)W_i + 2(R \cdot \text{Re}\{W_{ui}\} - X \cdot \text{Im}\{W_{ui}\})}{2kT_0 R} \ . \tag{B.93}$$

A short transformation yields

$$\begin{aligned}
-W_u = \ & R^2 W_i + 2R\left(kT_0\left(1 - F\right) + \text{Re}\{W_{ui}\}\right) \\
& + X^2 W_i - 2X \cdot \text{Im}\{W_{ui}\}
\end{aligned} \tag{B.94}$$

Further algebraic manipulations lead to

$$\begin{aligned}
-\frac{W_u}{W_i} &= \left(R^2 + 2R \cdot \frac{kT_0(1 - F) + \text{Re}\{W_{ui}\}}{W_i}\right) + \left(X^2 - 2X \cdot \frac{\text{Im}\{W_{ui}\}}{W_i}\right) \\
&= \left(R + \frac{kT_0(1 - F) + \text{Re}\{W_{ui}\}}{W_i}\right)^2 - \left(\frac{kT_0(1 - F) + \text{Re}\{W_{ui}\}}{W_i}\right)^2 \\
&\quad + \left(X - \frac{\text{Im}\{W_{ui}\}}{W_i}\right)^2 - \left(\frac{\text{Im}\{W_{ui}\}}{W_i}\right)^2
\end{aligned} \tag{B.95}$$

and

$$\begin{aligned}
&\left(R + \frac{kT_0(1 - F) + \text{Re}\{W_{ui}\}}{W_i}\right)^2 + \left(X - \frac{\text{Im}\{W_{ui}\}}{W_i}\right)^2 \\
&= \left(\frac{kT_0(1 - F) + \text{Re}\{W_{ui}\}}{W_i}\right)^2 + \left(\frac{\text{Im}\{W_{ui}\}}{W_i}\right)^2 - \frac{W_u}{W_i} \ ,
\end{aligned} \tag{B.96}$$

resulting in

$$C = (R - R_0)^2 + (X - X_0)^2 \ . \tag{B.97}$$

A further calculation shows that the constant C is greater or equal to zero for all cases which are physically possible. Thus the solution is the equation of a circle. Because the centers (R_0, X_0) of the circles depend on the noise figure F, the circles generally are not concentric.

Problem 2.19

Equating Eq. (2.126) with Eq. (2.136) leads to the following expression, which should be fulfilled:

$$F_{min} + \frac{R_n}{G_g} \cdot |Y_g - Y_{opt}|^2 = F_{min} + 4 \cdot \frac{R_n}{Z_0} \cdot \frac{|\Gamma_g - \Gamma_{opt}|^2}{|1 + \Gamma_{opt}|^2(1 - |\Gamma_g|^2)} \quad , \quad \text{(B.98)}$$

or, equivalently,

$$\frac{1}{G_g} \cdot |Y_g - Y_{opt}|^2 = 4 \cdot \frac{1}{Z_0} \cdot \frac{|\Gamma_g - \Gamma_{opt}|^2}{|1 + \Gamma_{opt}|^2(1 - |\Gamma_g|^2)} \quad .$$

A short transformation leads to

$$4G_g \cdot |\Gamma_g - \Gamma_{opt}|^2 = Z_0 \cdot |Y_g - Y_{opt}|^2 \cdot |1 + \Gamma_{opt}|^2(1 - |\Gamma_g|^2) \quad . \quad \text{(B.99)}$$

Inserting the equations:

$$\Gamma_g = \frac{1 - Y_g Z_0}{1 + Y_g Z_0} \quad , \qquad \Gamma_{opt} = \frac{1 - Y_{opt} Z_0}{1 - Y_{opt} Z_0} \quad \text{(B.100)}$$

and

$$G_g = \text{Re}\{Y_g\} \quad \text{(B.101)}$$

leads to:

$$4G_g \frac{|(1 - Y_g Z_0)(1 + Y_{opt} Z_0) - (1 + Y_g Z_0)(1 - Y_{opt} Z_0)|^2}{|(1 + Y_g Z_0) \cdot (1 + Y_{opt} Z_0)|^2}$$
$$= Z_0 \cdot |Y_g - Y_{opt}|^2 |\frac{1 + Y_{opt} Z_0}{1 + Y_{opt} Z_0} + \frac{1 - Y_{opt} Z_0}{1 + Y_{opt} Z_0}|^2$$
$$\cdot \left(\frac{|1 + Y_g Z_0|^2}{|1 + Y_g Z_0|^2} + \frac{|1 - Y_g Z_0|^2}{|1 + Y_g Z_0|^2} \right) \quad . \quad \text{(B.102)}$$

Simplifying this expression by using $|X|^2 = X \cdot X^*$ leads to the following equation:

$$G_g \cdot 4Z_0^2 |Y_g - Y_{opt}|^2 = Z_0 \cdot |Y_g - Y_{opt}|^2 \cdot 4G_g Z_0 \quad . \quad \text{(B.103)}$$

Thus it is proven that Eq. (2.126) is identical to Eq. (2.136).

Appendix C
Solutions to the
Problems of Chapter 3

Chapter 3

Problem 3.1

With

$$|V_L|^2 \;=\; c_i^2 \cdot \tau^2 \cdot \mathrm{si}^2(\pi f t) \qquad \text{and} \qquad \lim_{f \to 0} \mathrm{si}(\pi f t) = 1 \qquad (\text{C}.1)$$

we get

$$|V_L(0)|^2 \;=\; c_i^2 \cdot \tau^2 \qquad (\text{C}.2)$$

Thus Eq. (3.18) can be simplified as follows:

$$\Delta f_L = \frac{\int_0^\infty |V_L|^2 df}{c_i^2 \tau^2} = \frac{1}{c_i^2 \tau^2} \int\limits_0^\infty c_i^2 \frac{\sin^2(\pi f t)}{(\pi f)^2} df = \int\limits_0^\infty \mathrm{si}^2(\pi f t) df \;. \qquad (\text{C}.3)$$

The substitution $f' = \pi f \tau$ leads to

$$\Delta f = \frac{1}{\pi \tau} \int_0^\infty \mathrm{si}^2(f')df' = \frac{1}{2\tau} \ . \tag{C.4}$$

We will take an integration time of $\tau=1$s as a numerical example. A part of all measured values shall have an accuracy better than 0.1K, that means that the variance has to reach a given value. The measured values are normally distributed, thus the number X of measured values, which are located around T_m in an interval of $2\Delta T$ is given by

$$X = \frac{1}{\sqrt{2\pi}} \cdot \frac{1}{\Delta T_m} \int_{T_m - \Delta T}^{T_m + \Delta T} \exp\left(-\frac{(T - T_m)^2}{2\Delta T_m^2}\right) dT \ . \tag{C.5}$$

With the substitution

$$\frac{T - T_m}{\Delta T_m} = T' \tag{C.6}$$

and accounting for symmetry leads to

$$X = \frac{2}{\sqrt{2\pi}} \cdot \int_0^{\Delta T/\Delta T_m} \exp\left(-\frac{1}{2}T'^2\right) dT' \ . \tag{C.7}$$

The integral can be calculated numerically with different integration limits. With $X = 0.68$ we find

$$\frac{\Delta T}{\Delta T_m} = 1 \ . \tag{C.8}$$

With $\Delta T=0.1$K and Eq. (3.23) the necessary bandwidth is given by

$$\Delta f = \frac{1}{\tau}\left(\frac{T_m}{\Delta T_m}\right)^2 = 9\text{MHz} \ . \tag{C.9}$$

With $X = 0.95$ and $\Delta T/\Delta T_m = 2$ and $\Delta T =0.1$K we obtain

$$\Delta f = 36\text{MHz} \ . \tag{C.10}$$

The result clearly shows the dependence of the measurement precision on the bandwidth for a given measurement time.

Problem 3.2

With

$$1 - |\Gamma_l|^2 = \frac{|Z_l + Z_0|^2 - |Z_l - Z_0|^2}{|Z_l + Z_0|^2} = \frac{2Z_0(Z_l + Z_l^*)}{|Z_l + Z_0|^2} \qquad (C.11)$$

and

$$1 - |\Gamma_g|^2 = \frac{2Z_0(Z_g + Z_g^*)}{|Z_g + Z_0|^2} \qquad (C.12)$$

and

$$|1 - \Gamma_g\Gamma_l|^2 = \left| \frac{(Z_g + Z_0)(Z_l + Z_0) - (Z_g - Z_0)(Z_l - Z_0)}{(Z_g + Z_0)(Z_l + Z_0)} \right|^2$$

$$= \frac{|2Z_0(Z_g + Z_l)|^2}{|Z_g + Z_0|^2|Z_l + Z_0|^2} \qquad (C.13)$$

the fraction on the right-hand side of Eq. (3.48) is given by

$$\frac{(1 - |\Gamma_l|^2)(1 - |\Gamma_g|^2)}{|1 - \Gamma_g\Gamma_l|^2} = \frac{4Z_0^2(Z_g + Z_g^*)(Z_l + Z_l^*)}{4Z_0^2|Z_g + Z_l|^2}$$

$$= \frac{4\mathrm{Re}\{Z_g\} \cdot \mathrm{Re}\{Z_l\}}{|Z_g + Z_l|^2} . \qquad (C.14)$$

Using the definition of Eq. (3.50) leads to the following term for the right-hand side of Eq. (3.53):

$$1 - |\tilde{\rho}|^2 = \frac{|Z_l + Z_g|^2 - |Z_l - Z_g^*|^2}{|Z_l + Z_g|^2}$$

$$= \frac{Z_l Z_g^* + Z_l^* Z_g + Z_l^* Z_g^* + Z_l Z_g}{|Z_l + Z_g|^2}$$

$$= \frac{4\mathrm{Re}\{Z_g\} \cdot \mathrm{Re}\{Z_l\}}{|Z_l + Z_g|^2} . \qquad (C.15)$$

A comparison yields the identity of the equations (3.53) and (3.48):

$$1 - |\tilde{\rho}|^2 = \frac{(1 - |\Gamma_g|^2)(1 - |\Gamma_l|^2)}{|1 - \Gamma_l\Gamma_g|^2} . \qquad (C.16)$$

The power, absorbed by Z_l, can be calculated from the current I_l flowing through Z_l:

$$P_l = |I_l|^2 \cdot \mathrm{Re}\{Z_l\} . \qquad (C.17)$$

The current I_l is given by

$$|I_l|^2 = 4kT\Delta f \cdot \mathrm{Re}\{Z_g\} \cdot \frac{1}{|Z_l + Z_g|^2} = P_{av} \cdot \frac{4\mathrm{Re}\{Z_g\}}{|Z_l + Z_g|^2} . \qquad (C.18)$$

Then P_l can be written as follows:

$$P_l = P_{av} \cdot \frac{4\text{Re}\{Z_g\} \cdot \text{Re}\{Z_l\}}{|Z_l + Z_g|^2} \quad . \tag{C.19}$$

A comparison with Eq. (C.15) shows that

$$P_l = P_{av}(1 - |\tilde{\rho}|^2) \tag{C.20}$$

and thus the validity of Eq. (3.51).

Problem 3.3

As for the compensating radiometer in Fig. 3.26 the switching states I and II are treated separately. According to the dissipation theorem the power P_I reaching the amplifier in switching state I is given by

$$P_I = k\Delta f \cdot [T_0(1 - \kappa)\alpha + T_{ref}\kappa\alpha + (1 - \alpha)T_0] \quad . \tag{C.21}$$

The terms relate to Z_0, the reference, and the attenuator, respectively. Similarly, we obtain for switching state II:

$$
\begin{aligned}
P_{II} = \quad & k\Delta f \cdot \Big[T_0\kappa + T_{ref}\kappa(1 - \kappa)|\rho|^2 \\
& + T_m \cdot (1 - |\rho|^2)(1 - \kappa) + T_0(1 - \kappa)^2|\rho|^2 \Big]
\end{aligned}
\tag{C.22}
$$

Here, the noise powers derived from Z_0, the reference, the measurement object and from the isolator have been added. The noise wave emitted by the isolator passes through the coupler twice and is reflected by the measurement object. A balance of the measured powers, $P_I = P_{II}$, is achieved by a variation of the temperature of the reference noise source. Taking into account the losses of the attenuator and the coupler we get:

$$
\begin{aligned}
T_0(1 - \kappa)^2 + \kappa T_0 - T_0\kappa = \quad & T_m \cdot (1 - |\rho|^2)(1 - \kappa) + T_{ref}\kappa(1 - \kappa)|\rho|^2 \\
& + T_0(1 - \kappa)^2|\rho|^2 - T_{ref}\kappa(1 - \kappa) \quad .
\end{aligned}
\tag{C.23}
$$

Dividing both sides by $(1 - \kappa)$ leads to

$$T_0(1 - \kappa) = T_m \cdot (1 - |\rho|^2) + T_{ref}\kappa|\rho|^2 + T_0(1 - \kappa)|\rho|^2 - T_{ref}\kappa \quad . \tag{C.24}$$

Rearranging the terms results in

$$T_0(1 - \kappa)(1 - |\rho|^2) + T_{ref}\kappa(1 - |\rho|^2) = T_m \cdot (1 - |\rho|^2) \quad , \tag{C.25}$$

and, after eliminating $1 - |\rho|^2$,

$$T_m = T_0(1 - \kappa) + T_{ref}\kappa \quad . \tag{C.26}$$

This expression is independent of the reflection coefficient ρ of the measurement object. If the ambient temperature and the temperature of the reference are known, the temperature of the measurement object can be determined.

Problem 3.4

For the determination of the correlation between the input and output noise waves of the preamplifier with isolator, the powers P_I and P_{II} must be calculated at the output of the preamplifier.

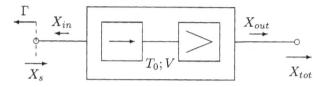

Fig. C.1   Noise waves at the amplifier with isolator.

Therefore, we must determine the squared magnitude of the total noise wave X_{tot} in terms of the quantities shown in Fig. (C.1).

Here, $X_{in/out}$ are the input and output equivalent noise waves of the preamplifier with isolator, Γ is the source reflection coefficient, X_s is the noise wave with contributions from the reference source and the measurement object, and V is the voltage gain of the amplifier. We obtain

$$P'_{I/II} = |X_{tot}|^2 = |X_{out}|^2 + |X_{in}|^2|\Gamma|^2|V|^2 + |X_s|^2|V|^2$$
$$+ \ 2 \cdot \operatorname{Re}\left\{\Gamma^* \cdot X_{in}^* \cdot X_{out} \cdot V^*\right\} \ . \tag{C.27}$$

The term $|X_{out}|^2$ is constant and will be neglected in the following. The last term in Eq. C.27 describes the effect of the correlation $X_{in}^* \cdot X_{out}$.

For the derivation of the balance condition of the circuit in Fig. 3.26 the switchable circulator was assumed to be lossless so that the term $|X_{in}|^2 \cdot |\Gamma|^2|V|^2$ did not need to be considered. In the case of a lossy isolator with the temperature T_0 the balance condition is given by

$$T_m = T_{ref} - T_0 \cdot \frac{|\rho|^2}{1 - |\rho|^2} \ . \tag{C.28}$$

If the combination of amplifier and isolator is completely de-correlated, the temperature of the measurement object T_m can be determined by using Eq. (C.28), on the basis of a known temperature of the isolator and a known reflection coefficient of the measurement object. If the amplifier and the isolator are not completely de-correlated, the balance condition reads:

$$T_m = T_{ref} - T_0 \cdot \frac{|\rho|^2}{1 - |\rho|^2} - \underbrace{\frac{2}{k\Delta f(1 - |\rho|^2)}\operatorname{Re}\left\{\rho^* \cdot \frac{X_{in}^* X_{out}}{V}\right\}}_{\text{correlation term}} \ . \tag{C.29}$$

The correlation term in Eq. (C.29) describes the measurement error resulting from a finite de-correlation i.e. $X_{in}^* \cdot X_{out} \neq 0$. For the circuit of Fig. 3.30 we have

$$T_m = T_0(1 - \kappa) + T_{ref}\kappa - \underbrace{\frac{2(1 - \kappa)}{k\Delta f(1 - |\rho|^2)} \mathrm{Re}\left\{\rho^* \cdot \frac{X_{in}^* X_{out}}{V}\right\}}_{\text{correlation term}} . \qquad \text{(C.30)}$$

A comparison of Eq. (C.30) with the result of problem 3.3 shows that the correlation term describes the resulting measurement error.

If the preamplifier and the isolator are at the same temperature T_0, the deviation $|\Delta T_m|$ of the measured temperature of the measurement object as compared to the correlation $X_{in}^* \cdot X_{out}$ can be evaluated. With the de-correlation Q $(0 \leq |Q| \leq 1)$ defined as

$$\frac{X_{in}^* X_{out}}{V} = kT_0\Delta f \cdot Q \qquad \text{(C.31)}$$

and $\kappa \to 0$ in Fig. 3.30 we obtain

$$|\Delta T_m|_{max} = \frac{2T_0}{1 - |\rho|^2} \cdot \mathrm{Re}\{\rho \cdot Q\} . \qquad \text{(C.32)}$$

With $\rho = 1/2 \stackrel{\wedge}{=} 6\mathrm{dB}$ the maximum measurement error is given by

$$|\Delta T_m|_{max} = \frac{4}{3} \cdot T_0 \cdot |Q| \qquad \text{(C.33)}$$

$|\Delta T_m|_{max} = 1\mathrm{K}$ and $T_0 = 290\mathrm{K}$ leads to $|Q| = 0.26\%$.

Problem 3.5

Figure C.2 shows the extended measurement circuit.

The powers P_I and P_{II} are obtained as

$$\frac{P_I}{k\Delta f} = T_{ref}\kappa_1(1 - \kappa_1)(1 - \kappa_2)$$
$$+ T_0\left[(1 - \kappa_1)^2(1 - \kappa_2) + (1 - \kappa_2)\kappa_1 + \kappa_2\right] \qquad \text{(C.34)}$$

$$\frac{P_{II}}{k\Delta f} = T_m(1 - |\rho|^2)(1 - \kappa_1)(1 - \kappa_2)$$
$$+ T_{ref}|\rho|^2\kappa_1(1 - \kappa_1)(1 - \kappa_2) + T_{aux}\kappa_2$$
$$+ T_0\left[|\rho|^2(1 - \kappa_1)^2(1 - \kappa_2)^2 \right.$$
$$\left. + \kappa_1(1 - \kappa_2) + |\rho|^2\kappa_2(1 - \kappa_1)^2(1 - \kappa_2)\right] . \qquad \text{(C.35)} \cdot$$

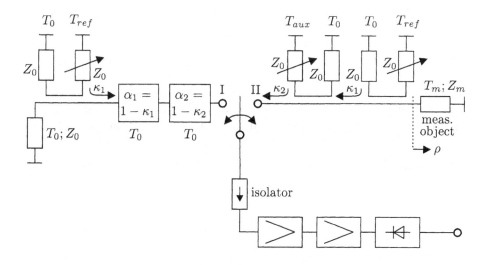

Fig. C.2 Compensated radiometer for measuring low noise temperatures.

After a procedure similar to problem 3.3, equating and solving the equation for T_m leads to

$$T_m = \kappa_1(T_{ref} - T_0) + T_0 - (T_{aux} - T_0) \cdot \frac{\kappa_2}{(1 - \kappa_1)(1 - \kappa_2)(1 - |\rho|^2)} \ .$$

$$(C.36)$$

This result is not sufficient to calculate T_m, because $|\rho|^2$ is unknown. The following procedure leads to a system of equations for T_m and $|\rho|^2$:

The first step is to adjust T_{ref1} to the ambient temperature ($T_{ref1} = T_0$). Then T_{aux} is raised until the balance condition is fulfilled. This leads to a first equation for T_m:

$$T_m = T_0 - (T_{aux1} - T_0) \cdot \frac{\kappa_2}{(1 - \kappa_1)(1 - \kappa_2)(1 - |\rho|^2)} \ .$$

$$(C.37)$$

The second step is to raise the excess temperature of the auxiliary noise source in a defined manner by a factor of n ($T_{aux2} - T_0 = n(T_{aux1} - T_0)$). The temperature of the reference noise source is raised to T_{ref2} until the balance condition is fulfilled. This leads to a second equation for T_m:

$$T_m = \kappa_1(T_{ref2} - T_0) + T_0 - n(T_{aux1} - T_0) \cdot \frac{\kappa_2}{(1 - \kappa_1)(1 - \kappa_2)(1 - |\rho|^2)} \ .$$

$$(C.38)$$

Multiplying the first equation by n and subtracting both equations leads to the following result:

$$T_m = T_0 - \frac{\kappa_1}{n-1}(T_{ref2} - T_0) \ . \tag{C.39}$$

With the circuit described above it is possible to measure temperatures in the range of $0K < T_m < T_0$. With $n = 2$ a fixed $3\,dB$ attenuator can be inserted in front of the auxiliary noise source in a first step. In a second step this attenuator is bypassed. There is no need to calibrate the auxiliary noise source. It just has to be variable and stable during the measuring time.

Problem 3.6

The lossless $3\,dB$-$90°$ coupler has the phase relations shown in Fig. C.3.

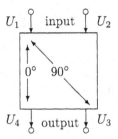

Fig. C.3

We get for the correlation between the output voltages:

$$
\begin{aligned}
U_4^* U_3 &= \frac{1}{\sqrt{2}}(U_1 + jU_2)^* \frac{1}{\sqrt{2}}(jU_1 + U_2) \\
&= \frac{1}{2}j|U_1|^2 - \frac{1}{2}j|U_2|^2 \ ,
\end{aligned}
\tag{C.40}
$$

because U_1 and U_2 are uncorrelated. This shows that for a correlation radiometer with a $90°$-coupler the real part of the correlation at the output is always zero. Thus the zero balance must be performed on the basis of the imaginary part of the correlation. As shown in the figure below this requires a further $90°$ phase shifter.

Problem 3.7

In Fig. 3.26 the circulator can be replaced by an isolator as shown in Fig. C.5:
    The isolator is assumed to have the same temperature as the reference noise source. A zero balance is obtained for

$$T_{ref} = T_{ref}|\rho|^2 + T_m(1 - |\rho|^2) \ , \tag{C.41}$$

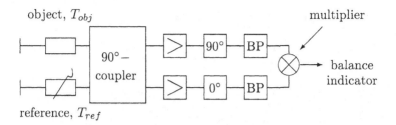

Fig. C.4

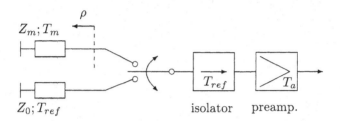

Fig. C.5

leading directly to:

$$T_m = T_{ref} \ . \tag{C.42}$$

The noise signals at the output of the preamplifier are uncorrelated for both switch positions. Thus the absolute error of the switching radiometer is given by

$$
\begin{aligned}
\Delta T_{sw} &= \sqrt{2} \cdot \sqrt{2} \Delta T_{ref} \\
&= 2\Delta T_{ref} = \frac{2}{\sqrt{\Delta f \cdot \tau}} (T_{ref} + T_a) \ .
\end{aligned} \tag{C.43}
$$

Inserting the balance equation C.41 into the error term ΔT_{sw} leads to

$$T_{ref} \pm \Delta T_{sw} = T_{ref}|\rho|^2 + T_m(1 - |\rho|^2) \ . \tag{C.44}$$

Solving this equation for T_m leads to

$$T_m = T_{ref} \pm \frac{\Delta T_{sw}}{1 - |\rho|^2} \ . \tag{C.45}$$

The temperature of the measurement object is measured with the following error:

$$\Delta \tilde{T}_m = \frac{\Delta T_{sw}}{1 - |\rho|^2} \ . \tag{C.46}$$

For the relative error we get

$$\frac{\Delta \tilde{T}_m}{T_m} = \frac{\Delta \tilde{T}_{sw}}{T_m} \cdot \frac{1}{1 - |\rho|^2} \; . \qquad (C.47)$$

For $|\rho| \to 1$ the relative error becomes arbitrarily large.

Problem 3.8

For the evaluation of a temperature error an equation for the correlation radiometer corresponding to Eq. (3.16) has to be determined. The following calculation is based on Fig. C.6.

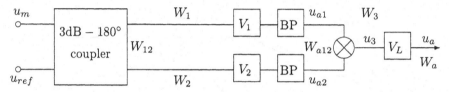

Fig. C.6   Principle circuit of the correlation radiometer.

The average squared output voltage is calculated according to Section 3.2.2 as follows:

$$\overline{u_a^2(t)} = \int\limits_{-\infty}^{\infty} W_a(f) df = \int\limits_{-\infty}^{\infty} W_3(f) |V_L(f)|^2 df \; . \qquad (C.48)$$

For the calculation of the spectrum $W_3(f)$ with

$$\begin{aligned} \rho_3(\theta) &= \overline{u_3(t) \cdot u_3(t+\theta)} \\ &= c^2 \, \overline{u_{a1}(t) \cdot u_{a1}(t+\theta) u_{a2}(t) \cdot u_{a2}(t+\theta)} \end{aligned} \qquad (C.49)$$

the results of Section 1.2.7 and accordingly problem 1.5 have to be extended to the form that is needed here. Under the assumptions made there, we have

$$\rho_3(\theta) = c^2 \left[\rho_{a12}^2(0) + \rho_{a1}(\theta)\rho_{a2}(\theta) + \rho_{a12}(\theta)\rho_{a21}(\theta) \right] \; . \qquad (C.50)$$

A Fourier transformation of Eq. (C.50) leads to an expression for $W_3(f)$:

$$\begin{aligned} W_3(f) &= c^2 \rho_{a12}^2(0)\delta(f) \\ &+ c^2 \int\limits_{-\infty}^{\infty} \left[W_{a1}(f') \cdot W_{a2}(f-f') + W_{a12}(f') \cdot W_{a21}(f-f') \right] df' \; . \end{aligned}$$

$$(C.51)$$

With $W(f) = W(-f)$ for the power spectra and $W_{xy}(f) = W_{yx}(-f)$ for the cross spectra, Eq. (C.51) can be written as follows:

$$
\begin{aligned}
W_3(f) &= c^2 \rho_{a12}^2(0)\delta(f) \\
&+ c^2 \int_{-\infty}^{\infty} [W_{a1}(f') \cdot W_{a2}(f' - f) + W_{a12}(f') \cdot W_{a12}(f' - f)]\, df' \ .
\end{aligned}
$$

(C.52)

With

$$
\begin{aligned}
\overline{u_a(t)}^2 &= \left(V_L(0) \cdot \overline{u_3(t)}\right)^2 = \left(c \cdot V_L(0) \cdot \overline{u_{a1}(t)u_{a2}(t)}\right)^2 \\
&= (c \cdot V_L(0) \cdot \rho_{a12}(0))^2
\end{aligned}
$$

(C.53)

the variance σ_a^2 can be calculated similar to the calculation in Eq. (3.15):

$$
\begin{aligned}
\sigma_a^2 &= \overline{u_a^2(t)} - \overline{u_a(t)}^2 = c^2 \rho_{a12}(0)^2 \int_{-\infty}^{\infty} |V_L(f)|^2 \delta(f) df \\
&+ c^2 \int_{-\infty}^{+\infty}\!\!\int |V_L(f)|^2 \left[W_{a1}(f')W_{a2}(f' - f) + W_{a12}(f')W_{a12}(f' - f)\right] df'df \\
&- (cV_L(0) \cdot \rho_{a12}(0))^2 \\
&= c^2 \int_{-\infty}^{+\infty}\!\!\int |V_L(f)|^2 \left[W_{a1}(f')W_{a2}(f' - f) + W_{a12}(f')W_{a12}(f' - f)\right] df'df \ .
\end{aligned}
$$

(C.54)

With the same approximations as made in Section 3.1.8 for the voltage gains V_1 and V_2 of both channels we get

$$
\sigma_a^2 \approx c^2 \int_{-\infty}^{\infty} |V_L(f)|^2 df
$$

(C.55)

$$
\cdot \left[\int_{-\infty}^{\infty} |V_1|^2 |V_2|^2 W_1(f)W_2(f)df + \int_{-\infty}^{\infty} (V_1^* V_2)^2 W_{12}^2(f)df \right] \ .
$$

Because of

$$
W_{12}^2(f) = W_{12}^{2*}(-f)
$$

(C.56)

we have

$$
\mathrm{Re}\left\{W_{12}^2(f)\right\} = \mathrm{Re}\left\{W_{12}^2(-f)\right\}
$$

(C.57)

and

$$\text{Im}\left\{W_{12}^2(f)\right\} = -\text{Im}\left\{W_{12}^2(-f)\right\} \tag{C.58}$$

and hence

$$
\begin{aligned}
\int\limits_{-\infty}^{\infty} W_{12}^2(f)df &= \int\limits_{-\infty}^{\infty} \left(\text{Re}\left\{W_{12}^2(f)\right\} + \text{Im}\left\{W_{12}^2(f)\right\}\right)df \\
&= 2\int\limits_{0}^{\infty} \text{Re}\left\{W_{12}^2(f)\right\}df \; .
\end{aligned} \tag{C.59}
$$

Then, the equation for σ_a^2 can be written as follows:

$$
\begin{aligned}
\sigma_a^2 &\approx 2 \cdot c^2 \int\limits_{0}^{\infty} |V_L(f)|^2 df \\
&\quad \cdot \left[2\int\limits_{0}^{\infty} |V_1|^2|V_2|^2 W_1(f)W_2(f)df + 2\int\limits_{0}^{\infty} \text{Re}\left\{(V_1^*V_2)^2 W_{12}^2(f)\right\}df\right] \; .
\end{aligned} \tag{C.60}
$$

With the assumption $V_1 = V_2 = V$ and rectangularly shaped band-pass filters Eq. (C.60) can be written as

$$
\begin{aligned}
\sigma_a^2 &\approx 2 \cdot c^2 \int\limits_{0}^{\infty} |V_L(f)|^2 df \\
&\quad \cdot \left[2|V|^4 \int\limits_{0}^{\infty} W_1(f)W_2(f)df + 2|V|^4 \int\limits_{0}^{\infty} \text{Re}\left\{W_{12}^2(f)\right\}df\right] \\
&= 4c^2 \int\limits_{0}^{\infty} |V_L(f)|^2 df \cdot |V|^4 \left[W_1(f_0)W_2(f_0) + \text{Re}\left\{W_{12}^2(f_0)\right\}\right]\Delta f \\
&= 4c^2 \int\limits_{0}^{\infty} |V_L(f)|^2 df \\
&\quad \cdot |V|^4 \left[W_1(f_0)W_2(f_0) + \text{Re}^2\left\{W_{12}(f_0)\right\} - \text{Im}^2\left\{W_{12}(f_0)\right\}\right]\Delta f \; .
\end{aligned} \tag{C.61}
$$

With the effective low-pass bandwidth Δf_L as given by Eq. (3.18) we obtain

$$\overline{u_a(t)}^2 = \left(cV_L(0)\rho_{a12}(0)\right)^2 = \left(cV_L(0)\int\limits_{-\infty}^{\infty} W_{a12}df\right)^2$$

$$= \left(2cV_L(0) \int\limits_0^\infty \mathrm{Re}\left\{ W_{a12}(f) \right\} df \right)^2$$

$$= \left(2cV_L(0) |V(f_0)|^2 \Delta f \, \mathrm{Re}\left\{ W_{12}(f_0) \right\} \right)^2 \; . \tag{C.62}$$

The normalized variance of $\overline{u_a(t)}^2$ is given by

$$\tilde{\sigma}_a^2 = \frac{\sigma_a^2}{\overline{u_a(t)}^2} \tag{C.63}$$

$$= \frac{W_1(f_0)W_2(f_0) + \mathrm{Re}^2\left\{ W_{12}(f_0) \right\} - \mathrm{Im}^2\left\{ W_{12}(f_0) \right\}}{\mathrm{Re}^2\left\{ W_{12}(f_0) \right\}} \cdot \frac{\Delta f_L}{\Delta f}$$

or, using the normalized cross spectrum k_{12} defined in Eq. (1.87):

$$\tilde{\sigma}_a^2 = \left(2 + \frac{1 - |k_{12}|^2}{\mathrm{Re}^2\left\{ k_{12} \right\}} \right) \frac{\Delta f_L}{\Delta f} \; . \tag{C.64}$$

The output signal of the correlation radiometer is proportional to the real part of the cross-spectrum W_{12} at the band-pass center frequency f_0:

$$\overline{u_a(t)} = C \cdot \mathrm{Re}^2\left\{ W_{12}(f_0) \right\} \; . \tag{C.65}$$

Inserting into (C.64) leads to

$$\sigma_a^2 = \tilde{\sigma}_a^2 \cdot \overline{u_a(t)}^2 \tag{C.66}$$

$$= C^2 \left[2\mathrm{Re}^2\left\{ W_{12}(f_0) \right\} W_1(f_0)W_2(f_0) - |W_{12}(f_0)|^2 \right] \frac{\Delta f_L}{\Delta f} \; .$$

The zero balance condition is

$$\overline{u_a(t)} = 0 \qquad \Rightarrow \qquad \mathrm{Re}^2\left\{ W_{12}(f_0) \right\} = 0 \; . \tag{C.67}$$

After adjusting for zero balance we get for the variance of the output voltage:

$$\sigma_a^2 = C^2 \left[W_1(f_0)W_2(f_0) - |W_{12}(f_0)|^2 \right] \frac{\Delta f_L}{\Delta f} \; . \tag{C.68}$$

Using the symbolic notation leads to the following expression for the spectra:

$$\begin{aligned}
W_1(f_0)\Delta f &= U_1 U_1^* = k\Delta f (T_m + T_{ref}) \tag{C.69} \\
&= U_2 U_2^* = W_2(f_0)\Delta f \\
&= 2k\, \Delta f\, T_m \qquad \text{for zero balance}
\end{aligned}$$

and

$$\begin{aligned}
W_{12}(f_0)\Delta f &= k\, \Delta f (T_m - T_{ref}) \tag{C.70} \\
&= 0 \qquad \text{for zero balance} \; .
\end{aligned}$$

The error of the output voltage can be interpreted as a temperature error:

$$
\begin{aligned}
\sigma_a^2 &= C^2 W_1(f_0) W_2(f_0) \frac{\Delta f_L}{\Delta f} = \overline{u_a(t)}^2 \Big|_{T_m - T_{ref} = \Delta T} \\
&= C^2 (k\Delta f)^2 (T_m - T_{ref})^2 = C^2 (k\Delta f)^2 \Delta T^2 \\
\Rightarrow\quad & 4 T_m^2 \frac{\Delta f_L}{\Delta f} = \Delta T^2 \\
\Rightarrow\quad & \frac{\Delta T}{\Delta T_m} = 2 \sqrt{\frac{\Delta f_L}{\Delta f}} \ .
\end{aligned}
\tag{C.71}
$$

Using an ideal integrator with the integration time τ as a low-pass filter we get with Eq. (3.22):

$$
\frac{\Delta T}{\Delta T_m} = \sqrt{2} \cdot \frac{1}{\sqrt{\Delta f \tau}} \ .
\tag{C.72}
$$

This result is identical to Eq. (3.70).

A direct calculation of the temperature error is explained in the following. The balance indicator signal BI of the correlator is proportional to the real part of the cross spectrum of its input variables:

$$
BI \sim \mathrm{Re} \left\{ \frac{1}{2} \left(U_m + U_{ref} \right) \left(U_m - U_{ref} \right)^* \right\} \ .
\tag{C.73}
$$

Using the identity

$$
\mathrm{Re}\{ab^*\} = \frac{1}{4} \left(|a+b|^2 - |a-b|^2 \right)
\tag{C.74}
$$

leads to

$$
\begin{aligned}
BI \ \sim \ & \frac{1}{8} |U_m + U_{ref} + U_m - U_{ref}|^2 \\
& - \frac{1}{8} |U_m + U_{ref} - U_m + U_{ref}|^2 \\
= \ & \frac{1}{2} \left(|U_m|^2 - |U_{ref}|^2 \right) \ .
\end{aligned}
\tag{C.75}
$$

Thus the expected proportionality

$$
BI \sim (T_m - T_{ref})
\tag{C.76}
$$

results. For the variance of the temperature error under balance conditions the Equations (3.69) and (3.70) apply.

Problem 3.9

A correlation radiometer as given in figure 3.31 with a non ideal multiplier does not provide a zero indication under a zero balance condition ($T_{ref} = T_m$) even if the measurement object and the reference are completely uncorrelated sources. The resulting direct voltage error shall be calculated in the following.

The output signal y of a non ideal multiplier can be described in general by a power series:

$$y = \sum_{m=0}^{\infty} \sum_{n=0}^{\infty} \beta_{mn} u_1^m u_2^n \ . \tag{C.77}$$

At the output of the low pass filter the time average is given by

$$\langle y \rangle = \sum_{m=0}^{\infty} \sum_{n=0}^{\infty} \beta_{mn} \langle u_1^m \cdot u_2^n \rangle \ . \tag{C.78}$$

For uncorrelated gaussian distributed input variables u_1 and u_2 Eqs. (3.69) and (3.72) lead to

$$\langle u_1^m \cdot u_2^n \rangle = \frac{1}{j^{m+n}} \left[\frac{d^m}{dv_1^m} e^{(-\frac{1}{2}\langle u_1^2 \rangle v_1^2)} \cdot \frac{d^n}{dv_2^n} e^{(-\frac{1}{2}\langle u_2^2 \rangle v_2^2)} \right]_{|v_1 = v_2 = 0} \ . \tag{C.79}$$

For the n-th derivative of the function $e^{a v^2}$ it holds:

$$\frac{d^n}{dv^n} e^{av^2} \cdot \sum_{k=0}^{\widehat{k}} \frac{a^k}{k!} \cdot \frac{n!}{(n-2k)!} \cdot (2av)^{n-2k} \tag{C.80}$$

$$\text{with} \quad \hat{k} = \begin{cases} \dfrac{n}{2} & \text{if } n \text{ is even} \\ \dfrac{n-1}{2} & \text{if } n \text{ is odd} \ . \end{cases} \tag{C.81}$$

As can be shown, the term $\langle u_1^m \cdot u_2^n \rangle$ becomes zero if either m or n are odd. If n and m are even, m can be replaced by $m = 2k$ and n by $n = 2l$. This leads to the following result for the output signal of the correlator:

$$\langle y \rangle = \sum_{k=0}^{\infty} \sum_{l=0}^{\infty} \beta_{2k2l} \frac{(-1)^{k+l} (2k)! \cdot (2l)!}{2^{k+l} \cdot k! \, l!} \cdot (\langle u_1^2 \rangle)^k (\langle u_2^2 \rangle)^l \ . \tag{C.82}$$

A dc offset appears at the output which depends on the power of the input variables.

Using a periodical phase shift as shown in figure 3.32, the coefficients $\langle u_1^m u_2^n \rangle$ also become periodical functions of time. However, the time average $\langle y \rangle$ remains constant. As in the unmodulated case the resulting error is a pure direct voltage error. Using a phase-sensitive detector at the switching frequency of e.g. $f_i = 10\text{kHz}$, only the correlation term makes a contribution at the frequency f_i. Then the display shows a zero for a vanishing correlation, independent of the type of multiplier.

Problem 3.10

Using Eqs. (3.75) and (3.68) we obtain for the Y-factor for small relative errors:

$$
Y_m = \frac{P_2' \pm \Delta P_2'}{P_2 \mp \Delta P_2} = \frac{P_2'}{P_2} \cdot \frac{1 \pm \dfrac{\Delta P_2'}{P_2'}}{1 \mp \dfrac{\Delta P_2}{P_2}}
$$

$$
= Y \cdot \frac{1 \pm \dfrac{1}{\sqrt{\Delta f \tau}}}{1 \mp \dfrac{1}{\sqrt{\Delta f \tau}}} \approx Y\left(1 \pm \frac{2}{\sqrt{\Delta f \tau}}\right) . \qquad (C.83)
$$

The error of the Y-factor causes an error of the noise figure:

$$
F_m = \frac{T_{ex}}{T_0} \cdot \frac{1}{Y\left(1 \pm \dfrac{2}{\sqrt{\Delta f \tau}}\right) - 1} \qquad (C.84)
$$

$$
\approx \frac{T_{ex}}{T_0} \frac{1}{Y-1}\left(1 \mp \frac{Y}{Y-1} \frac{2}{\sqrt{\Delta f \tau}}\right) = F \pm \Delta F . \qquad (C.85)
$$

For the given numerical example with

$$
\Delta f = 5\text{MHz}; \; \tau = 0.1\text{ns}; \; F = 6\text{dB} \stackrel{\wedge}{=} 4; \; \frac{T_{ex}}{T_0} = 16\text{dB} \stackrel{\wedge}{=} 40 \qquad (C.86)
$$

and

$$
Y = 1 + \frac{T_{ex}}{T_0 \cdot F} = 11 \qquad (C.87)
$$

we get for the relative error of the noise figure:

$$
\frac{\Delta F}{F} = 0.0031 . \qquad (C.88)
$$

Problem 3.11

With computer-controlled equipment for the measurement of noise figures of linear two-ports it is possible to measure the powers P_2 and P_2' both with and without a measurement object. If G_0 is the gain of the measurement circuit without a measurement object, the measured noise powers are

$$
\begin{aligned}
P_2 &= G_0(T_0 + T_a)k\Delta f \\
P_2' &= G_0(T_{g0} + T_a)k\Delta f \\
\Rightarrow \quad G_0 &= \frac{P_2' - P_2}{k\Delta f(T_{g0} - T_0)} .
\end{aligned}
\tag{C.89}
$$

Here $T_{g0} - T_0$ is the excess noise temperature of the noise generator and T_a is the system temperature of the preamplifier.

If $G_{tot} = G_{obj} \cdot G_0$ is the gain of the cascade of the measurement object and the measurement circuit, we get with the modified system temperature T_a':

$$
\begin{aligned}
\tilde{P}_2 &= G_{tot}(T_0 + T_a')k\Delta f \\
\tilde{P}_2' &= G_{tot}(T_{g0} + T_a')k\Delta f \\
\Rightarrow \quad G_{tot} &= \frac{\tilde{P}_2' - \tilde{P}_2}{k\Delta f(T_{g0} - T_0)} .
\end{aligned}
\tag{C.90}
$$

Thus the gain G_{obj} of the object under test is given by

$$
G_{obj} = \frac{G_{tot}}{G_0} = \frac{\tilde{P}_2' - \tilde{P}_2}{P_2' - P_2} .
\tag{C.91}
$$

Problem 3.12

For the determination of the input admittance Y_{in} the Eqs. (3.96) to (3.100) are combined in a matrix and vector form. With the matrix

$$
[K_1] =
\begin{bmatrix}
\sum a_{j5}a_{j1} & \sum a_{j5}a_{j2} & \sum a_{j5}a_{j3} & \sum a_{j5}a_{j4} \\
\sum a_{j6}a_{j1} & \sum a_{j6}a_{j2} & \sum a_{j6}a_{j3} & \sum a_{j6}a_{j4} \\
\sum a_{j7}a_{j1} & \sum a_{j7}a_{j2} & \sum a_{j7}a_{j3} & \sum a_{j7}a_{j4} \\
\sum a_{j8}a_{j1} & \sum a_{j8}a_{j2} & \sum a_{j8}a_{j3} & \sum a_{j8}a_{j4} \\
\sum a_{j9}a_{j1} & \sum a_{j9}a_{j2} & \sum a_{j9}a_{j3} & \sum a_{j9}a_{j4}
\end{bmatrix}
\tag{C.92}
$$

and the symmetrical matrix

$$
[K_2] =
\begin{bmatrix}
\sum a_{j5}^2 & \sum a_{j5}a_{j6} & \sum a_{j5}a_{j7} & \sum a_{j5}a_{j8} & \sum a_{j5}a_{j9} \\
\sum a_{j6}a_{j5} & \sum a_{j6}^2 & \sum a_{j6}a_{j7} & \sum a_{j6}a_{j8} & \sum a_{j6}a_{j9} \\
\sum a_{j7}a_{j5} & \sum a_{j7}a_{j6} & \sum a_{j7}^2 & \sum a_{j7}a_{j8} & \sum a_{j7}a_{j9} \\
\sum a_{j8}a_{j5} & \sum a_{j8}a_{j6} & \sum a_{j8}a_{j7} & \sum a_{j8}^2 & \sum a_{j8}a_{j9} \\
\sum a_{j9}a_{j5} & \sum a_{j9}a_{j6} & \sum a_{j9}a_{j7} & \sum a_{j9}a_{j8} & \sum a_{j9}^2
\end{bmatrix}
\tag{C.93}
$$

it follows:

$$
[K_1] \cdot \begin{bmatrix} 1 \\ G_{in} \\ B_{in} \\ G_{in}^2 + B_{in}^2 \end{bmatrix} = [K_2] \cdot \begin{bmatrix} m \\ \tilde{W}_i \\ \tilde{W}_u \\ \tilde{C}_r \\ \tilde{C}_i \end{bmatrix} \quad . \tag{C.94}
$$

For better clarity, the boundaries of the sum symbols have been omitted in this representation. By multiplication with the inverse of the symmetrical matrix $[K_2]$ a relation for the modified noise parameters and the factor m results:

$$
\begin{bmatrix} m \\ \tilde{W}_i \\ \tilde{W}_u \\ \tilde{C}_r \\ \tilde{C}_i \end{bmatrix} = [K_2]^{-1}[K_1] \cdot \begin{bmatrix} 1 \\ G_{in} \\ B_{in} \\ G_{in}^2 + B_{in}^2 \end{bmatrix} \quad . \tag{C.95}
$$

Equation (3.94) and (3.95) can also be converted into a matrix and vector form:

$$
\left([K_3] + \begin{bmatrix} G_{in} \\ B_{in} \end{bmatrix} \cdot \underline{v}_1 \right) \begin{bmatrix} m \\ \tilde{W}_i \\ \tilde{W}_u \\ \tilde{C}_r \\ \tilde{C}_i \end{bmatrix} = \left([K_4] + \begin{bmatrix} G_{in} \\ B_{in} \end{bmatrix} \cdot \underline{v}_2 \right) \begin{bmatrix} 1 \\ G_{in} \\ B_{in} \\ G_{in}^2 + B_{in}^2 \end{bmatrix} \quad , \tag{C.96}
$$

with

$$
[K_3] = \begin{bmatrix} \sum a_{j2}a_{j5} & \sum a_{j2}a_{j6} & \sum a_{j2}a_{j7} & \sum a_{j2}a_{j8} & \sum a_{j2}a_{j9} \\ \sum a_{j3}a_{j5} & \sum a_{j3}a_{j6} & \sum a_{j3}a_{j7} & \sum a_{j3}a_{j8} & \sum a_{j3}a_{j9} \end{bmatrix} \tag{C.97}
$$

$$
[K_4] = \begin{bmatrix} \sum a_{j2}a_{j1} & \sum a_{j2}^2 & \sum a_{j2}a_{j3} & \sum a_{j2}a_{j4} \\ \sum a_{j3}a_{j1} & \sum a_{j3}a_{j2} & \sum a_{j3}^2 & \sum a_{j3}a_{j4} \end{bmatrix} \tag{C.98}
$$

$$
\underline{v}_1 = 2\left[\sum a_{j4}a_{j5} \quad \sum a_{j4}a_{j6} \quad \sum a_{j4}a_{j7} \quad \sum a_{j4}a_{j8} \quad \sum a_{j4}a_{j9} \right] \tag{C.99}
$$

$$
\underline{v}_2 = 2\left[\sum a_{j4}a_{j1} \quad \sum a_{j4}a_{j2} \quad \sum a_{j4}a_{j3} \quad \sum a_{j4}^2 \right] \quad . \tag{C.100}
$$

With Eq. (C.95) and Eq. (C.96) the unknown noise parameters and the factor m can be eliminated. The following equations result for the determination of the real and imaginary part of the input admittance of the device under test.

$$
\left(\left([K_3][K_2]^{-1}[K_1] - [K_4] \right) + \begin{bmatrix} G_{in} \\ B_{in} \end{bmatrix} \left(\underline{v}_1[K_2]^{-1}[K_1] - \underline{v}_2 \right) \right) \cdot \begin{bmatrix} 1 \\ G_{in} \\ B_{in} \\ G_{in}^2 + B_{in}^2 \end{bmatrix} = \underline{0} \quad . \tag{C.101}
$$

With the terms

$$[K_5] = [K_3][K_2]^{-1}[K_1] - [K_4] \ , \tag{C.102}$$

$$\underline{v}_3 = \underline{v}_1[K_2]^{-1}[K_1] - \underline{v}_2 \ , \tag{C.103}$$

Equation (C.101) can be rewritten as follows:

$$G_{in}\left[\underline{v}_3(1,1)\cdot 1 + \underline{v}_3(1,2)\cdot G_{in} + \underline{v}_3(1,3)\cdot B_{in} + \underline{v}_3(1,4)\cdot\left(G_{in}^2 + B_{in}^2\right)\right]$$
$$[K_5](1,1)\cdot 1 + [K_5](1,2)\cdot G_{in} + [K_5](1,3)\cdot B_{in}$$
$$+ [K_5](1,4)\cdot\left(G_{in}^2 + B_{in}^2\right) = 0 \tag{C.104}$$
$$B_{in}\left[\underline{v}_3(1,1)\cdot 1 + \underline{v}_3(1,2)\cdot G_{in} + \underline{v}_3(1,3)\cdot B_{in} + \underline{v}_3(1,4)\cdot\left(G_{in}^2 + B_{in}^2\right)\right]$$
$$+ [K_5](2,1)\cdot 1 + [K_5](2,2)\cdot G_{in} + [K_5](2,3)\cdot B_{in}$$
$$+ [K_5](2,4)\cdot\left(G_{in}^2 + B_{in}^2\right) = 0 \tag{C.105}$$

The elimination of B_{in} and solving for G_{in} leads to a polynomial of 8th degree,

$$h_8 G_{in}^8 + h_7 G_{in}^7 + h_6 G_{in}^6 + h_5 G_{in}^5 + h_4 G_{in}^4 + h_3 G_{in}^3 + h_2 G_{in}^2 + h_1 G_{in} + h_0 = 0 \ , \tag{C.106}$$

which is unambiguously numerically solvable. Exemplarily, the coefficient h_8 is given by

$$\begin{aligned}
h_8 = \ & [\underline{v}_3(1,4)\left([K_5](2,2) - [K_5](1,3)\right) - \underline{v}_3(1,2)[K_5](2,4) \\
& + \underline{v}_3(1,3)[K_5](1,4)]^2 + [\underline{v}_3(1,4)\left([K_5](2,3) - [K_5](1,2)\right) \\
& - \underline{v}_3(1,3)[K_5](2,4) + \underline{v}_3(1,2)[K_5](1,4)]^2 \ . \tag{C.107}
\end{aligned}$$

The other coefficients h_0 to h_7 can be derived similarly.

Appendix D
Solutions to the
Problems of Chapter 4

Chapter 4

Problem 4.1

In order to derive the auto-correlation function of an irregular sequence of δ-impulses, the auto-correlation function of an irregular sequence of rectangular impulses is calculated first. Such a sequence $y(t)$ shall have the following properties:

$$y(t) = \sum_\nu u(t - t_\nu) \qquad u(t - t_\nu) = \frac{1}{\tau} \cdot \text{rect} \left\{ \frac{2(t - t_\nu)}{\tau} \right\} \qquad \text{(D.1)}$$

with

$$\text{rect}\{x\} = \begin{cases} 1 & \text{for } |x| \leq 1 \\ 0 & \text{for } |x| > 1 \end{cases} . \qquad \text{(D.2)}$$

Thus the rectangular impulses have the same momentum as the δ-impulses:

$$\int\limits_{-\infty}^{+\infty} u(t)dt = \int\limits_{-\infty}^{+\infty} \delta(t)dt = 1 \ . \tag{D.3}$$

The auto-correlation function of the sequence of rectangular functions,

$$P_{yy}(\theta) = \mathrm{E}\{y(t) \cdot y(t + \theta)\} \ , \tag{D.4}$$

will be calculated. Therefore the time intervals 1, 2 and 3 of Fig. D.1 are considered, in each of which a defined number m_i of impulses starts. With

$$y(t) = \frac{1}{\tau}(m_1 + m_2) \quad \text{and} \quad y(t + \theta) = \frac{1}{\tau}(m_2 + m_3) \tag{D.5}$$

the auto-correlation function is given by

$$P_{yy}(\theta) = \frac{1}{\tau^2}\mathrm{E}\{(m_1 + m_2) \cdot (m_2 + m_3)\} \ . \tag{D.6}$$

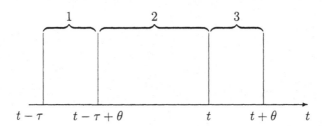

Fig. D.1

Because of the assumed independence of the impulses we have

$$\mathrm{E}\{m_1 m_2\} = \mathrm{E}\{m_1\} \cdot \mathrm{E}\{m_2\} \ , \tag{D.7}$$

and corresponding equations for the other products. Furthermore, with the impulse density z we have $\mathrm{E}\{m_1\} = \mathrm{E}\{m_3\} = z \cdot \theta$ and $\mathrm{E}\{m_2\} = z(\tau - \theta)$. Because of

$$\mathrm{E}\{m_2^2\} \ \ = \mathrm{E}\{m_2\} + \mathrm{E}\{m_2\}^2 \tag{D.8}$$

we obtain

$$\mathrm{E}\{m_2^2\} \ \ = z(\tau - \theta) + z^2(\tau - \theta)^2 \ . \tag{D.9}$$

For a positive θ and $\theta \leq \tau$, insertion of these variables into Eq. D.6 leads to

$$P_{yy}(\theta) = \frac{z}{\tau^2}(\tau - \theta) + z^2 \quad \text{for } \theta \leq \tau \ . \tag{D.10}$$

If $\theta \geq \tau$ only the intervals with the length τ before $y(t)$ and $y(t + \theta)$, respectively, have to be considered and thus

$$P_{yy}(\theta) = z^2 \quad \text{for } \theta > \tau \ . \tag{D.11}$$

Because the auto-correlation function of a stationary process is an even function, the following equation holds:

$$P_{yy}(\theta) = \begin{cases} \dfrac{z}{\tau^2} \cdot (\tau - |\theta|) + z^2 & \text{for } |\theta| \leq \tau \\ z^2 & \text{for } |\theta| > \tau \end{cases} \ . \tag{D.12}$$

Thus the auto-correlation function is constant for $|\theta| > \tau$ and has a triangular shape for $|\theta| \leq \tau$. The maximum value occurs at $\theta = 0$:

$$P_{yy}(\theta = 0) = \frac{z}{\tau} + z^2 \ . \tag{D.13}$$

The shape of this auto-correlation function is shown in Fig. D.2.

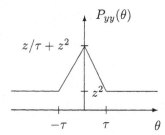

Fig. D.2 Auto-correlation function of rectangular impulses.

For the limit $\tau \to 0$ or $y(t) \to x(t)$, the auto-correlation function assumes the shape of a δ-function superimposed by the constant value z^2. Then the auto-correlation function of a sequence of rectangular impulses merges into the auto-correlation function of a sequence of δ-impulses:

$$P_{xx}(\theta) = z \cdot \delta(\theta) + z^2 \ . \tag{D.14}$$

Thus Eq. (4.14) has been derived. In Eq. (D.12) the constant component z^2, which is independent of θ, appears for both $|\theta| > \tau$ and $|\theta| \leq \tau$. Such a component can only result, if the considered process has a constant term. Thus the term z^2 becomes negligible if only the oscillating component of $y(t)$ is considered and we obtain

$$\tilde{P}_{sh}(\theta) = \begin{cases} \dfrac{z}{\tau^2} \cdot (\tau - |\theta|) & \text{for } |\theta| \leq \tau \\ 0 & \text{for } |\theta| > \tau \end{cases} \ . \tag{D.15}$$

In the limit for $\tau \to 0$, this function also approaches a Dirac-impulse with the weighting z:

$$P_{sh}(\theta) = z \cdot \delta(\theta) \ . \tag{D.16}$$

Thus Eq. (4.16) has been derived as well.

Problem 4.2

First the noisy diode will be replaced by an equivalent thermal noise source W_{UD} and an internal resistance R_i. For the elements of this source, according to Fig. 4.4, we have

$$R_i = R_b + \frac{1}{G_s} \tag{D.17}$$

and

$$W_{UD} = W_U + \frac{W_{is}}{G_s^2} = 4kTR_b + 4kT'_{ef} \cdot \frac{1}{G_s} \ . \tag{D.18}$$

T'_{ef} is the effective noise temperature of the Schottky diode with a negligible bulk resistance and with

$$\frac{T'_{ef}}{T} = \frac{1}{2} \tilde{n} \left(1 + \frac{I_{ss}}{I_0 + I_{ss}} \right) \tag{D.19}$$

as given by Eq. (4.31) we obtain, because of $W_{UD} = 4kT_{ef}R_i$:

$$\frac{T_{ef}}{T} = \frac{\dfrac{T'_{ef}}{T} + R_bG_s}{1 + R_bG_s} \ . \tag{D.20}$$

With G_s as given by Eq. (4.28) and the given values, the numbers in the following table are calculated for $T = 300\text{K}$. For $I_0 > 0, 2\text{mA}$ the results well agree with measured values.

I_0/mA	0.1	0.2	0.4	0.8	1.2	1.6
T_{ef}/T	0.61	0.62	0.64	0.68	0.71	0.73

$$\tag{D.21}$$

Problem 4.3

The noise figure of a two-port is defined as the ratio of the output noise spectrum of the noisy two-port and that of the noise-free two-port. First, the two spectra are calculated.

Because of Eq. (4.48) we have for the spectrum related to U^e

$$|U^e|^2 = 2kTR_{e0} \ . \tag{D.22}$$

With the currents $\tilde{I}_b$ and $\tilde{I}_c$ taken as mesh currents it follows for the voltage u_1 in the left mesh as induced by the noise sources:

$$u_1 = (R_g + R_b + R_{e0}) \cdot \tilde{I}_b + R_{e0}\tilde{I}_c \; . \tag{D.23}$$

Here $\tilde{I}_c$ is given by:

$$\tilde{I}_c = -\left(I^a + \alpha_0\tilde{I}_e\right) = -\left(I^a - \alpha_0(\tilde{I}_b + \tilde{I}_c)\right) \; . \tag{D.24}$$

Thus with $R = R_g + R_b + R_{e0}$ we have

$$
\begin{aligned}
u_1 &= R\tilde{I}_b + R_{e0}\tilde{I}_c \\
\tilde{I}_c &= \frac{\alpha_0\tilde{I}_b - I^a}{1 - \alpha_0} \; .
\end{aligned} \tag{D.25}
$$

Solving the first equation for $\tilde{I}_b$, inserting the result into the second equation and solving the second equation for $\tilde{I}_c$ leads to

$$\tilde{I}_c = \frac{1}{R_{e0} + R\left(\dfrac{1}{\alpha_0} - 1\right)} \cdot \left(u_1 - \frac{R}{\alpha_0}I^a\right) \; . \tag{D.26}$$

Using a symbolic notation we get

$$|\tilde{I}_c|^2 = W_{i2} = \frac{1}{\left(R_{e0} + R\left(\dfrac{1}{\alpha_0} - 1\right)\right)^2} \cdot \left(|u_1|^2 + \frac{R^2}{\alpha_0^2}|I^a|^2\right) \; . \tag{D.27}$$

Here the mixed terms are omitted because the sources are uncorrelated. Writing the parts of $|u_1|^2$ separately, leads to

$$
\begin{aligned}
|\tilde{I}_c|^2 &= W_{i2} \tag{D.28} \\
&= \frac{1}{\left(R_{e0} + R\left(\dfrac{1}{\alpha_0} - 1\right)\right)^2} \cdot \left[4kT\left(R_g + R_b + \frac{R_{e0}}{2}\right) + \frac{R^2}{\alpha_0^2}W_i^a\right] \; .
\end{aligned}
$$

The output spectrum for the noisy transistor is thus calculated. In order to get the spectrum of a noise-free transistor, the spectra W_i^a, $|U^b|^2$ and $|U^e|^2$ just have to be set to zero:

$$W_{i20} = \frac{1}{\left(R_{e0} + R\left(\dfrac{1}{\alpha_0} - 1\right)\right)^2} \cdot [4kTR_g] \; . \tag{D.29}$$

Taking the ratio of both spectra leads to the noise figure:

$$
\begin{aligned}
F = \frac{W_{i2}}{W_{i20}} &= \frac{4kT\left(R_g + R_b + \dfrac{R_{e0}}{2}\right) + \dfrac{R^2}{\alpha_0^2}W_i^a}{4kTR_g} \\
&= 1 + \frac{R_b}{R_g} + \frac{R_{e0}}{2R_g} + \frac{(R_g + R_b + R_{e0})^2}{4\alpha_0^2 kTR_g} \cdot W_i^a \; . \tag{D.30}
\end{aligned}
$$

With Eq. (4.49) and Eq. (4.47) the result of Eq. (4.52) is confirmed. Thus the noise figure depends on the internal generator resistance as expected. For the value $R_{g\,opt}$ the noise figure F has its minimum value. In order to calculate $R_{g\,opt}$ the derivative dF/dR_g of Eq. (4.52) has to be calculated and set to zero. We thus obtain:

$$R_{g\,opt}^2 = (R_b + R_{e0})^2 + (2R_b + R_{e0}) \cdot \frac{R_{e0}\alpha_0^2}{\alpha_0(1 - \alpha_0) + \dfrac{(\alpha_0^2 I_{ee} + I_{cc})}{I_e + I_{ee}}} \quad . \tag{D.31}$$

After inserting this result into Eq. (4.52), a short calculation leads to the following expression for the minimum noise figure:

$$F_{min} = 1 + \frac{R_{g\,opt} + R_b + R_{e0}}{R_{e0}\alpha_0^2} \left[\alpha_0(1 - \alpha_0) + \frac{(\alpha_0^2 I_{ee} + I_{cc})}{I_e + I_{ee}} \right] \quad . \tag{D.32}$$

For the given numerical values the results are

$$R_{g,opt} = 222\,\Omega \ , \tag{D.33}$$

and

$$F_{min} = 1.57 \quad \text{or} \quad F_{min} = 2\,\text{dB} \ . \tag{D.34}$$

Problem 4.4

The calculation of the noise figure of a common-base circuit starts from the equivalent circuit shown below. The calculation is performed in analogy to problem 4.3.

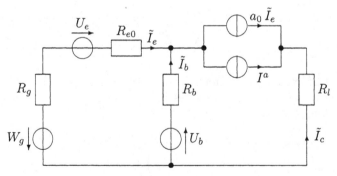

Fig. D.3

$$u_1 = (R_g + R_{e0} + R_b)\tilde{I}_e + R_b \cdot \tilde{I}_c = R\tilde{I}_e + R_b\tilde{I}_c \tag{D.35}$$

$$\tilde{I}_c = -(I^a + \alpha_0 \cdot \tilde{I}_e) \ . \tag{D.36}$$

Solving Eq. (D.35) for $\tilde{I}_e$ and inserting into Eq. (D.36) leads to:

$$\tilde{I}_c = \frac{-u_1 - \dfrac{R}{\alpha_0}I^a}{\dfrac{R}{\alpha_0} - R_b} \tag{D.37}$$

and

$$|\tilde{I}_c|^2 = \frac{1}{\left(\dfrac{R}{\alpha_0} - R_b\right)^2}\left[|u_1|^2 + \frac{R^2}{\alpha_0^2}W_i^a\right] \tag{D.38}$$

for the squared magnitude. If W_{i2} and W_{i20} are determined, the same relation for the noise figure is obtained as in problem 4.3.

Thus the noise figure and the optimum noise figure for the common-emitter circuit and the common-base circuit are equal under the assumptions made.

Problem 4.5

Equations (4.115), (4.117), (4.119), and (4.121) yield

$$-j\tilde{C} = \frac{-j\omega C_g \tilde{Q}(V_g)}{\sqrt{\dfrac{(\omega C_g)^2}{g_m}\tilde{R}(V_g)\cdot g_m \cdot \tilde{P}(V_g)}} \tag{D.39}$$

and hence

$$\tilde{C}(V_g) = \frac{\tilde{Q}(V_g)}{\sqrt{\tilde{P}(V_g)\tilde{R}(V_g)}} . \tag{D.40}$$

Inserting Eq. 4.116, 4.118 and 4.120 leads to

$$
\begin{aligned}
\tilde{C}(V_g) &= \frac{\dfrac{1}{10}\dfrac{1+3\sqrt{V_g}(2+\sqrt{V_g})}{(1+\sqrt{V_g})(1+2\sqrt{V_g})}}{\sqrt{\dfrac{1}{2}\dfrac{1+3\sqrt{V_g}}{1+2\sqrt{V_g}}\dfrac{1}{10}\dfrac{1+7\sqrt{V_g}}{1+2\sqrt{V_g}}}} \\[2em]
&= \frac{1+3\sqrt{V_g}(2+\sqrt{V_g})}{(1+\sqrt{V_g})\sqrt{5(1+3\sqrt{V_g})(1+7\sqrt{V_g})}} .
\end{aligned} \tag{D.41}
$$

For $V_g = 0$ the function $\tilde{C}(V_g)$ has the numerical value 0.447 and continuously decreases to the numerical value 0.395 at $V_g = 1$. Thus, the normalized cross-spectrum is almost independent of the operating point.

Problem 4.6

The differentiation of Eq. (4.127) with respect to X_0 leads to

$$\frac{\partial F}{\partial X_0} = -\frac{2\omega C_g}{g_m R_0}(\tilde{P} - \tilde{Q}) + 2X_0 \frac{(\omega C_g)^2}{g_m R_0}(\tilde{P} + \tilde{R} - 2\tilde{Q}) \ . \qquad (D.42)$$

From this the reactive part of the optimum source impedance follows as

$$X_{opt} = \frac{1}{\omega C_g} \cdot \frac{\tilde{P} - \tilde{Q}}{\tilde{P} + \tilde{R} - 2\tilde{Q}} \ . \qquad (D.43)$$

By inserting Eq. (4.127) we obtain

$$F(X_{opt}) = 1 + \frac{R_g + R_s}{R_0} + \frac{1}{g_m R_0}\left[\frac{\tilde{P}\tilde{R} - \tilde{Q}^2}{\tilde{P} + \tilde{R} - 2\tilde{Q}}\right.$$
$$\left. + (\omega C_g)^2 (R_0 + R_g + R_s)^2 (\tilde{P} + \tilde{R} - 2\tilde{Q})\right] \ . \qquad (D.44)$$

The differentiation of this function with respect to R_0 leads to

$$\frac{dF(X_{opt})}{dR_0} = -\frac{R_g + R_s}{R_0^2}$$
$$+ \frac{1}{(g_m R_0)^2}\left[2g_m R_0 (\omega C_g)^2 (R_0 + R_g + R_s)(\tilde{P} + \tilde{R} - 2\tilde{Q})\right.$$
$$- g_m \frac{\tilde{P}\tilde{R} - \tilde{Q}^2}{\tilde{P} + \tilde{R} - 2\tilde{Q}}$$
$$\left. - g_m (\omega C_g)^2 (R_0 + R_g + R_s)^2 (\tilde{P} + \tilde{R} - 2\tilde{Q})\right] \ . \qquad (D.45)$$

From the condition $dF(X_{opt})/dR_0 = 0$ for the real part of the optimum source impedance we get

$$(\omega C_g)^2 (R_{opt} + R_g + R_s)(R_{opt} - R_g - R_s)\left(\tilde{P} + \tilde{R} - 2\tilde{Q}\right)$$
$$- g_m(R_g + R_s) - \frac{\tilde{P}\tilde{R} - \tilde{Q}^2}{\tilde{P} + \tilde{R} - 2\tilde{Q}} = 0 \ . \qquad (D.46)$$

Using $\tilde{Q}^2 = \tilde{C}^2 \tilde{P}\tilde{R}$ from Eq. 4.121 the result for R_{opt} is

$$R_{opt} = \sqrt{(R_g + R_s)^2 + \frac{\tilde{P}\tilde{R}(1 - \tilde{C}^2) + g_m(R_g + R_s)(\tilde{P} + \tilde{R} - 2\tilde{Q})}{(\omega C_g)^2(\tilde{P} + \tilde{R} - 2\tilde{Q})^2}} \ . \quad (D.47)$$

Insertion of

$$\frac{\tilde{P}\tilde{R} - \tilde{Q}^2}{(\tilde{P} + \tilde{R} - 2\tilde{Q})} + g_m(R_g + R_s) + (\omega C_g)^2(\tilde{P} + \tilde{R} - 2\tilde{Q})(R_g + R_s)^2$$
$$= (\omega C_g)^2 R_{opt}^2(\tilde{P} + \tilde{R} - 2\tilde{Q}) \qquad (D.48)$$

into the equation for $F(X_{opt})$ leads to

$$F_{min} = 1 + 2 \cdot \frac{(\omega C_g)^2}{g_m}(R_{opt} + R_g + R_s)(\tilde{P} + \tilde{R} - 2\tilde{Q}) , \tag{D.49}$$

and with the result for R_{opt}, Eq. (4.131) follows directly.

Problem 4.7

Inserting Eq. (4.82) and Eq. (4.84) into Eq. (4.134) leads to the minimum noise figure:

$$F_{min} = 1 + 3K\omega C_0 \sqrt{\frac{R_g + R_s}{G_0}} \cdot \frac{1 + \sqrt{V_g}}{(1 + 2\sqrt{V_g})^2\sqrt{1 - \sqrt{V_g}}} . \tag{D.50}$$

Setting $x = \sqrt{V_g}$, the evaluation of the optimum operating point is equivalent to finding the minimum of the following function:

$$\frac{1 + x}{(1 + 2x)^2\sqrt{1 - x}} . \tag{D.51}$$

Setting the derivative to zero leads to

$$\sqrt{1 - x}(1 + 2x) - (1 + x)\left(-\frac{1 + 2x}{2 \cdot \sqrt{1 - x}} + 4 \cdot \sqrt{1 - x}\right) = 0 \tag{D.52}$$

and after some algebraic conversions we obtain the following quadratic equation:

$$x^2 + \frac{5}{6}(x - 1) = 0 . \tag{D.53}$$

The solution to this equation is

$$x = \frac{1}{12}(\sqrt{145} - 5) \approx \frac{7}{12} , \tag{D.54}$$

from which the optimum value of the normalized voltage V_g follows as

$$V_{gopt} \approx \left(\frac{7}{12}\right)^2 \approx \frac{1}{3} . \tag{D.55}$$

For this calculation the factor K in Eq. (4.134) was taken as constant. Accounting for the V_g-dependence of K as described by Eq. (4.135) results in a marginally different value for the normalized voltage:

$$V_{gopt} = \left(\frac{\sqrt{7} - 1}{3}\right)^2 \approx 0.3 . \tag{D.56}$$

Appendix E
Solutions to the
Problems of Chapter 5

Chapter 5

Problem 5.1

If the resistors R_b are interpreted as the external circuit of the two-port with the admittance matrix $[G]$, a short calculation with

$$\begin{bmatrix} I_s \\ I_i \end{bmatrix} = [G] \begin{bmatrix} U_s' \\ U_i' \end{bmatrix} \quad \text{with} \quad U_s' = U_s - I_s R_b \quad \text{and} \quad U_i' = U_i - I_i R_b \quad \text{(E.1)}$$

leads to

$$\begin{aligned} \begin{bmatrix} I_s \\ I_i \end{bmatrix} &= \frac{1}{(1 + G_0 R_b)^2 - G_1^2 R_b^2} \begin{bmatrix} R_b(G_0^2 - G_1^2) + G_0 & G_1 \\ G_1 & R_b(G_0^2 - G_1^2) + G_0 \end{bmatrix} \begin{bmatrix} U_s \\ U_i \end{bmatrix} \\ &= [G'] \begin{bmatrix} U_s \\ U_i \end{bmatrix} . \end{aligned} \quad \text{(E.2)}$$

Extending $[G']$ by Y_s and Y_i leads to the matrix $[\tilde{G}']$, from which the ratio U_i/I_{sg} can be calculated:

$$\frac{U_i}{I_{sg}} = -\frac{G_1}{\det[\tilde{G}']} \quad . \tag{E.3}$$

Thus the gain is given by

$$G_p = \frac{4\mathrm{Re}\{Y_s\} \cdot \mathrm{Re}\{Y_i\} \cdot G_1^2}{|\det[\tilde{G}']|^2} \quad . \tag{E.4}$$

With the matrix $[G_e']$, extended by the generator admittance only, we obtain for the input admittance:

$$Y_{in} = \frac{I_i}{U_i} = \frac{\det[G_e']}{[G_e']_{11}} \quad . \tag{E.5}$$

Inserting $Y_i = Y_{in}^*$ in Eq. (E.4) yields an expression for the available gain. For a power match at the input and output we have

$$Y_s = \frac{\det[G_e']}{[G_e']_{11}} \quad . \tag{E.6}$$

A short calculation leads to

$$Y_s^2 = \frac{[R_b(G_0^2 - G_1^2) + G_0]^2 - G_1^2}{[(1 + G_0 R_b)^2 - G_1^2 R_b^2]^2} \quad . \tag{E.7}$$

By inserting this relation into Eq. (E.4), an expression for the maximum available gain results after some manipulations:

$$G_m = \frac{G_1^2}{(R_b(G_0^2 - G_1^2) + G_0)^2} \left(\frac{1}{1 + \sqrt{1 - \frac{G_1^2}{(R_b(G_0^2 - G_1^2) + G_0)^2}}} \right)^2 \quad . \tag{E.8}$$

This result can also be derived directly by a comparison of Eq. (E.2) and Eq. (5.9). Replacing G_0 in Eq. (5.32) by

$$R_b(G_0^2 - G_1^2) + G_0$$

leads to (E.8).

Problem 5.2

The ratio U_i/I_{sg} as needed for the determination of the gain, can be calculated with the matrix $[\tilde{G}]$ extended by Y_s, Y_i and Y_{im}. Because of

$$I_{sg} = Y_s U_s + I_s , \qquad 0 = Y_i U_i + I_i , \qquad 0 = Y_{im}^* U_{im}^* + I_{im}^* , \qquad (E.9)$$

we have

$$\begin{bmatrix} I_{sg} \\ 0 \\ 0 \end{bmatrix} = \begin{bmatrix} G_0 + Y_s & G_1 & G_2 \\ G_1 & G_0 + Y_i & G_1 \\ G_2 & G_1 & G_0 + Y_{im}^* \end{bmatrix} \begin{bmatrix} U_s \\ U_i \\ U_{im}^* \end{bmatrix} \qquad (E.10)$$

and thus

$$\frac{U_i}{I_{sg}} = [\tilde{G}]_{21}^{-1} = -\frac{G_1(Y_{im}^* + G_0) - G_1 G_2}{\det[\tilde{G}]} . \qquad (E.11)$$

With Eq. (5.20) we obtain for the gain

$$G_p = \left| \frac{G_1(Y_{im}^* + G_0) - G_1 G_2}{\det[\tilde{G}]} \right|^2 \cdot 4\mathrm{Re}\{Y_s\} \cdot \mathrm{Re}\{Y_i\} . \qquad (E.12)$$

For the calculation of the available gain, the input admittance of the intermediate frequency port has to be determined. Therefore, the matrix $[G]$ is extended by Y_s and Y_{im}^*:

$$\begin{bmatrix} 0 \\ I_i \\ 0 \end{bmatrix} = \begin{bmatrix} G_0 + Y_s & G_1 & G_2 \\ G_1 & G_0 & G_1 \\ G_2 & G_1 & G_0 + Y_{im}^* \end{bmatrix} \begin{bmatrix} U_s \\ U_i \\ U_{im}^* \end{bmatrix} . \qquad (E.13)$$

Thus the input admittance Y_{in} is given by

$$Y_{in} = \frac{1}{[G_e]_{22}^{-1}} = \frac{\det[G_e]}{(G_0 + Y_s)(G_0 + Y_{im}^*) - G_2^2} . \qquad (E.14)$$

Inserting $Y_i = Y_{in}^*$ into Eq. (E.12) leads to an expression for the available gain.

Problem 5.3

For a 180° coupler we observe the following signals at diode I and diode II, respectively:

diode I : $\quad \dfrac{1}{\sqrt{2}} \left(\hat{U}_s \cdot \cos\left(\omega_s t + 180° + \phi_s\right) + \hat{U}_p \cdot \cos\left(\omega_p t\right) \right)$

diode II : $\quad \dfrac{1}{\sqrt{2}} \left(\hat{U}_s \cdot \cos\left(\omega_s t + \phi_s\right) + \hat{U}_p \cdot \cos\left(\omega_p t\right) \right) . \qquad (E.15)$

The intermediate frequency signal of diode I is:

$$\text{diode I}: \quad u_i^I \sim G_1 \, \hat{U}_s \cos{(\omega_i \, t + 180° + \phi_s)} \quad \text{for } \omega_s > \omega_p \, , \qquad \text{(E.16)}$$

and because the polarity has been changed for the diode II ,

$$\begin{aligned} \text{diode II}: \quad u_i^{II} \quad &\sim \quad -G_1 \, \hat{U}_s \cos{(\omega_i \, t + \phi_s)} \\ &= \quad G_1 \, \hat{U}_s \cos{(\omega_i \, t + 180° + \phi_s)} \, . \end{aligned} \qquad \text{(E.17)}$$

Thus the intermediate frequency signals at both diodes are in phase also for a 180° coupler. The difference between the 90° and the 180° coupler is the following: For similar mismatched diodes and a 90° coupler the signal path and the pump oscillator path are not isolated, but they are both matched. In contrast both paths are isolated, if a 180° coupler is used, but the reflection might have increased in this case. Thus it has to be decided which coupler is more advantageous for the given application.

Problem 5.4

With rectangularly shaped band-pass filters the signals $X_1(t)$ at the frequency f_1 and $X_2(t)$ at the frequency f_2 are filtered from unmodulated white Gaussian noise. The bandwidth of the band-pass filters is assumed to be small with respect to the frequency offset $f_2 - f_1$. The bandpass filtered signals can be written as the convolution of the unmodulated white Gaussian noise signal $s(t)$ with the corresponding impulse responses $h_1(t)$ and $h_2(t)$, respectively:

$$\begin{aligned} X_1(t) &= \int_{-\infty}^{\infty} h_1(t') \cdot s(t - t') dt' \\ X_2(t) &= \int_{-\infty}^{\infty} h_2(t'') \cdot s(t - t'') dt'' \, . \end{aligned} \qquad \text{(E.18)}$$

The signal $X_2(t)$ shall be shifted in frequency by $f_2 - f_1$. This is possible, for example, by an ideal multiplication with $2 \cdot \cos[2\pi(f_2 - f_1)t]$. Thus the resulting signal $\tilde{X}_2(t)$ has frequency components at f_1. In addition to $X_1(t)$, it can be considered as an input signal of a correlator as shown in Fig. 3.15. If integration and averaging in time are interchanged, then the correlation is given by

$$\langle X_1(t) \cdot \tilde{X}_2(t) \rangle = \qquad \text{(E.19)}$$

$$\int\!\!\int_{-\infty}^{+\infty} h_1(t') h_2(t'') \langle 2 \cdot \cos(2\pi(f_2 - f_1)t) \cdot s(t - t') s(t - t'') \rangle dt' dt'' \, .$$

The average over $s(t - t') s(t - t'')$ yields a non-zero contribution only for $t' = t''$, because $s(t)$ is a white Gaussian noise process. Thus, the expression in angular brackets has to be evaluated for $t' = t''$ only:

$$\langle 2 \cdot \cos(\omega_p t) \cdot s(t - t')s(t - t'') \rangle|_{t'=t''} = \langle 2 \cdot \cos(\omega_p t) \cdot s^2(t - t') \rangle . \quad \text{(E.20)}$$

The time average of a function multiplied with $\cos(\omega t)$ is always zero. Thus, the real part of the cross-spectrum of $X_1(t)$ and $\tilde{X}_2(t)$ vanishes. In the same way, it can be shown that the imaginary part also disappears.

Problem 5.5

First, it will be shown that the imaginary part of the cross spectrum disappears for an even pump drive signal. For the measurement of the imaginary part with a circuit as shown in Fig. 3.15 a phase shift by 90° has to be performed in one path. In the case discussed here, this 90° phase shift may be achieved by means of an ideal frequency shifter and by multiplying with $2 \sin(\omega_p t)$ instead of $2 \cos(\omega_p t)$. Then an expression, abbreviated by AT_1, similar to Eq. (5.47) results:

$$AT_1 = \frac{2G_1}{G_0} \sin(\omega_p t') \rho_0 \delta(t' - t'') . \quad \text{(E.21)}$$

For the ratio of the cross-spectrum and the power spectrum we obtain

$$\frac{\langle X_i(t) \cdot \tilde{X}_s(t) \rangle}{\langle X_i^2(t) \rangle} = \frac{\text{Im}\{I_{n1}^* I_{n2}\}}{|I_{n1}|^2}$$

$$= \frac{\rho_0 \frac{2G_1}{G_0} \cdot \int_{-\infty}^{\infty} h_i(t') h_s(t'') \cdot \sin(\omega_p t') dt'}{\rho_0 \cdot \int_{-\infty}^{\infty} h_i^2(t') dt'} . \quad \text{(E.22)}$$

As can easily be shown, the integral in the numerator is zero because the expression

$$\cos(\omega_i t') \cdot \cos(\omega_s t') \cdot \sin(\omega_p t')$$

$$= \frac{1}{4}[\sin(2\omega_p t') + \sin(2\omega_s t') + \sin(2\omega_i t')] \quad \text{(E.23)}$$

has no constant term.

For a mixed even-odd pump drive signal the following relation results instead of Eq. (5.42):

$$s_m(t) = s(t)\sqrt{1 + \frac{2\text{Re}\{G_1\}}{G_0} \cdot \cos(\omega_p t) - \frac{2\text{Im}\{G_1\}}{G_0} \cdot \sin(\omega_p t)} . \quad \text{(E.24)}$$

Obviously, the integration in Eq. (5.48) yields a contribution different from zero only, if Eq. (5.47) includes a cosine function. For the calculation of the real part, i.e. by multiplying with $2\cos(\omega_p t)$, an expression, abbreviated by AT_2, similar to Eq. (5.47) results as follows:

$$AT_2 = \rho_0 \cdot \delta(t' - t'') \left[\frac{2\text{Re}\{G_1\}}{G_0} \cdot \cos(\omega_p t') + \frac{2\text{Im}\{G_1\}}{G_0} \cdot \sin(\omega_p t') \right] .$$

(E.25)

Only the first part in the bracket yields a contribution. A calculation similar to the one for an even pump drive leads to

$$\frac{\langle X_i(t) \cdot \tilde{X}_s(t) \rangle}{\langle X_i^2(t) \rangle} = \frac{\text{Re}\{I_{n1}^* I_{n2}\}}{|I_{n1}|^2} = \frac{\text{Re}\{G_1\}}{G_0} .$$

(E.26)

For the calculation of the imaginary part, i.e. the multiplication with $2\sin(\omega_p t)$, we obtain an expression corresponding to Eq. (5.47) :

$$AT_3 = \rho_0 \cdot \delta(t' - t'') \left[\frac{2\text{Re}\{G_1\}}{G_0} \cdot \sin(\omega_p t') - \frac{2\text{Im}\{G_1\}}{G_0} \cdot \cos(\omega_p t') \right] . \quad (E.27)$$

In this case, only the second part in the bracket results in a contribution and we have

$$\frac{\langle X_i(t) \cdot \tilde{X}_s(t) \rangle}{\langle X_i^2(t) \rangle} = \frac{\text{Im}\{I_{n1}^* I_{n2}\}}{|I_{n1}|^2} = -\frac{\text{Im}\{G_1\}}{G_0} .$$

(E.28)

Inserting Eq. (5.39) leads to the wanted result:

$$
\begin{aligned}
I_{n1}^* I_{n2} &= \text{Re}\{I_{n1}^* I_{n2}\} + j\text{Im}\{I_{n1}^* I_{n2}\} \\
&= 2k \left(\frac{\tilde{n}}{2} T \right) [\text{Re}\{G_1\} - j\text{Im}\{G_1\}] \\
&= 2k \left(\frac{\tilde{n}}{2} T \right) G_1^*
\end{aligned}
$$

(E.29)

. For the other matrix elements we obtain with Eq. (5.54):

$$
\begin{bmatrix}
I_{ns}^* I_{ns} & I_{ns}^* I_{ni} & I_{ns}^* I_{nim}^* \\
I_{ni}^* I_{ns} & I_{ni}^* I_{ni} & I_{ni}^* I_{nim}^* \\
I_{nim} I_{ns} & I_{nim}^* I_{ni} & I_{nim} I_{nim}^*
\end{bmatrix}
= 2k \cdot \frac{\tilde{n}}{2} \cdot T
\begin{bmatrix}
G_0 & G_1^* & G_2^* \\
G_1 & G_0 & G_1^* \\
G_2 & G_1 & G_0
\end{bmatrix} . \quad (E.30)
$$

If the correlation matrix of the Schottky diode mixer for a mixed even-odd pump drive is compared with the correlation matrix of a passive thermally noisy N-port network at a homogeneous temperature (Eq. (2.45)), i.e.

$$
k \cdot T
\begin{bmatrix}
2 \cdot \text{Re}\{Y_{11}\} & Y_{12}^* + Y_{21} & Y_{13}^* + Y_{31} \\
Y_{21}^* + Y_{12} & \cdots & \cdots \\
\cdots & \cdots & \cdots
\end{bmatrix} ,
$$

(E.31)

with

$$Y_{12}^* + Y_{21} \hat{=} G_1^* + G_1^* = 2 \cdot G_1^* \ , \tag{E.32}$$

the proportionality of both matrices is obvious.

Problem 5.6

The down converter to be analyzed is shown in Fig. E.1.

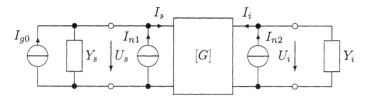

Fig. E.1

With

$$\begin{bmatrix} I_s \\ I_i \end{bmatrix} = \begin{bmatrix} G_0 & G_1 \\ G_1 & G_0 \end{bmatrix} \begin{bmatrix} U_s \\ U_i \end{bmatrix} \tag{E.33}$$

for a real G_1 and

$$I_s = I_{n1} + I_{g0} - U_s Y_s \tag{E.34}$$

$$I_i = I_{n2} - U_i Y_i \tag{E.35}$$

we have

$$\begin{bmatrix} I_{g0} + I_{n1} \\ I_{n2} \end{bmatrix} = \begin{bmatrix} G_0 + Y_s & G_1 \\ G_1 & G_0 + Y_i \end{bmatrix} \begin{bmatrix} U_s \\ U_i \end{bmatrix} = [\tilde{G}] \cdot \begin{bmatrix} U_s \\ U_i \end{bmatrix} \ . \tag{E.36}$$

Solving for U_i leads to

$$\begin{aligned} U_i &= [\tilde{G}]_{22}^{-1} \cdot I_{n2} + [\tilde{G}]_{21}^{-1} \cdot (I_{g0} + I_{n1}) \\ &= \frac{(G_0 + Y_s)I_{n2} - G_1(I_{g0} + I_{n1})}{\det[\tilde{G}]} \ . \end{aligned} \tag{E.37}$$

Thus the noise figure F is given by

$$F = \frac{|U_i|^2}{|U_{i0}|^2} = \frac{|(G_0 + Y_s)I_{n2} - G_1(I_{g0} + I_{n1})|^2}{|G_1 I_{g0}|^2} \ . \tag{E.38}$$

With Eqs. (5.39) and (5.52) a short calculation with a real Y_s leads to

$$F = 1 + \frac{\tilde{n}}{2} \frac{T}{T_0} \frac{G_1^2 G_0 - 2G_1^2(G_0 + Y_s) + G_0(G_0 + Y_s)^2}{G_1^2 \cdot Y_s} \ . \tag{E.39}$$

With Eqs. (5.25) and (5.24) the last equation can be rearranged to

$$F = 1 + \frac{\tilde{n}}{2} \frac{T}{T_0} \left(\frac{1}{G_{av}} - 1 \right) . \tag{E.40}$$

From this, the wanted result, Eq. (5.55), follows directly. A similar calculation with complex G_1 and Y_s leads to the same result.

Problem 5.7

The circuit in Fig. 5.16 can be interpreted as a cascade connection of three noisy two-ports. Two two-ports consist of a noisy series resistor R_b at the temperature T. The mixer at the temperature $\tilde{n}T/2$ without a series resistance with the noise figure F and the available gain G_{av} is embedded between these two-ports. With the noise figures F_{b1}, F_{b2} and the available gains G_{av1} and G_{av2} of the series resistors, the total noise factor F_t follows by means of the cascade formula Eq. (2.100):

$$F_t = F_{b1} + \frac{F - 1}{G_{av1}} + \frac{F_{b2} - 1}{G_{av1} G_{av}} . \tag{E.41}$$

For the noise figures F_{b1} and F_{b2} of the two-ports, formed by the series resistors, Eq. (2.90) yields

$$F_{b1} = 1 + \frac{T}{T_0} \frac{1 - G_{av1}}{G_{av1}} , \qquad F_{b2} = 1 + \frac{T}{T_0} \frac{1 - G_{av2}}{G_{av2}} . \tag{E.42}$$

Inserting Eq. (5.55) into (E.41) and manipulating the expression leads to

$$F_t = 1 + \frac{T}{T_0} \left(\frac{1 - G_{av2} + G_{av2}G_{av}(1 - G_{av1}) + \frac{\tilde{n}}{2}G_{av2}(1 - G_{av})}{G_{av1} G_{av2} G_{av}} \right) . \tag{E.43}$$

The available gain G_{av} of the mixer without losses from the series resistance can be calculated by means of Eq. (5.25). For the available gain of the circuits with series resistors as a function of the load resistance at the input we obtain

$$G_{av1} = \frac{R_1}{R_b + R_1} \qquad G_{av2} = \frac{R_2}{R_b + R_2} . \tag{E.44}$$

For the first series resistor we have $R_1 = Z_s$, i.e. R_1 is equal to the source resistance on the signal side. For the second series resistor we have $R_2 = Z_{in}$, where Z_{in} is the real input resistance of the mixer at the intermediate frequency side.

Problem 5.8

The mixer is interpreted as a cascade connection of three two-ports as in problem 5.7. The dissipation theorem yields for the cascade connection of N two-ports:

$$F = \frac{\sum\limits_{i=1}^{N} \beta_i T_i}{\beta_0 T_0} \ . \tag{E.45}$$

Here, the coefficients β_j denote the normalized dissipated power in the different temperature regions.

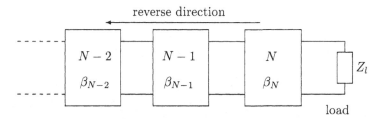

Fig. E.2

For convenience, as the noise figure is independent of the load resistance, a power match is assumed for Z_l at the output. For the normalized real power, dissipated by the N-th two-port if the feeding is performed from the output side, we have

$$\beta_N = \frac{P_g \cdot (1 - G'_{KN})}{P_g} = 1 - G'_{KN} \ . \tag{E.46}$$

Here, G'_{KN} is the power gain of the N-th two-port in reverse direction and P_g is the available source power. Furthermore, we obtain for the $(N-1)$-th two-port:

$$\beta_{N-1} = G'_{KN}(1 - G'_{(KN-1)}) \ , \tag{E.47}$$

and, in general,

$$\beta_j = (1 - G'_{Kj}) \cdot \prod_{i=j+1}^{N} G'_{Ki} \ . \tag{E.48}$$

In order to calculate β_j as a function of the available gains in forward direction, we will utilize the fact that the gain is independent of the direction (problem 2.14).

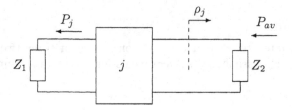

Fig. E.3

Taking any of the cascaded two-ports Fig. E.3, the power transferred in reverse direction by this two-port is given by

$$P_j = G'_{Kj} \cdot (1 - |\rho_j|^2) \cdot P_{av} \; . \tag{E.49}$$

Here, P_{av} is the available power and ρ_j is the reflection coefficient according to a mismatch. Eq. (E.49) can also be written as

$$P_j = G'_{pj} \cdot P_{av} \; , \tag{E.50}$$

where G'_{pj} is the gain in reverse direction. With Eqs. (E.49) and (E.50) we obtain

$$G'_{Kj} = \frac{G'_{pj}}{1 - |\rho_j|^2} \; . \tag{E.51}$$

For power match on the right hand side ($Z_2 = Z_{opt}$, $\rho_j = 0$) we have $G'_{Kj} = G'_{pj}$. The complex conjugate match on the generator side, which is assumed for the definition of the gain, is considered in Eq. (E.51). In a similar way a relation between the gain and the available gain in forward direction can be derived:

$$G''_{pj} = (1 - |\rho_j|^2) \cdot G''_{avj} \; . \tag{E.52}$$

With the gain being independent of the direction ($G'_p = G''_p$), Eqs. (E.51) and (E.52) yield

$$G'_{Kj} = G''_{avj} \; . \tag{E.53}$$

Thus, for every reciprocal two-port, the available gain G_{avj} is equal in forward and reverse direction.

Inserting (E.53) into (E.48) leads to the β_j of the series connection of three two-ports. With the relations of problem 5.7 we have

$$\beta_3 = 1 - G_{av2} \tag{E.54}$$

$$\beta_2 = G_{av2}(1 - G_{av}) \tag{E.55}$$

$$\beta_1 = G_{av} \cdot G_{av2}(1 - G_{av1}) \tag{E.56}$$

$$\beta_0 = G_{av} \cdot G_{av1} \cdot G_{av2} = \frac{1}{L} \; . \tag{E.57}$$

Insertion into Eq. (E.45) leads to the result for the noise figure of a down converter with a series resistance, already known from problem 5.7:

$$F_t = 1 + \frac{T}{T_0} \left(\frac{1 - G_{av2} + G_{av2}G_{av}(1 - G_{av1}) + \frac{\tilde{n}}{2}G_{av2}(1 - G_{av})}{G_{av1}G_{av2}G_{av}} \right) .$$

(E.58)

Problem 5.9

With

$$s(t) = \sum_{n=-\infty}^{\infty} S_n \exp(jn\omega_p t)$$

(E.59)

and

$$S_{-n} = S_n^* \qquad (s(t) \text{ real})$$

(E.60)

we get according to Eq. (5.66):

$$\begin{bmatrix} U_u \\ U_i \\ U_d^* \end{bmatrix} = \begin{bmatrix} S_0 & S_1 & S_2 \\ S_1^* & S_0 & S_1 \\ S_2^* & S_1^* & S_0 \end{bmatrix} \begin{bmatrix} Q_u \\ Q_i \\ Q_d^* \end{bmatrix} .$$

(E.61)

Problem 5.10

The gain in Eq. (5.97) becomes maximum if Z_i approaches the negative reference impedance $-Z_0$. Z_0 is assumed to be real and, for example, equal to R_g. Thus, the imaginary part of Z_i must vanish. This can be achieved by a match at both the input and the output. If the gain is very high, then $-Z_i$ is close to $Z_0 = R_g$. Inserting $-Z_i \approx R_g = R_l$ into Eq. (5.106) leads to the noise figure:

$$F = 1 + \frac{T}{T_0} \left(\frac{1 + \dfrac{S_1^2}{\omega_i \omega_d \cdot R_b^2} \dfrac{\omega_i}{\omega_d}}{\dfrac{S_1^2}{\omega_i \omega_d \cdot R_b^2} - 1} \right) .$$

(E.62)

As can be seen by Eqs. (5.96) and (5.97), we obtain for a high gain, i.e.

$$\frac{S_1^2}{\omega_i \omega_d \cdot R_b^2} \gg 1 ,$$

(E.63)

for the noise figure

$$F = 1 + \frac{T}{T_0}\left(\frac{\omega_i\omega_d \cdot R_b^2}{S_1^2} + \frac{\omega_i}{\omega_d}\right) \quad . \tag{E.64}$$

Differentiating the expression in brackets with respect to ω_d leads to the optimum frequency $\omega_{d\,opt}$:

$$\omega_{d\,opt} = \frac{S_1}{R_b} \tag{E.65}$$

Inserting $\omega_{d\,opt}$ into Eq. (E.64) leads to the optimum noise figure:

$$F_{opt} = 1 + \frac{2 \cdot T}{T_0}\frac{\omega_i}{\omega_{d\,opt}} \quad . \tag{E.66}$$

Problem 5.11

In order to determine the available and the maximum gain, the input resistances at the signal and at the load side have to be calculated. With a compensation of the inductive reactances we have

$$\begin{bmatrix} 0 \\ U_u \end{bmatrix} = \begin{bmatrix} R_b + R_g & \dfrac{S_1}{j\omega_u} \\ \dfrac{S_1}{j\omega_i} & R_b \end{bmatrix}\begin{bmatrix} I_i \\ I_u \end{bmatrix} = [Z]\begin{bmatrix} I_i \\ I_u \end{bmatrix} \quad . \tag{E.67}$$

For a power match at the output, R_u has to be equal to the real input impedance Z_{in} on the load side. We have

$$Z_{in} = R_u = \frac{U_u}{I_u} = \frac{1}{[Z]_{22}^{-1}} = R_b + \frac{S_1^2}{\omega_i\omega_u(R_b + R_g)} \quad . \tag{E.68}$$

Inserting R_u into Eq. (5.108) leads to the following expression for the available gain:

$$G_{av} = \frac{1}{\dfrac{\omega_i}{\omega_u}\dfrac{R_g + R_b}{R_g} + \dfrac{R_b\omega_i^2}{S_1^2}\dfrac{(R_g + R_b)^2}{R_g}} \quad . \tag{E.69}$$

Assuming symmetry and a match at both ends,

$$R_g = R_u = R_b + \frac{S_1^2}{\omega_i\omega_u(R_g + R_b)} \quad , \tag{E.70}$$

we get

$$R_u = R_g = \sqrt{R_b^2 + \frac{S_1^2}{\omega_i\omega_u}} \quad . \tag{E.71}$$

Inserting Eq. (E.70) into Eq. (E.69) leads to the maximum gain:

$$G_m = \frac{S_1^2}{\omega_i^2 R_b^2} \cdot \frac{1}{\left[1 + \sqrt{1 + \dfrac{S_1^2}{\omega_i \omega_u R_b^2}} \right]^2} \cdot \tag{E.72}$$

For the calculation of the noise figure, a power match at the output is assumed. The noise contribution of the series resistance is taken into account by two noise sources $|U_b|^2 = 4kTR_b\Delta f$ connected in series. The sources are uncorrelated because they operate at different frequencies. The noise of the resistor at the signal side is transmitted to the load side according to the squared magnitude of the corresponding transfer function. For the available power at the load resistance caused by the noise of the series resistance we obtain

$$\Delta P_2 = \frac{kT\Delta f \cdot R_b}{R_u} \left(\left| \frac{S_1}{\omega_i (R_g + R_b)} \right|^2 + 1 \right) . \tag{E.73}$$

With R_u from Eq. (E.68) and the available gain from Eq. (E.71) some calculations lead to the noise figure of the up-converter:

$$\begin{aligned} F &= 1 + \frac{\Delta W_2}{W_{20}} = 1 + \frac{\Delta W_2}{G_{av} k T_0} \\ &= 1 + \frac{T}{T_0} \frac{R_b}{R_g} \left(1 + \frac{\omega_i^2}{S_1^2} (R_g + R_b)^2 \right) . \end{aligned} \tag{E.74}$$

Appendix F
Solutions to the
Problems of Chapter 6

Chapter 6

Problem 6.1

At the output of the sideband filter either the lower sideband $x_l(t)$ or the upper sideband $x_u(t)$ appears:

$$
\begin{aligned}
x_l(t) &= \text{Re}\left\{X_l \cdot \exp[j(\Omega_0 - \omega)t]\right\} \\
&= \frac{1}{2}\left\{X_l \cdot \exp[j(\Omega_0 - \omega)t] + X_l^* \cdot \exp[-j(\Omega_0 - \omega)t]\right\}, \quad \text{(F.1)} \\
x_u(t) &= \text{Re}\left\{X_u \cdot \exp[j(\Omega_0 + \omega)t]\right\} \\
&= \frac{1}{2}\left\{X_u \cdot \exp[j(\Omega_0 + \omega)t] + X_u^* \cdot \exp[-j(2\Omega_0 + \omega)t]\right\}. \quad \text{(F.2)}
\end{aligned}
$$

These signals are multiplied by $2 \cdot \cos\Omega_0 t = \exp(j\Omega_0 t) + \exp(-j\Omega_0 t)$ in the mixer. The mixing product for the lower sideband results as

$$
x_l'(t) = \frac{1}{2}\left\{X_l \cdot \exp[j(2\Omega_0 - \omega)t] + X_l^* \cdot \exp(j\omega t)\right.
$$

$$+X_l \cdot \exp(-j\omega t) + X_l^* \cdot \exp[-j(\Omega_0 - \omega)t] \Big\} \ . \qquad \text{(F.3)}$$

A similar expression follows for the upper sideband:

$$x_u'(t) = \frac{1}{2} \Big\{ X_u \cdot \exp\left[j(2\Omega_0 + \omega)t\right] + X_u^* \cdot \exp(-j\omega t) $$
$$+ X_u \cdot \exp(j\omega t) + X_u^* \cdot \exp\left[-(2\Omega_0 + \omega)t\right] \Big\} \ . \qquad \text{(F.4)}$$

The low-pass filter suppresses the frequency components at $2\Omega_0 \pm \omega$. Thus the output signal is either given by

$$
\begin{aligned}
x_{lb}(t) &= \frac{1}{2} \left[X_l^* \exp(j\omega t) + X_l \exp(-j\omega)t \right] \\
&= \mathrm{Re}\{X_l^* \exp(j\omega t)\} = \mathrm{Re}\{X_{lb} \exp(j\omega t)\} \qquad \text{(F.5)}
\end{aligned}
$$

or

$$
\begin{aligned}
x_{ub}(t) &= \frac{1}{2} \left[X_u \exp(j\omega t) + X_u^* \exp(-j\omega t) \right] \\
&= \mathrm{Re}\{X_u \exp(j\omega t)\} = \mathrm{Re}\{X_{ub} \exp(j\omega t)\} \ . \qquad \text{(F.6)}
\end{aligned}
$$

The circuit can be calculated in a similar way, if used as a single sideband modulator.

Problem 6.2

With Eq. (6.19) and Eq. (6.20), Eq. (6.14) yields

$$
\begin{bmatrix} X_{lb} \\ X_{ub} \end{bmatrix} = \frac{X_0}{2} \begin{bmatrix} \exp(-j\Phi_0) & -j \cdot \exp(-j\Phi_0) \\ \exp(j\Phi_0) & j \cdot \exp(j\Phi_0) \end{bmatrix} \begin{bmatrix} \dfrac{\Delta X}{X_0} \\ \Delta\Phi \end{bmatrix} \ , \quad \text{(F.7)}
$$

$$
\begin{aligned}
|X_{lb}|^2 &= \frac{X_0^2}{4} \left[\left| \frac{\Delta X}{X_0} \right|^2 + |\Delta\Phi|^2 - j \cdot \frac{\Delta X^*}{X_0} \Delta\Phi + j \frac{\Delta X}{X_0} \Delta\Phi^* \right] \\
&= \frac{X_0^2}{4} \left[\left| \frac{\Delta X}{X_0} \right|^2 + |\Delta\Phi|^2 + 2\mathrm{Im}\{ \frac{\Delta X^*}{X_0} \Delta\Phi \} \right] \ , \qquad \text{(F.8)}
\end{aligned}
$$

$$
\begin{aligned}
|X_{ub}|^2 &= \frac{X_0^2}{4} \left[\left| \frac{\Delta X}{X_0} \right|^2 + |\Delta\Phi|^2 + j \cdot \frac{\Delta X^*}{X_0} \Delta\Phi - j \frac{\Delta X}{X_0} \Delta\Phi^* \right] \\
&= \frac{X_0^2}{4} \left[\left| \frac{\Delta X}{X_0} \right|^2 + |\Delta\Phi|^2 - 2\mathrm{Im}\{ \frac{\Delta X^*}{X_0} \Delta\Phi \} \right] \ . \qquad \text{(F.9)}
\end{aligned}
$$

Replacing the phasor products by the corresponding spectra leads to Eq. (6.21) and Eq. (6.22).

With Eq. (6.19), Eq. (6.20) and Eq. (6.15) we get

$$\begin{bmatrix} \dfrac{\Delta X}{X_0} \\ \Delta \Phi \end{bmatrix} = \dfrac{1}{X_0} \begin{bmatrix} \exp(j\Phi_0) & \exp(-j\Phi_0) \\ j \cdot \exp(j\Phi_0) & -j \cdot \exp(-j\Phi_0) \end{bmatrix} \begin{bmatrix} X_{lb} \\ X_{ub} \end{bmatrix} , \qquad (F.10)$$

$$\left| \dfrac{\Delta X}{X_0} \right|^2 = \dfrac{1}{X_0^2} \left[|X_{lb}|^2 + |X_{ub}|^2 + X_{lb}^* X_{ub} \cdot \exp(-2j\Phi_0) + X_{lb} X_{ub}^* \cdot \exp(2j\Phi_0) \right]$$

$$= \dfrac{1}{X_0^2} \left[|X_{lb}|^2 + |X_{ub}|^2 + 2\mathrm{Re}\{X_{lb}^* X_{ub} \cdot \exp(-2j\Phi_0)\} \right] , \qquad (F.11)$$

$$|\Delta \Phi|^2 = \dfrac{1}{X_0^2} \left[|X_{lb}|^2 + |X_{ub}|^2 - X_{lb}^* X_{ub} \cdot \exp(-2j\Phi_0) - X_{lb} X_{ub}^* \cdot \exp(2j\Phi_0) \right]$$

$$= \dfrac{1}{X_0^2} \left[|X_{lb}|^2 + |X_{ub}|^2 - 2\mathrm{Re}\{X_{lb}^* X_{ub} \cdot \exp(-2j\Phi_0)\} \right] . \qquad (F.12)$$

We get Eq. (6.23) and Eq. (6.24) by using the corresponding spectra.

Problem 6.3

For the squared magnitude of the normalized cross-spectrum we obtain:

$$\left| \dfrac{W_{lub}(\omega)}{\sqrt{W_n(\Omega_0 - \omega) \cdot W_n(\Omega_0 + \omega)}} \right|^2$$

$$= \dfrac{\left(|m_\alpha|^2 - |m_\Phi|^2 \right)^2 + 4\mathrm{Re}^2\{m_\alpha^* m_\Phi\}}{\left(|m_\alpha|^2 + |m_\Phi|^2 \right)^2 - 4\mathrm{Im}^2\{m_\alpha^* m_\Phi\}}$$

$$= \dfrac{\left(|m_\alpha|^2 + |m_\Phi|^2 \right)^2 - 4|m_\alpha|^2 |m_\Phi|^2 + 4\mathrm{Re}^2\{m_\alpha^* m_\Phi\}}{\left(|m_\alpha|^2 + |m_\Phi|^2 \right)^2 - 4\mathrm{Im}^2\{m_\alpha^* m_\Phi\}} . \qquad (F.13)$$

With

$$\begin{aligned} |m_\alpha|^2 |m_\Phi|^2 &= |m_\alpha^*|^2 |m_\Phi|^2 \\ &= |m_\alpha^* m_\Phi|^2 \\ &= \mathrm{Re}^2\{m_\alpha^* m_\Phi\} + \mathrm{Im}^2\{m_\alpha^* m_\Phi\} , \qquad (F.14) \end{aligned}$$

we conclude that the squared magnitude is equal to one.

Problem 6.4

We have $u_{in}(t) = \hat{u}_{in} \cdot \cos \omega t$. Then the output voltage is given by

$$u_{out}(t) = \begin{cases} V_0 \hat{u}_{in} \cos \omega t\ , & |u_{in}| \leq u_0 \\ u_m\ , & u_{in} > u_0 \\ -u_m\ , & u_{in} < -u_0 \end{cases} \tag{F.15}$$

with the small-signal gain

$$V_0 = \frac{u_m}{u_0}\ . \tag{F.16}$$

For $\hat{u}_{in} \leq u_0$ the amplifier operates in a linear range and the describing function D is equal to the small-signal gain V_0:

$$D = V_0 \quad \text{for} \quad \hat{u}_{in} \leq u_0\ . \tag{F.17}$$

For $\hat{u}_{in} > u_0$ the waveform of $u_{out}(t)$ is shown in the figure below.

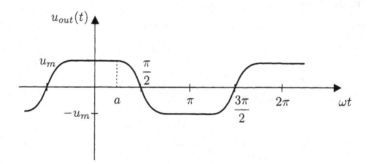

Fig. F.1

The boundary point a between the linear range and the saturation range follows from

$$V_0 \hat{u}_{in} \cdot \cos a = u_m \tag{F.18}$$

as

$$a = \arccos \frac{u_m}{V_0 \hat{u}_{in}} = \arccos \frac{u_0}{\hat{u}_{in}}\ . \tag{F.19}$$

The output voltage can be written as a Fourier series:

$$u_{out}(t) = \sum_{n=1}^{\infty} u_{out\,n} \cdot \cos n\omega t\ . \tag{F.20}$$

The amplitude of the fundamental wave is given by

$$
\begin{aligned}
u_{out\,1} &= \frac{1}{\pi} \int_0^{2\pi} u_{out}(t) \cdot \cos \omega t \, d\omega t \\
&= \frac{4}{\pi} \int_0^{\pi/2} u_{out}(t) \cdot \cos \omega t \, d\omega t \\
&= \frac{4}{\pi} \left[\int_0^a u_m \cos \omega t \, d\omega t + \int_a^{\pi/2} V_0 \hat{u}_{in} \cos^2 \omega t \, d\omega t \right] \\
&= \frac{4}{\pi} \left[u_m \cdot \sin a + V_0 \hat{u}_{in} \left(\frac{\pi}{4} - \frac{a}{2} \right) - \frac{1}{4} V_0 \hat{u}_{in} \cdot \sin 2a \right] \\
&= \frac{4}{\pi} u_m \left[\left(1 - \frac{1}{2} \frac{\hat{u}_{in}}{u_0} \cdot \cos a \right) \cdot \sin a + \frac{1}{2} \frac{\hat{u}_{in}}{u_0} \left(\frac{\pi}{2} - a \right) \right] \;. \quad \text{(F.21)}
\end{aligned}
$$

With

$$
\cos a = \frac{u_0}{\hat{u}_{in}} \tag{F.22}
$$

and

$$
\sin a = \sin \left[\arcsin \sqrt{1 - \left(\frac{u_0}{\hat{u}_{in}} \right)^2} \right] = \sqrt{1 - \left(\frac{u_0}{\hat{u}_{in}} \right)^2} \tag{F.23}
$$

we obtain

$$
\begin{aligned}
u_{out1} &= \frac{4}{\pi} u_m \left[\frac{1}{2} \sqrt{1 - \left(\frac{u_0}{\hat{u}_{in}} \right)^2} + \frac{1}{2} \frac{\hat{u}_{in}}{u_0} \left(\frac{\pi}{2} - \arccos \frac{u_0}{\hat{u}_{in}} \right) \right] \\
&= \frac{2}{\pi} u_m \left[\sqrt{1 - \left(\frac{u_0}{\hat{u}_{in}} \right)^2} + \frac{\hat{u}_{in}}{u_0} \arcsin \frac{u_0}{\hat{u}_{in}} \right] \;. \quad \text{(F.24)}
\end{aligned}
$$

The describing function finally is given by

$$
D = V_0 \frac{2}{\pi} \left[\frac{u_0}{\hat{u}_{in}} \sqrt{1 - \left(\frac{u_0}{\hat{u}_{in}} \right)^2} + \arcsin \frac{u_0}{\hat{u}_{in}} \right] \;. \tag{F.25}
$$

At the transition to the linear range, i.e. if $\hat{u}_{in} = u_0$, we get $D = V_0$, as expected. For $\hat{u}_{in} \to \infty$, D approaches zero and u_{out1} approaches $4/\pi \cdot u_m$. In summary, the describing function depends on the input drive level as shown in Fig. F.2. D is continuously differentiable, even at the transition to the

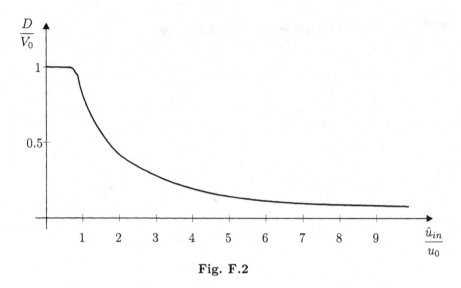

Fig. F.2

saturation range.

Problem 6.5

Using the notation of problem 6.4 we obtain

$$k_\alpha = -\frac{\hat{u}_{in}}{D(\hat{u}_{in})} \cdot \frac{d\,D(\hat{u}_{in})}{d\,\hat{u}_{in}} \quad . \tag{F.26}$$

The derivative of the describing function is given by

$$\frac{dD}{d\hat{u}_{in}} = V_0 \frac{2}{\pi} \left[-\frac{u_0}{\hat{u}_{in}^2}\sqrt{1-\left(\frac{u_0}{\hat{u}_{in}}\right)^2} + \frac{u_0}{\hat{u}_{in}} \frac{-2\frac{u_0}{\hat{u}_{in}} \cdot \left(-\frac{u_0}{\hat{u}_{in}^2}\right)}{2 \cdot \sqrt{1-\left(\frac{u_0}{\hat{u}_{in}}\right)^2}} + \frac{-\frac{u_0}{\hat{u}_{in}^2}}{\sqrt{1-\left(\frac{u_0}{\hat{u}_{in}}\right)^2}} \right]$$

$$= -V_0 \frac{2}{\pi} \frac{u_0}{\hat{u}_{in}^2} \left[\sqrt{1-\left(\frac{u_0}{\hat{u}_{in}}\right)^2} + \frac{1-\left(\frac{u_0}{\hat{u}_{in}}\right)^2}{\sqrt{1-\left(\frac{u_0}{\hat{u}_{in}}\right)^2}} \right]$$

$$= -V_0 \frac{4}{\pi} \frac{u_0}{\hat{u}_{in}^2} \sqrt{1-\left(\frac{u_0}{\hat{u}_{in}}\right)^2} \quad . \tag{F.27}$$

Thus we get

$$k_\alpha = \frac{V_0 \dfrac{4}{\pi} \dfrac{u_0}{\hat{u}_{in}} \cdot \sqrt{1 - \left(\dfrac{u_0}{\hat{u}_{in}}\right)^2}}{V_0 \dfrac{2}{\pi} \left[\dfrac{u_0}{\hat{u}_{in}} \sqrt{1 - \left(\dfrac{u_0}{\hat{u}_{in}}\right)^2} + \arcsin \dfrac{u_0}{\hat{u}_{in}}\right]} ,$$

$$= \frac{2\dfrac{u_0}{\hat{u}_{in}} \cdot \sqrt{1 - \left(\dfrac{u_0}{\hat{u}_{in}}\right)^2}}{\dfrac{u_0}{\hat{u}_{in}} \sqrt{1 - \left(\dfrac{u_0}{\hat{u}_{in}}\right)^2} + \arcsin \dfrac{u_0}{\hat{u}_{in}}} . \qquad (F.28)$$

At the boundary to the linear range, that means $\hat{u}_{in} = u_0$, we have $k_\alpha = 0$. For $\hat{u}_{in} \gg u_0$ and thus $\arcsin u_0/\hat{u}_{in} \approx u_0/\hat{u}_{in}$ and with

$$\sqrt{1 - \left(\frac{u_0}{\hat{u}_{in}}\right)^2} \approx 1 \qquad (F.29)$$

we obtain $k_\alpha \approx 1$. In the linear range we have $D = V_0 =$const and accordingly $k_\alpha = 0$. In summary, the curve of the amplitude compression coefficient is shown in Fig. F.3. If the linear range is exceeded, k_α rapidly rises to values close to one.

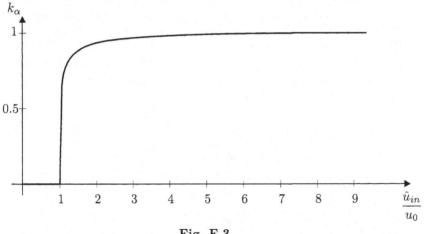

Fig. F.3

In this example, there is no phase shift between the input signal and the output signal, independent of the amplitude. Thus the phase ϱ of the describing function is $\varrho = $ const $= 0$ and consequently $k_\phi = 0$.

Problem 6.6

The input signal can be considered as a combination of a carrier signal at f_1 and an upper sideband at $f_1 + \Delta f$. The sideband causes both an amplitude modulation and a phase modulation of the carrier with the modulation frequency $\Delta f = f_2 - f_1$. Both types of modulation can be treated separately, if the upper sideband is split into two in-phase components with the amplitude $A/2$ and if two antiphase signals with the same amplitudes are added at the lower sideband frequency $f_1 - \Delta f$ (Fig. F.5).

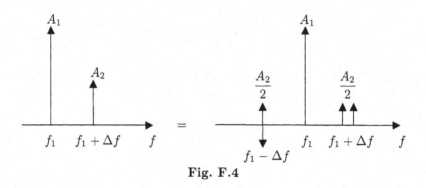

Fig. F.4

Since $A_2/A_1 \gg 1$, the in-phase pair of sideband signals causes a pure amplitude modulation, the other pair a pure phase modulation. In a non-linear system with hard amplitude clipping the amplitude modulation is almost completely suppressed so that at the output only the carrier and the sideband signals of the phase modulation will appear. If the peak phase deviation of the modulation remains constant, the relation of the sideband to the carrier amplitudes does not change and with an amplification factor V the following spectrum is observed at the output:

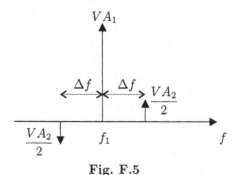

Fig. F.5

If for a frequency multiplier or frequency divider the phase deviations are multiplied or divided by the factor N, the sidebands amplitudes are changed by the same factor relative to the carrier. The resulting spectra are shown in Fig. F.6.

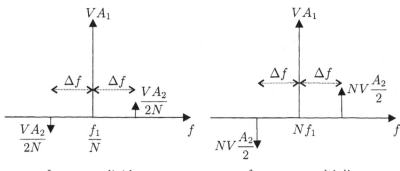

frequency divider   frequency multiplier

Fig. F.6

Note that by a multiplication or division the phase or frequency deviations change but not the modulation frequency. Thus the sidebands of the output signals have the same offset Δf to the carrier as the component A_2 of the input signal.

Problem 6.7

We will assume that the signal $x(t)$ is not directly fed into the mixer but via a phase shifter with the phase shift Φ_0. Then we have

$$x(t) \ = \ X_0 \cdot \cos\left[\Omega_0 t + \Phi_0 + \Delta\Phi(t)\right] \ , \qquad (\text{F}.30)$$

$$y(t) \ = \ Y_0 \cdot \cos\left[\Omega_0 t + \Delta\Psi(t)\right] \ . \qquad (\text{F}.31)$$

If the mixer is treated as a multiplier with the multiplier constant K_M, the output signal is given by

$$
\begin{aligned}
u(t) \ &= \ K_M \cdot x(t) \cdot y(t) \\
&= \ K_M \cdot X_0 Y_0 \cos\left[\Omega_0 t + \Phi_0 + \Delta\Phi(t)\right] \cdot \cos\left[\Omega_0 t + \Delta\Psi(t)\right] \\
&= \ \frac{1}{2} K_M X_0 Y_0 \big\{ \cos\left[\Delta\Psi(t) - \Delta\Phi(t) - \Phi_0\right] \\
&\quad + \ \cos\left[2\Omega_0 t + \Phi_0 + \Delta\Phi(t) + \Delta\Psi(t)\right] \big\}
\end{aligned}
\qquad (\text{F}.32)
$$

The second term, which is a high-frequency component, is suppressed by a low-pass filter. With $|\Delta\Psi(t) - \Delta\Phi(t)| \ll 1$ we obtain

$$u(t) \ = \ \frac{1}{2} K_M X_0 Y_0 \big\{ \cos\Phi_0 \cos[\Delta\Psi(t) - \Delta\Phi(t)]$$

$$+ \sin \Phi_0 \sin[\Delta\Psi(t) - \Delta\Phi(t)]\}$$

$$\approx \frac{1}{2}K_M X_0 Y_0 \{\cos \Phi_0 + \sin \Phi_0 \cdot [\Delta\Psi(t) - \Delta\Phi(t)]\} \; . \qquad (\text{F.33})$$

The output signal consists of a dc and an alternating voltage which is proportional to the difference of the phase fluctuations of both input signals. The phase detector constant K_{PD} thus follows as

$$K_{PD} = \frac{1}{2}K_M X_0 Y_0 \cdot \sin \Phi_0 \; . \qquad (\text{F.34})$$

For $\Phi_0 = 0$ we get $K_{PD} = 0$. The highest sensitivity is obtained if both input signals have a phase difference of 90°. In this case, the d.c. voltage of the output signal is zero. By monitoring the d.c. voltage while tuning a variable phase shifter, a balanced mixer can be adjusted for maximum phase detector sensitivity.

Problem 6.8

The three measurement objects shall have the noise spectra $W_{\psi n1}$, $W_{\psi n2}$ and $W_{\psi n3}$. If for all pairs of objects the phase jitter is measured with the circuit shown in Fig. 6.10, the following noise spectra of the output voltage $u(t)$ are obtained:

$$\begin{aligned}
W_{u1} &= K_{PD}^2(W_{\psi n1} + W_{\psi n2}) \; , & (\text{F.35}) \\
W_{u2} &= K_{PD}^2(W_{\psi n1} + W_{\psi n3}) \; , & (\text{F.36}) \\
W_{u3} &= K_{PD}^2(W_{\psi n2} + W_{\psi n3}) \; . & (\text{F.37})
\end{aligned}$$

This linear system of equations can be solved for the unknown spectra $W_{\psi ni}$, $i = 1, 2, 3$:

$$W_{\Psi n1} = \frac{1}{2K_{PD}^2}(W_{u1} + W_{u2} - W_{u3}) \; , \qquad (\text{F.38})$$

$$W_{\Psi n2} = \frac{1}{2K_{PD}^2}(W_{u1} - W_{u2} + W_{u3}) \; , \qquad (\text{F.39})$$

$$W_{\Psi n3} = \frac{1}{2K_{PD}^2}(-W_{u1} + W_{u2} + W_{u3}) \; . \qquad (\text{F.40})$$

Appendix G
Solutions to the
Problems of Chapter 7

Chapter 7

Problem 7.1

The optimum signal power can be determined with Eq. (6.88).

$$P_{sopt} = \frac{P_{sat}}{G_0} \cdot \ln G_0 = (5\,\text{dBm})\ln(15\,\text{dB}) = 10.4\,\text{dBm} \ . \qquad (G.1)$$

The gain for this signal power is

$$G(P_{sopt}) = \frac{G_0 - 1}{\ln G_0} = 9.5\,\text{dB} \ . \qquad (G.2)$$

Therefore, the power at the output is 19.9 dBm. Due to the signal divider the output power of the oscillator is reduced to 16.9 dBm ≈ 50 mW.

Equation (5.91) leads to

$$G(P_{sopt}) \cdot 2\,\beta^2 = (1 + 2\,\beta)^2 = 4\,\beta^2 + 4\,\beta + 1 \qquad (G.3)$$

or

$$\beta^2 - \frac{2\,\beta}{G(P_{sopt}) - 2} - \frac{1}{2\,G(P_{sopt}) - 4} = 0 \ . \tag{G.4}$$

This quadratic equation has the solution

$$\beta = \frac{1 + \sqrt{G(\frac{P_{sopt})}{2}}}{G(P_{sopt}) - 2} = 0.45 \ . \tag{G.5}$$

Problem 7.2

According to Eq. (6.90) the amplitude compression factor is

$$k_\alpha = 1 - \frac{\ln G_0}{G_0 - 1} = 0.887 \ . \tag{G.6}$$

With $F_r = \Omega_r/(2\pi)$, the corner frequencies $f_{t1} = \omega_{t1}/2\pi$ and $f_{t2} = \omega_{t2}/2\pi$ are given by

$$
\begin{aligned}
f_{t1} &= k_\alpha(1 + 2\,\beta)\frac{F_r}{2\,Q_0} = 8.43\,\text{MHz} \ , \\
f_{t2} &= \frac{1}{k_\alpha} f_{t1} = 9.50\,\text{MHz} \ .
\end{aligned}
\tag{G.7}
$$

The spectral power density W_0 is calculated as

$$W_0 = \frac{F_{ef}\,k\,T_0}{2\,P_{in}} = (-174 + 20 - 3 - 10.4)\,\text{dB/Hz} = -167.4\,\text{dB/Hz} \ . \tag{G.8}$$

Without consideration of the $1/f$-noise, the amplitude noise at frequencies below f_{t1} is

$$W_\alpha = \left(\frac{1}{k_\alpha} - 1\right)^2 W_0 = -185.3\,\text{dB/Hz} \tag{G.9}$$

and above f_{t2} it is

$$W_\alpha = (1 - k_\alpha)^2 \, W_0 = -186.3\,\text{dB/Hz} \ . \tag{G.10}$$

For strong amplitude compression both corner frequencies are close to each other. Then, the drop in amplitude noise is not very pronounced.

Without $1/f$-noise the phase noise is given by

$$W_\phi(f) \left[1 + \left(\frac{1 + 2\,\beta}{2\,Q_0\,f/F_0}\right)^2\right] W_0 \ . \tag{G.11}$$

At $f = 1\,\text{kHz}$ and in $1\,\text{Hz}$ bandwidth we obtain the numerical value $W = -87.8$ dB/Hz. With the $1/f$-noise included all numerical values for the spectra are

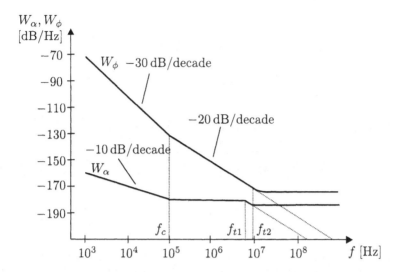

Fig. G.1 Quantitative spectra of the amplitude and phase noise of a two-port oscillator with $1/f$-noise included.

multiplied by $(1 + f_b/f)$. The result is shown in Fig. G.1.

Problem 7.3

Interchanging the positions of the resonator and the signal divider does not change the function $H(\Omega)$. Thus, also the oscillation condition remains unchanged and therefore the values for the oscillation amplitude, the coupling factor β and the amplitude compression factor k_α are the same as those of the problems 7.1 and 7.2. However, by modifying the output coupling network, the function $A(\Omega)$ is no longer frequency independent but is given by

$$A(\Omega) = H(\Omega) . \tag{G.12}$$

With the Eq. (7.14) we obtain for the amplitude and phase fluctuations of the output signal:

$$
\begin{bmatrix} \dfrac{\Delta Z}{Z_0} \\ \Delta\theta \end{bmatrix} =
\begin{bmatrix} H_\Sigma & j\,H_\Delta \\ -j\,H_\Delta & H_\Sigma \end{bmatrix}
\begin{bmatrix} \dfrac{\Delta Y}{Y_0} \\ \Delta\Psi \end{bmatrix} \tag{G.13}
$$

with

$$H_\Sigma = \frac{1 + 2\,\beta}{1 + 2\,\beta + j\,2\,Q_0\,\omega/\Omega_r} \tag{G.14}$$

and

$$H_\Delta = 0 . \tag{G.15}$$

Hence we get

$$\frac{\Delta Z}{Z_0} = \frac{1 + 2\beta}{k_\alpha (1 + 2\beta) + j\, 2\, Q_0\, \omega/\Omega_r} \cdot \frac{\Delta Y_n}{Y_0} \qquad (G.16)$$

$$\Delta\theta = \frac{1 + 2\beta}{j\, 2\, Q_0\, \omega/\Omega_0} \cdot \Delta Y_n \;, \qquad (G.17)$$

with the spectra

$$W_\alpha(\omega) = (1 - k_\alpha)^2\, \frac{(1 + 2\beta)^2}{k_\alpha^2 (1 + 2\beta)^2 + (2\, Q_0\, \omega/\Omega_0)^2}\, W_0 \;, \qquad (G.18)$$

$$W_\Phi(\omega) = \frac{(1 + 2\beta)^2}{(2\, Q_0\, \omega/\Omega_r)^2}\, W_0 \;. \qquad (G.19)$$

A comparison with the Eqs. (7.41) and (7.42) shows that there is no corner frequency ω_{t2} any more. The phase noise decreases steadily by 20 dB/decade. The same holds for the amplitude noise at offset frequencies above ω_{t1}. Below the corner frequency $\omega_{t1} = 8.43$ MHz we obtain the same quantitative values for the spectra W_α and W_Φ as in 7.2. The constant decrease by 20 dB/decade for $f \gg f_{t1}$ leads to the dotted curves in Fig. G.1.

Problem 7.4

With U and I real, we have $R = -U/I$ and

$$k_\alpha = -\frac{I}{R(I)} \cdot \frac{dR}{dI} = -\frac{I^2}{U} \frac{d}{dI}\left(\frac{U}{I}\right)$$

$$= -\frac{I^2}{U} \cdot \frac{I\dfrac{dU}{dI} - U}{I^2} = -\frac{I}{U} \cdot \frac{dU}{dI} + 1 \;. \qquad (G.20)$$

The output power is given by

$$P = \frac{1}{2} U I \;. \qquad (G.21)$$

For the maximum power the derivative of the power with respect to the current amplitude I equals zero:

$$\frac{dP}{dI} = \frac{1}{2}\left(I\frac{dU}{dI} + U\right) = 0 \;. \qquad (G.22)$$

Hence

$$\frac{dU}{dI} = -\frac{U}{I} < 0 \;. \qquad (G.23)$$

This means that the maximum of the output power occurs in the descending part of the U/I characteristic of Fig. 7.7a. We obtain for the compression factor at this operating point

$$k_\alpha = 2 \ . \tag{G.24}$$

Problem 7.5

The oscillation condition follows from Eq. (7.54):

$$\Omega_r/2\pi = \frac{1}{2}\pi\sqrt{LC} = 503.3 \,\mathrm{MHz} \ . \tag{G.25}$$

With $kT_0 = -174$ dBm/Hz and Eq. (7.12) we obtain for W_0:

$$W_0 = \frac{F_{ef} \cdot kT_0}{2\,P_{in}} = \frac{(-174 + 20)\,\mathrm{dBm/Hz}}{3\,\mathrm{dBm}} = -157\,\mathrm{dB/Hz} \ . \tag{G.26}$$

The corner frequency $f_t = \omega_t/2\pi$ is given by

$$f_t = \frac{k_\alpha\,R_0}{4\pi\,L} = \frac{2 \cdot 50}{4\pi \cdot 10^{-6}}\,\mathrm{Hz} = 7.96\,\mathrm{MHz} \ . \tag{G.27}$$

For $f \ll f_t$ the amplitude noise has the constant value

$$W_\alpha = \frac{1}{k_\alpha^2}\,W_0 = (-157 - 6)\,\mathrm{dB/Hz} = -163\,\mathrm{dB/Hz} \ . \tag{G.28}$$

The phase noise at an offset frequency of 1 kHz is calculated as

$$W_\phi = \frac{R_0^2}{(2\,\omega\,L)^2}\,W_0 = \frac{2500}{(4\pi \cdot 10^3 \cdot 10^{-6})^2}\,(-157\,\mathrm{dB/Hz}) = -85\,\mathrm{dB/Hz} \ . \tag{G.29}$$

The corresponding spectra W_α and W_ϕ are shown in Fig. G.2 .

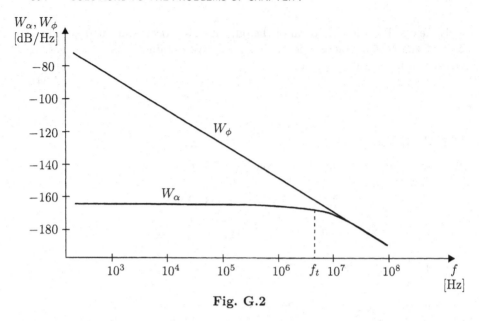

Fig. G.2

Problem 7.6

The function $H(\Omega)$ of Eq. (7.50) is given by

$$H(\Omega) = \frac{1}{R_0} + j\,\Omega\,C + \frac{1}{j\,\Omega\,L} \ . \tag{G.30}$$

By inserting this relation into the oscillation condition we obtain the same expressions as for the series resonance circuit given by the Eqs. (7.53) and (7.54). The oscillation frequency and amplitude are equal for both circuits. The voltage U is identical to the voltage at the load resistance, hence $A(\Omega) \equiv 1$. With $\Omega = \Omega_0 + \omega$ and $\omega \ll \Omega_0$ we obtain

$$H(\Omega) = \frac{1}{R_0} + j\,2\,\omega\,C \ , \tag{G.31}$$

and

$$H_\Sigma = 1 + j\,2\,\omega\,R_0\,C \ , \quad H_\Delta = 0 \ . \tag{G.32}$$

The results for the amplitude and phase fluctuations are

$$\frac{\Delta U_R}{U_{R0}} = \frac{\Delta U}{U_0} = \frac{1}{k_\alpha - j\,2\,\omega R_0\,C(1 - k_\alpha)}\,\frac{\Delta U_n}{U_0} \ ,$$

$$\Delta\theta = \Delta\Psi = -\frac{1}{j\,2\,\omega R_0\,C}\,\Delta\Psi_n \ . \tag{G.33}$$

The stability can be checked with the function

$$De(p) = k_\alpha - 2\,R_0\,C(1 - k_\alpha)p \ . \tag{G.34}$$

From $De(p_1) = 0$ we get

$$p_1 = \frac{k_\alpha}{2\,R_0\,C(1 - k_\alpha)} \ . \tag{G.35}$$

For $0 < k_\alpha < 1$ we have $p_1 > 0$. Thus, in general, the circuit is not stable. However, one-port oscillators are stable if the active one-port has a characteristic of the form shown in Fig. G.3

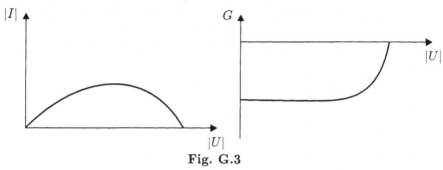

Fig. G.3

Compared to Fig. 7.7 current and voltage have been exchanged. Then, the circuit with the parallel resonance circuit can be calculated in exactly the same way as the circuit with the series resonance circuit by exchanging the quantities current and voltage.

Problem 7.7

The calculation of the input admittance of the circuit in Fig. 7.11 can be performed as follows. The input admittance at the emitter is given by:

$$Y_{ei} = \frac{i_e}{u_e} \ . \tag{G.36}$$

For the currents i_b and i_c we have

$$i_b = (g_{be} + j\omega C_{be})u_{be} \ , \tag{G.37}$$

$$i_c = g_m u_{be} \ . \tag{G.38}$$

The voltage u_e and the current i_e at the emitter port are given by

$$u_e = -(u_{be} + i_b Z_B) \ , \tag{G.39}$$

$$i_e = -(i_b + i_c) \ , \tag{G.40}$$

with $Z_B = 1/Y_B$. Inserting Eq. (G.37) and Eq. (G.38) leads to

$$u_e = -(1 + Z_B(g_{be} + j\omega C_{be}))u_{be} \ , \tag{G.41}$$

$$i_e = -(g_{be} + j\omega C_{be} + g_m)u_{be} \ . \tag{G.42}$$

Thus the input admittance at the emitter can be calculated as

$$Y_{ei} = \frac{i_e}{u_e} = \frac{g_{be} + j\omega C_{be} + g_m}{1 + Z_B(g_{be} + j\omega C_{be})} \quad . \tag{G.43}$$

Considering Z_B as a pure reactance with $Z_B(j\omega) = jX_B(\omega)$ and separating Y_{ei} into its real and imaginary parts leads to

$$\begin{aligned} Y_{ei} &= \frac{g_{be} + g_m(1 - \omega C_{be} X_B)}{(1 - \omega C_{be} X_B)^2 + (g_{be} X_B)^2} \\ &+ j\frac{\omega C_{be} - \omega^2 C_{be}^2 X_B - g_{be} X_B (g_{be} + g_m)}{(1 - \omega C_{be} X_B)^2 + (g_{be} X_B)^2} \quad . \end{aligned} \tag{G.44}$$

The requirement of a negative real part of the input admittance can be met if $X_B(\omega) > 0$. The base of the transistor should be connected to an inductive circuit with $Z_B(j\omega) = jX_B(\omega)$.

Problem 7.8

The transfer function of the symmetrical transmission resonator was already given by Eq. (7.28). With $\beta_1 = \beta_2 = \beta$ as the coupling factor, Ω_r as the angular resonance frequency, Ω_1 as the angular input carrier frequency and Q_0 as the unloaded Q-factor we get for the transmission coefficient S_{21} or transfer function H:

$$S_{12} = S_{21} = H(\Omega_1) = \frac{2\beta}{1 + 2\beta + jQ_0\left(\dfrac{\Omega_1}{\Omega_r} - \dfrac{\Omega_r}{\Omega_1}\right)} \quad . \tag{G.45}$$

The input angular carrier frequency Ω_1 differs from the angular resonance frequency Ω_r by the angular displacement frequency $\Delta\Omega_r$, i.e. $\Omega_1 = \Omega_r + \Delta\Omega_r$. With ω as the angular offset frequency of the noise sidebands and with the assumption that $Q_0\omega/\Omega_r \ll 1$ and $\Delta\Omega_r/\Omega_r \ll 1$ we can write for Eq. (7.16):

$$\begin{aligned} H_\Delta &= \frac{1}{2}\left(\frac{H_u}{H_c} - \frac{H_l^*}{H_c^*}\right) \\ &= \frac{1}{2}\left(\frac{H(\Omega_1 + \omega)}{H(\Omega_1)} - \frac{H^*(\Omega_1 - \omega)}{H^*(\Omega_1)}\right) \\ &= \frac{1}{2}\left[\frac{1 + 2\beta + j\,2\,Q_0\cdot\dfrac{\Delta\Omega_r}{\Omega_r}}{1 + 2\beta + j\,2\,Q_0\dfrac{\Delta\Omega_r}{\Omega_r} + j\,2\,Q_0\dfrac{\omega}{\Omega_r}}\right. \\ &\quad \left. - \frac{1 + 2\beta - j\,2\,Q_0\cdot\dfrac{\Delta\Omega_r}{\Omega_r}}{1 + 2\beta - j\,2\,Q_0\cdot\dfrac{\Delta\Omega_r}{\Omega_r} + j\,2\,Q_0\dfrac{\omega}{\Omega_r}}\right] \end{aligned} \tag{G.46}$$

$$\approx \frac{-2\,Q_0\dfrac{\Delta\Omega_r}{\Omega_r}\cdot 2\,Q_0\dfrac{\omega}{\Omega_r}}{(1+2\,\beta)^2+(2\,Q_0\dfrac{\Delta\Omega_r}{\Omega_r})^2} \tag{G.47}$$

There is no useful optimum choice for the coupling factor β. Although the magnitude of H_Δ has a maximum at $\beta=0$, this choice for β is not practical. However, $2\,Q_0\Delta\Omega_r/\Omega_r = 1+2\beta$ leads to an optimum value with respect to the angular displacement frequency $\Delta\Omega_r$. Then, we get for the spectrum the expression:

$$W_u(\omega) = \left(K_{ad}\,Y_0\,\frac{Q_0}{1+2\,\beta}\,\frac{\omega}{\Omega_r}\right)^2 W_\phi(\omega)\ . \tag{G.48}$$

For a practical choice of the coupling factor of $\beta=0.5$ leading to a transmission loss of the transmission resonator of $6\,\mathrm{dB}$, we get for the spectra:

$$W_u(\omega) = \left(K_{ad}\,Y_0\,\frac{Q_0}{2}\,\frac{\omega}{\Omega_r}\right)^2 W_\phi(\omega)\ . \tag{G.49}$$

Comparing this result with the circuit of Fig. 7.27, i.e. Eq. (7.102), one may argue that the transmission resonator type discriminator shows a degradation of the discrimination efficiency by 6dB as compared to the circuit with the reflection type resonator and the by-pass phase shifter.

Problem 7.9

The delay line has the transfer function

$$H(\Omega) = \exp\left[-\left(\alpha'+j\,\frac{\Omega}{v}\right)l\right] = \exp\left[-\left(\alpha'+j\,\frac{\Omega_0+\omega}{v}\right)l\right]\ , \tag{G.50}$$

where v is the phase velocity of the line. Because l is assumed to be an integer multiple n of the wavelength at the oscillation frequency Ω_0, we have $\Omega l/v = n2\pi$ and thus

$$H(\Omega_0+\omega) = \exp\left[-\left(\alpha'+j\,\frac{\omega}{v}\right)l\right]\ . \tag{G.51}$$

In conjunction with $G(\Omega) = j$ we obtain from Eq. (7.95):

$$W_u(\omega) = \left(\frac{K_{ad}\,Y_0}{2}\right)^2 \left| \frac{\exp\left[-\left(\alpha'+j\,\dfrac{\omega}{v}\right)l\right]+j}{\exp(-\alpha'l)+j} \right.$$

$$\left. -\frac{\exp\left[-\left(\alpha'+j\,\dfrac{\omega}{v}\right)l\right]-j}{\exp(-\alpha'l)-j} \right|^2 W_\Phi(\omega)$$

$$
= \left(\frac{K_{ad}\,Y_0}{2}\right)^2 \left|\frac{j\,2\exp(-\alpha'l)\left[1-\exp\left(-\dfrac{j\,\omega}{v}\,l\right)\right]}{1+\exp(-2\alpha'l)}\right|^2 W_\Phi(\omega)
$$

$$
= \left(K_{ad}\,Y_0\,\frac{\exp(-\alpha'l)}{1+\exp(-2\alpha'l)}\right)^2 \left[\left(1-\cos\frac{\omega l}{v}\right)^2 + \sin^2\frac{\omega l}{v}\right] W_\Phi(\omega) \;.
$$

$$
\text{(G.52)}
$$

After some further manipulations we finally obtain

$$
W_u(\omega) = (K_{ad}\,Y_0)^2 \left[\frac{\sin\dfrac{\omega l}{2v}}{\cosh\alpha'l}\right]^2 W_\Phi(\omega) \;. \tag{G.53}
$$

The optimum length l_{opt} can be determined by means of a differentiation of the expression within the brackets with respect to the length l. With the approximation $\sin\omega l/2v \approx \omega l/2v$ for low offset frequencies the optimum length l_{opt} follows from the condition

$$
\alpha'l_{opt}\tanh(\alpha'l_{opt}) = 1 \;. \tag{G.54}
$$

This transcendental equation has the approximate solution

$$
\alpha'l_{opt} \approx 1.2 \qquad \text{or} \qquad l_{opt} \approx 1.2/\alpha'_{opt} \;. \tag{G.55}
$$

Problem 7.10

The resonator is assumed to be critically coupled so that the frequency discriminator has its maximum sensitivity. With $\beta = 1$ Eq. (7.97) yields

$$
H(\Omega_0) = 0, \qquad H(\Omega_0 + \omega) = H^*(\Omega_0 - \omega) \;. \tag{G.56}
$$

If a 3 dB-180° coupler is employed, then at one output the sum and at the other output the difference of the input signals is obtained. The amplitude fluctuations of both output signals are calculated with Eq. (7.14):

$$
\begin{aligned}
\frac{\Delta Y_1}{Y_0} &= \frac{1}{2}\left[\frac{H(\Omega_0+\omega)+\exp(j\gamma)}{\exp(j\gamma)} + \frac{H(\Omega_0+\omega)+\exp(-j\gamma)}{\exp(-j\gamma)}\right]\frac{\Delta X}{X_0} \\
&\quad + \frac{j}{2}\left[\frac{H(\Omega_0+\omega)+\exp(j\gamma)}{\exp(j\gamma)} - \frac{H(\Omega_0+\omega)+\exp(-j\gamma)}{\exp(-j\gamma)}\right]\Delta\Phi \\
&= \left[1+H(\Omega_0+\omega)\cdot\cos\gamma\right]\frac{\Delta X}{X_0} + H(\Omega_0+\omega)\cdot\sin\gamma\cdot\Delta\Phi \;,
\end{aligned}
$$

$$
\text{(G.57)}
$$

$$\frac{\Delta Y_2}{Y_0} = \frac{1}{2} \left[\frac{H(\Omega_0 + \omega) - \exp(j\gamma)}{-\exp(j\gamma)} + \frac{H(\Omega_0 + \omega) - \exp(-j\gamma)}{-\exp(-j\gamma)} \right] \frac{\Delta X}{X_0}$$

$$+ \frac{j}{2} \left[\frac{H(\Omega_0 + \omega) - \exp(j\gamma)}{-\exp(j\gamma)} - \frac{H(\Omega_0 + \omega) - \exp(-j\gamma)}{-\exp(-j\gamma)} \right] \Delta\Phi$$

$$= [1 - H(\Omega_0 + \omega) \cdot \cos\gamma] \frac{\Delta X}{X_0} - H(\Omega_0 + \omega) \cdot \sin\gamma \cdot \Delta\Phi \ . \tag{G.58}$$

The output voltage $u(t)$ is the difference of the voltages of both detectors, which are proportional to the amplitude fluctuations $\Delta Y_1/Y_0$ and $\Delta Y_2/Y_0$. We thus obtain

$$u(t) \sim \frac{\Delta Y_1 - \Delta Y_2}{Y_0} = 2 H(\Omega_0 + \omega) \left(\cos\gamma \frac{\Delta X}{X_0} + \sin\gamma \Delta\Phi \right) \ . \tag{G.59}$$

We note that the contribution of the amplitude fluctuations of the input signal vanishes for $\gamma = 90°$ while the conversion efficiency for the detection of phase fluctuations reaches its maximum. If a 3 dB-90° coupler is employed, the same result is achieved for $\gamma = 0$.

Problem 7.11

For the output signal of the amplifier, the 3 dB coupler has the same effect as the signal divider in Fig. 7.3. Therefore, the transfer functions $H(\Omega)$ and $A(\Omega)$ do not change. The transfer function $E(\Omega)$ of the input coupling network of Fig. 7.32 has the constant value $1/\sqrt{2}$ as $A(\Omega)$. With $Y_i = Z_i/\sqrt{2}$ and $Z = Y/\sqrt{2}$ we have

$$\frac{Y_i}{Y} = \frac{1}{2} \frac{Z_i}{Z} = q_i \exp(j\vartheta) \tag{G.60}$$

and

$$q_i = \frac{1}{2} \left| \frac{Z_i}{Z} \right| = \frac{1}{2} \sqrt{\frac{P_i}{P_0}} \ . \tag{G.61}$$

Since according to Eq. (7.27) $D(X)$ is real, the imaginary part of the oscillation condition, Eq. (7.105), leads to

$$(1 + 2\beta) q_i \sin\vartheta - 2 Q_0 \frac{\Omega_i - \Omega_0}{\Omega_0} (1 + q_i \cos\vartheta) = 0 \tag{G.62}$$

At the boundaries of the synchronization range we have $\Omega_i - \Omega_0 = \pm\Delta\Omega_m$ and $\vartheta = \pm 90°$. This results in

$$(1 + 2\beta) q_i = 2 Q_0 \frac{\Delta\Omega_m}{\Delta\Omega_0} \tag{G.63}$$

and

$$\Delta\Omega_m = \frac{\Omega_0}{4 Q_0} (1 + 2\beta) \sqrt{\frac{P_i}{P_0}} \ . \tag{G.64}$$

Problem 7.12

With Eq. (7.145) the standard deviation of the measured velocity is given by

$$\sigma_v = \frac{c}{\Omega_0}\, \sigma_\Omega \ , \tag{G.65}$$

with σ_Ω as the standard deviation of the angular frequency difference. Similar to the calculation of σ_Φ, the standard deviation σ_Ω can be determined by an integral of the corresponding spectrum:

$$\sigma_\Omega^2 = 2 \int\limits_{-\infty}^{+\infty} W_\Omega(f)\,[1 - \cos(2\pi f\tau)]\, df \ . \tag{G.66}$$

Because of

$$\Delta\Omega(t) = \frac{d}{dt}\left[\Delta\Phi(t)\right] \tag{G.67}$$

the relation between the spectra W_Ω and W_Φ is given by

$$W_\Omega = (2\pi f)^2\, W_\Phi \ . \tag{G.68}$$

This leads to

$$W_\Omega = 2 \int\limits_{-\infty}^{+\infty} (2\pi f)^2\, W_\Phi(f)\,[1 - \cos(2\pi f\tau)]\, df \ . \tag{G.69}$$

Appendix H
Solutions to the
Problems of Chapter 8

Chapter 8

Problem 8.1

The cascade circuit generates impulse like signals behind the 1-bit quantizer at the positions y_1, y_2 and y_3 with amplitude values of zero or one. A signal with the amplitude value one for one time step shall be denoted as unity impulse or simply impulse. For the first stage a unity impulse (y_1 in Fig. 8.7), is directly transmitted to the output y. After the second and third stage the unity impulses (y_2 and y_3 in Fig. 8.7) are differentiated once or twice, respectively, and also transmitted to the output (y in Fig. 8.7), where all three signals (y_1, y_2' and y_3'' in Fig. 8.7) are summed up.

The Z-transformed transfer functions of the three differentiation networks are given in the table. By a reverse Z-transformation into the time domain one obtains the corresponding finite impulse responses $h(k)$, also given in the table.

stage	$H(z)$	$h(n)$	
1	$(1 - z^{-1})^0 = 1$	$\delta(k)$	(H.1)
2	$(1 - z^{-1})^1 = 1 - z^{-1}$	$\delta(k) - \delta(k - 1)$	
3	$(1 - z^{-1})^2 = 1 - 2z^{-1} + z^{-2}$	$\delta(k) - 2\delta(k - 1) + \delta(k - 2)$	

The contribution to the total deviation of the division factor ΔN for the different stages is shown in Fig. H.1.

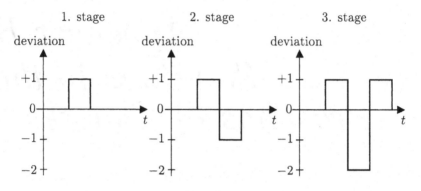

Fig. H.1

For instance, with the appearance of an impulse at the position y_3, an increase of the division factor N behind the differentiator of second order by +1 is induced, a decrease in the succeeding clock cycle by -2 and once again an increase by +1 in the following clock cycle. Because the impulses of the different stages can arrive either simultaneously or at earlier or later clock cycles and because the differentiation networks are linear circuits, the maximum division factor deviation can attain the value +4 for a three-stage configuration. This situation occurs if an impulse is produced by all three stages simultaneously and in addition an impulse signal already appeared two clock cycles earlier. The minimum deviation of the division factor is obtained when at least the second and the third stage produce impulses simultaneously in front of the differentiation networks, but not the first stage. The sum of -1 and -2 leads to the maximum negative deviation of -3. Similar reasoning results in division factor deviations of -1 to +2 for the two-stage and 0 to +1 for the one-stage configuration.

Problem 8.2

The transfer function $H(z)$ is given by

$$H(z) = \frac{Y(z)}{F} = \frac{\kappa_1}{D^3 + (D^2 \cdot \kappa_3 + D \cdot \kappa_2 + \kappa_1) \cdot z^{-1}}$$

$$= \frac{\dfrac{1}{4}}{(1 - z^{-1})^3 + \left[(1 - z^{-1})^2 + \dfrac{1}{2}(1 - z^{-1}) + \dfrac{1}{4}\right] z^{-1}}$$

$$= \frac{\dfrac{1}{4} z^2}{z^2 - \dfrac{5}{4} z + \dfrac{1}{2}} \ . \tag{H.2}$$

Thus the zeros $z_{1,2}$ of the denominator are

$$z_{1,2} = \frac{1}{8}(-5 \pm j\sqrt{7}) \tag{H.3}$$

or

$$|z_1| = |z_2| = \frac{\sqrt{2}}{2} \ . \tag{H.4}$$

We conclude that the zeros lie within the unit circle and stability is guaranteed.

Problem 8.3

The transfer function $H(z)$ is given by

$$H(z) = \frac{Y(z)}{F}$$

$$= \frac{z^{-1}}{D^4 + D^3 \cdot z^{-1} + D^2 \cdot z^{-1} + \dfrac{1}{2} D \cdot z^{-2} + \dfrac{3}{16} \cdot z^{-2}}$$

$$= \frac{z^2}{z^3 - 2z^2 + \dfrac{27}{16} z - \dfrac{1}{2}} \ . \tag{H.5}$$

The zeros of the denominator are

$$|z_1| = 0.5772\ldots$$
$$|z_2| = |z_3| = 0.9307\ldots \ . \tag{H.6}$$

Again we conclude that the zeros lie within the unit circle and stability is guaranteed.

References

1. R. Adler, A Study of Locking Phenomena in Oscillators, *Proceedings of the IEEE*, Vol. 61, pp. 1380-1385, October, 1973.

2. A. R. Bennett, *Electrical Noise*, McGraw Hill, New York, 1960.

3. Heinz Bittel, L. Storm, *Rauschen*, Springer, New York, 1971.

4. Alfons Blum, *Elektronisches Rauschen*, B.G. Teubner, Stuttgart, 1996.

5. Alain Cappy, Noise Modelling and Measurement Techniques, *IEEE Transactions on Microwave Theory and Techniques*, Vol. 36, pp. 1-10, January, 1988.

6. N.R. Campbell, V.J. Francis, A Theory of Valve and Circuit Noise, *Proceedings of the IRE*, Vol. 93, pp. 45-62, 1946.

7. Wilbur B. Davenport, William L. Root, *An Introduction to the Theory of Random Signals and Noise*, McGraw-Hill, New York, 1958.

8. H.T. Friis, Noise Figures of Radio Receivers, *Proceedings of the IRE*, Vol. 32, July, 1944.

9. F.N. Hooge, The Relation between $1/f$ Noise and the Number of Electrons, *Physica*, vol. 162, pp. 344-352, 1990.

10. K. Kurokawa, Noise in Synchronized Oscillators, *IEEE Transactions on Microwave Theory and Techniques*, Vol. 16, pp. 234-240, April, 1968.

11. K. Kurokawa, Some Basic Characteristics of Broadband Negative Resistance Oscillator Circuits, *Bell Systems Technology Journal*, pp. 1937-1955, July-August, 1969.

12. D.B. Leeson, A Simple Model of Feedback Oscillator Noise Spectrum, *Proceedings of the IEEE*, Vol. 54, pp. 329-330, 1966.

13. Stephen A. Maas, *Noise in linear and nonlinear circuits*, Artech House Publishers, Boston, 2005.

14. C. D. Motchenbacher, J. A.Connelly, *Low-Noise Electronic System Design*, John Wiley & Sons, New York, 1993.

15. Alan V. Oppenheim, Ronald W. Schafer, *Discrete-Time Signal Processing*, Prentice Hall, Englewood Cliffs, 1989.

16. Henry W. Ott, *Noise Reduction Techniques in Electronic Systems*, John Wiley & Sons, New York, 1988.

17. Athanasios Papoulis, *Probability, Random Variables and Stochastic Processes*, McGraw Hill, New York, 1965.

18. Athanasios Papoulis, *Signal Analysis*, McGraw Hill, New York, 1977.

19. S.O. Rice, Mathematical Analysis of Random Noise, *Bell Systems, Technical Journal*, Vol. 23, pp. 282-292, 1944, Vol. 24, pp. 146-156, 1945.

20. Ulrich L. Rohde, Ajay K. Poddar, Georg Böck, *The Design of Modern Microwave Oscillators for Wireless Applications*, John Wiley & Sons, New York, 2005.

21. Peter Russer, Noise Analysis of Linear Microwave Circuits with General Topology, *Review of Radio Science 1993-1996*, pp. 361-393, 1996.

22. Burkhard Schiek, *Grundlagen der Hochfrequenzmesstechnik*, Springer, New York, 1999.

23. W. Schottky, Über spontane Stromschwankungen in verschiedenen Elektrizitätsleitern, *Analen der Physik*, Bd. 57, pp. 541-567, 1918.

24. Mischa Schwartz, *Information, Transmission, Modulation and Noise*, McGraw Hill, New York, 1970.

25. Simon M. Sze, *Physics of Semiconductor Devices*, John Wiley & Sons, New York, 1981.

26. Simon M. Sze, *Modern Semiconductor Device Physics*, John Wiley & Sons, New York, 1997.

27. B. Van der Pol, The Nonlinear Theory of Electrical Oscillators, *Proceedings of the IRE*, Vol. 22, No. 9, pp. 1051-1086, September, 1934.

28. Aldert Van der Ziel, *Noise*, Chapman and Hall, London, 1955.

29. Aldert Van der Ziel, *Noise, Sources, Characterisation, Measurement*, Prentice-Hall, Englewood Cliffs, 1970.

30. Aldert Van der Ziel, *Noise in Measurements*, John Wiley & Sons, New York, 1976.

31. Aldert Van der Ziel, *Noise in Solid State Devices and Circuits*, John Wiley & Sons, New York, 1986.

32. Edgar Voges, *Hochfrequenztechnik*, Hüthig, Heidelberg, 2004.

Index